Meteors in the Earth's Atmosphere
Meteoroids and Cosmic Dust and their Interactions with the Earth's Upper Atmosphere

A huge amount of extraterrestrial matter enters the Earth's atmosphere every year and eventually settles on the ground. The two main sources of this matter are cosmic dust and meteoroid streams. Meteorites form only a very small fraction of the total mass that is captured by the Earth's atmosphere. Most of the matter is in the form of very fine dust particles. Because of the temperatures reached during entry, a large proportion of these particles evaporates at high altitudes, giving rise to radar signatures and the visual phenomenon of shooting stars.

This book integrates astronomical observations and theories with geophysical studies to present a comprehensive overview of the extraterrestrial matter that falls to Earth from space. Meteoroids are the main topic of the book, although cosmic dust, interplanetary matter and meteorites are also discussed. This work will be of great value to researchers involved in the study of meteor phenomena.

Meteors in the Earth's Atmosphere

Meteoroids and Cosmic Dust and their Interactions
with the Earth's Upper Atmosphere

Edited by

Edmond Murad
Air Force Research Laboratory, Hanscom, USA

Iwan P. Williams
Queen Mary University of London, UK

CAMBRIDGE
UNIVERSITY PRESS

University Printing House, Cambridge CB2 8BS, United Kingdom

One Liberty Plaza, 20th Floor, New York, NY 10006, USA

477 Williamstown Road, Port Melbourne, VIC 3207, Australia

314-321, 3rd Floor, Plot 3, Splendor Forum, Jasola District Centre, New Delhi - 110025, India

79 Anson Road, #06-04/06, Singapore 079906

Cambridge University Press is part of the University of Cambridge.

It furthers the University's mission by disseminating knowledge in the pursuit of education, learning and research at the highest international levels of excellence.

www.cambridge.org
Information on this title: www.cambridge.org/9780521804318

First published 2002

A catalogue record for this publication is available from the British Library

ISBN 978-0-521-80431-8 Hardback

Contents

Colour plates between pages 150 and 151

Contributors

Arthur C. Aikin, NASA, Goddard Space Flight Center, Greenbelt, MD 20771, USA.

W. Jack Baggaley, Physics and Astronomy Department, University of Canterbury, Christchurch, New Zealand.

Valeri Dikarev, Max-Planck-Institut für Kernphysik, Heidelberg, Germany, and Astronomical Institute of St Petersburg University, Russia.

George J. Flynn, Departments of Physics & Mathematics, State University of New York–Plattsburgh, Plattsburgh, 12901 NY, USA.

Luigi Foschini, Instituti TeSRE–CNR, Via Gobetti 101, I-40129 Bologna, Italy.

Joseph M. Grebowsky, NASA, Goddard Space Flight Center, Greenbelt, MD 20771, USA.

Eberhard Grün, Max-Planck-Institut für Kernphysik, Heidelberg, Germany.

Robert L. Hawkes, Department of Physics, Mount Allison University, Sackville, NB E4L 1E6, Canada.

J. Höffner, Leibniz-Institute of Atmospheric Physics, Schloss-Str. 6, 18225 Kühlungsborn, Germany.

Harald Krüger, Max-Planck-Institut für Kernphysik, Heidelberg, Germany.

Markus Landgraf, ESOC, Darmstadt, Germany.

William J. McNeil, Radex, Inc., 3 Preston Court, Bedford, MA 01731, USA.

Edmond Murad, Space Weather Center of Excellence, Space Vehicles Directorate, Air Force Research Laboratory, Hanscom AFB, MA 01731, USA.

John M. C. Plane, School of Environmental Sciences, University of East Anglia, Norwich NR4 7TJ, UK.

Frans J. M. Rietmejer, Institute of Meteoritics, Department of Earth and Planetary Sciences, University of New Mexico, Albuquerque, NM 87131, USA.

U. von Zahn, Leibniz-Institute of Atmospheric Physics, Schloss-Str. 6, 18225 Kühlungsborn, Germany.

Iwan P. Williams, Astronomy Unit, Queen Mary, University of London, London E1 4NS, UK.

Preface

Many new advances have been made since the standard textbooks on meteoric phenomena such as Bronshten (1983), McKinley (1961) and Öpik (1958) were published. New observational techniques, such as lidar and wide-aperture radar, have come into regular use, data from space, such as from the Ulysses and Gallileo spacecraft, have become available, the collection and analysis of interplanetary dust particles have become almost routine, and new techniques have made the analysis of surviving meteorites more quantifiable. In addition, new theoretical understanding of orbital mechanics and the physics of meteoroid ablation, coupled with very good new data from mass spectrometric and spectroscopic observations and measurements of accurate and precise chemical kinetic data together with the increase in available computing power, have made possible the development of comprehensive models that integrate all these phenomena into a uniform framework. Our aim in this book has been to introduce information from a variety of disparate fields and to act as facilitators in the crosslinking of data and ideas from these disciplines to the understanding of the meteoric phenomena itself.

We thank all the contributors of individual chapters that have added a depth of expertise that we could not possibly have equalled ourselves. We thank a number of scientists for acting as reviewers of the manuscripts, in particular, Rainer A. Dressler, Michael Horanyi, John Mathews, and Asta Pellinen-Wannberg for expending much time on their reviews. Of course, we thank the editors at Cambridge University Press for having made this book possible and for help on the production details.

REFERENCES

BRONSHTEN V. A. 1983. *Physics of Meteoric Phenomena*, Kluwer, Dordrecht, Holland, 356 pp.

MCKINLEY D. W. R. 1961. *Meteor Science and Engineering* McGraw-Hill, New York, 309 pp.

ÖPIK, E. J. 1958. *Physics of Meteor Flight in the Atmosphere*, Interscience Publishers, New York, 174 pp.

1
INTRODUCTION

By IWAN P. WILLIAMS[1†] AND EDMOND MURAD[2‡]

[1] Astronomy Unit, Queen Mary, University of London, London, E1 4NS, UK

[2] Space Weather Center of Excellence, Space Vehicles Directorate, Air Force Research
Laboratory, Hanscom AFB, MA 01731, USA

A brief review of previous work on meteors is presented, and reference is made to most published
texts on the subject. We present a summary of the evolution of the field over several millenia
from a tool for prognostication and forecasting to one useful for understanding the origin of
meteorites and their impact on Earth.

Approximately 4×10^4 metric tons (40×10^6 kg) per year of extraterrestrial matter
enters the Earth's atmosphere and eventually settles on the ground (see discussion by
Flynn in Chapter 4). This extraterrestrial matter originates from two components, cosmic dust background and meteoroid streams. The first is the general cosmic dust that
pervades the universe; what arrives on Earth is mostly dust from within our solar system,
although a small fraction arises from interstellar space. The second source of extraterrestrial matter is the meteoroids generated from the breakup of comets and asteroids. It is
the meteoroids that are the subject of most of this book, although cosmic dust, interplanetary matter, and fragments of meteoroids that survive transit through the atmosphere
(i.e. meteorites) will also be discussed. We now know that the larger meteoroids (such as
those that survive transit through the atmosphere) are but a small fraction of the total
extraterrestrial mass that is captured by Earth. While there is a large uncertainty in the
absolute fluxes, there seems to be agreement that mass influx of particles having masses
in the range 10^{-5} to 10^{-6} g accounts for about 20% of the total flux with the remaining
80% arising from objects in the mass range 10^{-6} to 10^{+15} g, as shown in Figure 1.1 which
is taken from Rietmeijer (2000).

The micrometeorites and fine dust have a size distribution that shows a peak for
particles having a diameter of $\sim$200 μm [Love & Brownlee (1993)]. It is worthwhile, at
the risk being slightly pedantic, to define terms here. The term meteor generally refers
to the luminous phenomenon generated by the entry of a meteoroid into the Earth's
atmosphere. The term meteoroid refers to the solid particle of extraterrestrial dust that
enter the Earth's atmosphere. The term meteorite refers to a meteoroid that is large
enough to survive (in a reduced size) transit through the Earth's atmosphere. It is also
useful in the introduction to discuss in general the orbits of meteors with reference to the
ecliptic and to the celestial sphere. Generally, most meteor streams arise from comets,
although one, the Geminids, arises from the breakup of the asteroid 3200 Phaethon.
Whether Phaethon is a true asteroid or a dormant cometary nucleus is still a matter of
active debate. Cometary orbits occur on a plane that is different from the ecliptic, as
shown in Figure 1.2.

These planes are related to the galactic plane on the celestial sphere as illustrated in
Figure 1.3. This figure shows a coarse sketch of two constellations, Ursa Minor and Leo.
The figure would be too busy if we were to show all the constellations that have meteor
streams associated with them. The velocity of a meteoroid is obtained from the sum of the

† Email:I.P.Williams@qmul.ac.uk
‡ Email:ed.murad@hanscom.af.mil

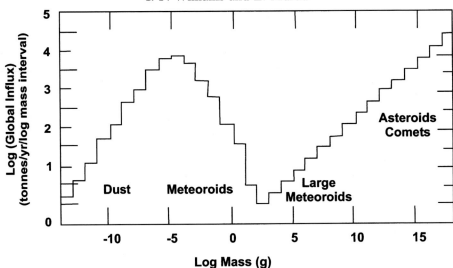

FIGURE 1.1. Total mass accumulation. Figure reproduced from [Rietmeijer (2000)] with permission from *Meteoritics and Planetary Sciences*.

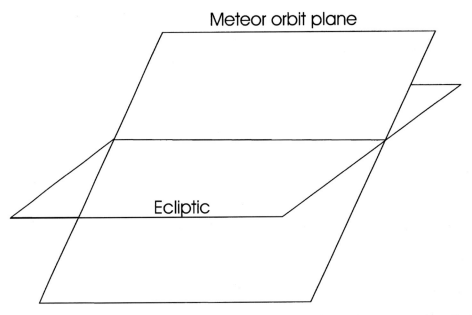

FIGURE 1.2. Simple sketch showing the plane of a cometary orbit with respect to the ecliptic plane.

velocity of the meteoroid and that of Earth. For a prograde meteoroid stream (inclination less than 90°), the meteors follow the Earth and the net velocity is the difference between the two velocities. For a retrograde stream (inclination 90–180°), the net velocity is the sum of the two velocities. These points will be discussed in detail in the next chapter.

Meteoroid streams will be discussed by Williams in the next chapter and the various components of extraterrestrial matter will be discussed by Grün, and Flynn, in Chapters 3 and 4, respectively. Observation of meteoroids in the Earth's atmosphere can be done visibly, by radar, by lidar, and by space-borne mass spectrometers. These observational

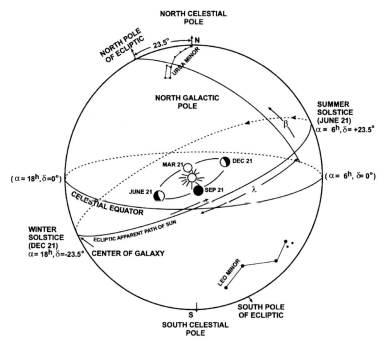

FIGURE 1.3. Simple sketch showing the plane of a cometary orbit with respect to the ecliptic plane.

methods will be discussed by Hawkes, Baggaley, von Zahn, and Grebowsky, in Chapters 5, 6, 7, and 8, respectively. Meteoroids can survive atmospheric heating and can land on the ground; these survivors (meteorites and interplanetary dust particles) will be discussed by Rietmeijer in Chapter 9. Finally, a natural question arises as to the meaning of it all and how the scientific method (model, theory, independent laboratory measurements, and validation of the model) is used to develop an understanding of the interaction of meteoroids and cosmic dust with the Earth's atmosphere. This topic will be the subject of Chapters 10, 11, and 12: models of the impact of extraterrestrial matter on spacecraft in Earth orbit (Foschini), modeling of vaporization and atmospheric interactions (by McNeil, Murad and Plane), and crucial laboratory measurements needed to develop the necessary models (Plane). It is worth noting at this point that we will not consider in this book the much rarer, but related, phenomenon, namely that of bolides or fireballs that are caused by large meteoroids (100 g to 1 kg) entering the atmosphere and in the process of transit to ground generating dense clouds of metallic and other vapor species. This phenomenon is a sub-specialty of its own.

In the remainder of this introductory chapter we will give a summary of historical material, some of the popular theories, manifestation of meteoroids, and what we know about meteoroids at present. It is worth mentioning at this point several older texts that cover various aspects of meteoric phenomena namely Porter (1952), Öpik (1958), Fedynsky (1959), the compendium by Bronshten (1983), and the review by Bone (1993).

1.1. Historical Survey

Two excellent source books for the history of the study of meteors and meteorites are by Krinov (1960) and Burke (1986). The former is particularly valuable as it refers to publications on the subject of meteors published in the (then) Soviet Union. The relative rarity of the phenomenon of shooting stars made it an ideal candidate for myth and magic before the relatively recent emergence of astronomy from its association with astrology. Shooting stars seem to have been associated in ancient times with omens. In Mesopotamian civilizations the observation of meteors sometimes suggested a victory in an impending battle or an attack by any enemy encamped nearby [Bjorkman (1973)]. In contrast to the omens, Spartan writings dating to c. 1200 BC [Leach & Fried (1984)] suggest that the observation of shooting stars by wise men on particular nights was used to conclude that the king had committed a sin; hence the need to depose him. What a pity this valuable tool of political theory became extinct! Aside from the folklore and mythological applications, there were notable attempts in Asia to systematize the observations. For example, Chinese, Japanese, and Korean archives contain detailed accounts of ancient observations of meteors and meteor showers. Detailed historical descriptions of the work on meteors and meteorites are given by Olivier (1925) and Burke (1986).

The first scientific analysis seems to have been initiated by the eminent astronomer Chladni (1798a), who attributed the phenomenon of fireballs to extraterrestrial bodies entering the Earth's atmosphere and burning in the process. This explanation was not universally accepted even though it was based in part on crude triangulation experiments by two German students, Brandes and Benzenberg, who derived an altitude of $\sim$ 95 km for the height of the meteors. Chladni (1798b) obtained further evidence for his hypothesis by analysis of two fallen meteorites. Interest in the subject was aided by what we now know as the great Leonid storm of November 13, 1833, when for a short period of time as many as 100,000 meteors per hour appeared. This Leonid storm gave rise to the most famous picture of a meteor storm, which showed a village scene near Basel, Switzerland, that seems engulfed by streamers of light (shooting stars) [Hughes (1995)]. Indeed, the Leonid meteor storm of 1833 was responsible for much of the work aimed at identifying the sources of the Leonids and the Perseids. Following the Leonid storm of 1833, and the next one in 1866, Giovanni Schiaparelli is credited with having determined the orbits and origin of the Leonid meteor shower and storm and of pointing out the conjection between these orbits with those of Comet 1862 III (now known as comet 55P/Tempel–Tuttle). A good historical description of the scientific development of meteor research is given in the reviews cited earlier [Burke (1986); Olivier (1925)].

Once the hypothesis that meteors represented the fiery entry of extraterrestrial matter into the Earth's atmosphere was accepted, meteor observation and analysis became an established sub-field of astronomy. Meteor astronomy really flowered with the advent of radio communication and with the intense scientific activity taking place around the understanding of radio wave propagation and disruption. For example, Nagaoka (1929) was first to note coincidence between meteor altitudes and the E-layer (then called the Kennelly–Heaviside layer); he also suggested that meteoric dust (particle sizes varying between 0.01 and 0.1 µm) leads to reduced ionization (because the meteoroids sweep away electrons that are in their path), and hence disturbances in the E-layer. The correct explanation, i.e. enhanced ionization, was made by Skellet (1931, 1935). Rather than review this exciting history, the reader is referred to two standard references by Lovell (1954) and McKinley (1961). Following World War II, the introduction of radar into astronomical observations confirmed prior studies that meteoroids entering the atmosphere lead to enhanced ionization and that the phenomenon of shooting stars is correlated with ionization trails. Thus, it became possible to obtain important parameters for meteoroids entering the Earth's atmosphere: velocities (initial and terminal), electron densities in the trail, and heights (initial altitude at which the ionization patch appears and the terminal altitude). When these observations were combined with visual observations, it became possible to draw a reasonable picture of the nature of the interaction. Identification of enhanced ionization with meteor trails even made possible the operation of a communication system, Meteor Burst Communication, based on the ionization trails generated by meteoroids entering the Earth's atmosphere [Dennis (1993); Desourdis (1993); Weitzen (1993)].

1.2. Meteoroid Observations

1.2.1. *Visual*

As alluded above, historically, meteoroids were seen as visible bright streaks (meteors) whose luminosity and longevity depended on various properties of the particular meteoroid giving rise to the particular streak. In standardizing reports of the visual observations, several quantities evolved for the characterization of meteors: visual magnitude (M), radiant, velocity ($\mathbf{v}$), height, the population index (r), the solar longitude (λ), and the Zenithal Hourly Rate (ZHR). An other important quantity is the orbit of the meteoroids. This will be discussed in detail in the next chapter as it is related to the origin of the stream, i.e. whether cometary or asteroidal, and which comet. The visual magnitude, M, is expressed in a logarithmic scale related to the luminous intensity, I, via the relationship

$$M = 6.8 - 2.5 \log_{10} I \qquad (1.2.1)$$

where I is expressed in watts. Visual observations of meteors (shooting stars) generally refer to meteoroids having diameters ≥ 2 mm. The recent use of low light level television to study meteors has increased the detection limit considerably and now meteoroids of magnitude $+7$ corresponding approximately to meteoroids having diameters of ~ 1 mm can be observed [Fujiwara *et al.* (1998)]. In comparison, radar ionization trails are seen for particles having diameters ≥ 80 μm, equivalent to visual magnitude $+10$ [Hughes (1978); Pellinen-Wannberg *et al.* (1998)].

Radiant: The radiant is another quantity that specifies the direction from which the meteor appears to be coming. This direction is obtained by observing a meteor and extending the direction of its visible path to a point on the celestial sphere (i.e. to a constellation). Thus meteors observed in August seem to originated from a direction that connects with the constellation Perseus. Two quantities which are specified along with the radiant are α and δ, which are coordinates for a shower's radiant position, usually at maximum. α is right ascension, while δ is declination.

Velocity: A third quantity that is used to characterize meteors is the velocity. Here, it is important to distinguish between the initial velocity, i.e. that of the meteor when it is first seen, and terminal velocity, which is the velocity when the meteor seems to vanish. In general, the velocity is different for different meteor streams, as will be discussed in the next chapter. As an example, however, we might mention that the Leonids, which occur in mid-November and are associated with comet 55P/Tempel–Tuttle, have a velocity of ~ 71 km/s. On the other hand, the Draconids or Giacobinids (which occur in October and are associated with Comet Giacobini–Zinner) have a velocity of 20 km/s. The fourth quantity mentioned above is the height; here, again, it is important to distinguish between the initial height at which the meteor is observed and the terminal height at which the meteor appears to vanish.

Zenithal Hourly Rate, or ZHR: This a calculated maximum number of meteors that an observer would see in perfectly clear skies with the shower radiant overhead.

Population Index, r: Population index is a measure of the quality of the meteor shower. It is related to the magnitude and to the ZHR.

TABLE 1.1. Regular meteor showers

Shower	v(km/s)	r	ZHR	Dates	Date of Max Activity
Quadrantids	41	2.1	120	1–5 January	03 January
η-Aquarids	66	2.7	60	19 April–28 May	5 May
Perseids	59	2.6	200	17 July–24 August	12 August
Orionids	66	2.9	20	2 October–27 November	21 October
Leonids	71	2.5	40+	14–21 November	17 November
Geminids	35	2.6	110	7–17 December	13 December
Ursids	33	3.0	10	17–26 December	22 December

Data taken from the Internation Meteor Organization website (http://www.imo.org)

Much useful and timely information about meteors, meteor streams, and their observation is available at the we site of the International Meteor Organization, http://www.imo.net/. Another helpful web site that contains additional information is that of Commission 22 (meteors and interplanetary dust) of the International Astronomical Union, http://www.mta.ca/~rhawkes/IAU22/. Finally, a valuable printed reference is the compilation of meteor showers by Kronk (1988). Table 1.1 presents a summary of the most common showers along with their properties.

The visual observations will be discussed in detail by Hawkes in Chapter 5. The visual observation of meteors, of course, indicates not only a hot body emitting continuum radiation but also the presence of highly excited atomic or molecular species that radiate in the visible region of the spectrum. Even though most visible observations are broad-band observations in the visible region (400–700 nm), a few spectra have been obtained at moderate spectral resolution, and these show emissions from metal atoms, metal ions, and atmospheric species [Millman & McKinley (1963)]. The observations of ions is very interesting because it connects the visible observations with the radar signatures.

1.2.2. *Radar Observation*

Another way in which meteoroids manifest themselves is through generation of ionization trails that are studied by radar. When meteoroids collide with the atmosphere at the entry velocities, they give rise to ionization trails that contain a large number of ions and electrons in small, confined space. The ionization results from shock heating and high energy collisions between meteoroids and the atmosphere; spectra taken of such trails show the presence of metal ions as well as atmospheric ions [this subject will be discussed in some detail by Hawkes in Chapter 5, but reference can be made here to an earlier work by Millman & McKinley (1963) that shows clearly the presence of these ions (and electrons)]. This method, which had its genesis in radio transmission work, consists of reflecting radio or radar signals from the electrons in a meteor trail. Two ways can be used to observe the reflected radiation: back-scatter or forward scatter. In the former, the transmitter of the radio waves is located at the same position as the receiver, and the technique was called radio echo in its early days. In the second method, the transmitter and receiver are separated by great distances; this technique has the advantage of being able to observe fainter and higher meteor trails. Both techniques are described in detail by McKinley (1961) and will be considered in more detail in Chapter 6 by Baggaley. The detailed analysis of these trails requires a differentiation between two types, underdense trails and overdense trails [Greenhow (1963)]. The former refers to ionization trails where the number of reflecting electrons is $< 10^{14}$ m^{-1}, and where diffusion is the primary electron loss mechanism; the latter term refers to those trails where the number of reflecting electrons is $> 10^{14}$ m^{-1}, and where other processes such as electron attachment become important electron loss mechanisms. As will be discussed in Chapter 6 by Baggaley, there is tradeoff between sensitivity, power requirements, radar frequency, trail altitude, and electron density in the trail. Suffice it to say here that meteoroids of diameter ≥ 80 µm can now be observed by radar [Pellinen-Wannberg *et al.* (1998)]. This technique makes possible the precise determination of trail altitude and meteor velocity. The relationship between radar and visible observations has been the subject of much discussion [Linblad (1961); Millman & McKinley (1956)]; in the latter study Millman and McKinley found a direct relationship between the absolute visual magnitude and the duration of a meteor echo (i.e. the number of electrons in a trail) for a given meteor stream. This relationship holds for visual meteors having magnitudes in the range +5 to −5. For example, it was found that a Perseid meteor of visual magnitude +5 corresponded to a trail containing 2×10^{14} electrons m^{-1}.

1.2.3. *Metallic Deposits in the Atmosphere: Mass spectrometry and Airglow*

Slipher (1929), in his studies of the nightglow, established tentatively the presence of sodium, Na, at high altitudes. Following that early work, spectroscopic studies at higher resolutions [Cabannes (1938a); Cabannes & Dufay (1938b); Bernard (1938a); Bernard (1939)] clearly established that the yellow light in the nightglow was due to Na. The attribution of sodium in the nightglow as arising from meteorites seems to have been first made by Cabannes & Dufay (1938b), who also concluded that the emission arose at an altitude of 130 km. By contrast, Bernard (1938b) calculated that the altitude of emission to be near 60 km and he suggested that Na in the dayglow spectrum arose from updraft of sea water followed by chemical reactions. The identification of the source of the Na nightglow as being due to chemical reactions between Na from meteor ablation and atmospheric constituents was first made by Chapman (1939), who suggested his famous mechanism:

$$\text{Na} + \text{O}_3 \rightarrow \text{NaO} + \text{O}_2 \qquad (1.2.2)$$

$$\text{NaO} + \text{O} \rightarrow \text{Na}^* + \text{O}_2 \qquad (1.2.3)$$

$$\text{Na}^* \rightarrow \text{Na} + h\nu \qquad (1.2.4)$$

Even though later work has introduced more complex processes into the explanation for the Na nightglow, in the main this simple mechanism has been proven to be correct. This mechanism and others that detail the morphology of meteoric metals from their deposition in the upper atmosphere to their transport to the ground will be discussed in much more detail by McNeil, Murad, and Plane in Chapter 11.

Following World War II rockets became available for atmospheric research, and an early application was the release of compact mass spectrometers, such as the compact Bennett rf mass spectrometer [Johnson & Meadows (1955)]. The first detection of metal ions was made by Istomin & Pokhunkov (1963) using an rf mass spectrometer. They had sufficient mass resolution to identify Mg$^+$, Ca$^+$, and Fe$^+$ at an altitude of 105 km with Fe$^+$ being ten times larger than Mg$^+$. The presence of metal ions in the D- and E-regions of the ionosphere was firmly established when a compact quadrupole mass filter was built and flown by Narcisi and his colleagues [Narcisi & Bailey (1965)]. They observed these metals to be present in well-defined layers; they also established that the metal ions

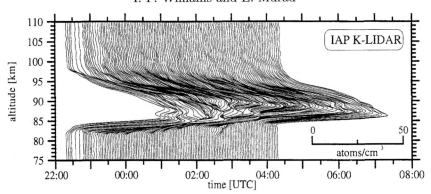

FIGURE 1.4. Profile of a lidar observation of K during a Perseid shower. Data taken from Eska *et al.* (1999) and reproduced with permission of authors and the *Journal of Geophysical Research.*

are formed by rapid charge transfer between the ambient ions, O_2^+ and NO^+, and the metals:

$$NO^+(O_2^+) + M \rightarrow M^+ + NO(O_2) \qquad (1.2.5)$$

where M is any metal atom [Narcisi (1968)]. This process was established when it was realized that the abundance of NO^+ and O_2^+ is reduced greatly when that of the metal ions increases. This subject will be discussed in detail by Grebowsky in Chapter 8.

1.2.4. *Lidar*

The availability of high power lasers provided an impetus to the application of radar techniques to optical methods for the study of atmospheric phenomena, such as composition, winds, temperature. Fiocco & Smullin (1963) first suggested use of this technique for atmospheric research and then applied it to study Sporadic-E layer formation [Fiocco & Colombo (1964)] and meteoroid fragmentation [Fiocco (1965)]. Lidar can be used to study a wide range of atmospheric properties, such as temperature, winds, and gravity waves [Gardner (1989)]. Briefly, a pulse from a laser is aimed at the upper atmosphere, the laser wavelength being chosen to correspond with an allowed transition for the atom or molecules that is the subject of study. Following the initial pulse, fluorescence signals are received on the ground at different times corresponding to the altitude of the atom. In this way one can obtain altitude profiles for a given species. In terms of its application to meteor research, lidar is now used to follow the metal atoms and metal neutral components of the D- and E-regions of the ionosphere at altitudes varying between 80 and 100 km, as will be described in detail by von Zahn in Chapter 7. Careful work and the use of narrow-band excitation has enabled von Zahn and his colleagues [Eska *et al.* (1998, 1999)] to detect extremely low abundances, as in the case of K, where densities as low as 5 cm^{-3} were measured at 85 km altitude, where the normal atmospheric density is 10^{14} cm^{-3}. An example of this is shown in Figure 1.4 which displays the measured abundance of K in the altitude range 80–105 km; it is adapted from the data presented by Eska *et al.* (1999). Lidar can also be used to follow the vapor trail left by an individual meteoroid entering the atmosphere, as was recently done for the observation of the Leonid showers of 1998 and 1999 [von Zahn *et al.* (1999)].

1.2.5. *Satellite Impact*

The advent of space exploration and especially manned space travel made it necessary to evaluate the likelihood and severity of meteoroid (and debris in general) impact upon spacecraft. This general topic will be treated in detail by Foschini in Chapter 10. Suffice it to state that current assessments indicate that a meteoroid 1 mm in diameter can penetrate the space suit of an astronaut [Cevolani & Foschini (1998); Gleghorn *et al.* (1995)]. In addition to direct impact and penetration of surfaces, other effects of direct impact include the generation of plasma that can lead to discharge of power sources on a spacecraft [Foschini (1998)].

1.3. Sources of Meteoroids

Meteoroids can come from many sources, in fact from any situation where the production of solid particles can take place. Many of the cases are discussed more fully in the following chapters, but it is useful to obtain an overview of the possibilities here. In a strict sense, material can only come from two distinct locations, either from Earth orbit, or from the Interplanetary dust Complex. Giving such an answer, however, rather begs the question; the interesting aspect of the question is where did the particles originate from. Considering first the material in Earth orbit, a recent source is debris resulting from the exploration and exploitation of space by humans. For example, a majority of samples gathered by NASA's Long Term Exposure Facility experiment (LDEF) was debris and waste from other spacecraft and shuttle flights. A second source for material in Earth orbit is the Moon. The impact of any body on the Moon, and the cratering indicates that this is a common occurence, will eject miriads of small particles and some will undoubtedly escape the Moon's gravity and be captured in Earth orbit. It has even been suggested [Hartung (1993)] that the Corvid meteor shower resulted from the lunar impact allegedly observed by Giordano Bruno in 1178.

Turning now to the Interplanetary Dust Complex, collision with any significant solid body in the Solar System will release dust particles and these, if originally on orbits exterior to the Earth, will eventually decay and become Earth crossers. Hence these particles in due course have the potential to become meteors and meteorites. In general, the history of the original orbit will have been lost, but potential strong source regions here are the Edgeworth–Kuiper belt and the asteroid belt. Because of its relative proximity, Mars is also a potential source and indeed, some meteorites are known to originate from there [McSween (1994)]. If the formation of the meteoroids was not too long in the past, then some memory of the parent may remain within the orbital elements which manifests itself as a cluster of meteors being observed at the same time. The two most obvious sources of such activity are the comets and the Near Earth asteroids. These will be discussed in other chapters in detail and so no further discussion is included here.

Acknowledgements: We wish to thank *Meteoritics and Planetary Sciences* and F. J. M. Rietmeijer for permission to include in this chapter Figure 1.1, which is taken from Rietmeijer (2000). We also thank the *Journal of Geophysical Research* and U. von Zahn for several helpful comments and for permission to include Figure 1.4, which is taken from Eska *et al.* (1999).

REFERENCES

BERNARD R. 1938a. Das Vorhandensein von Natrium in der Atmosphäre auf Grund von interferometrischen Untersuchungen der D-linie im Abend- und Nachlhimmellicht, *Z. Physik*, **110**, 291–302.

BERNARD R. 1938b. Sur la formation d'atomes libres de sodium dans la haute atmosphère, *Compte Rendus*, **206**, 1669–1672.

BERNARD R. 1939. The identification and the origin of atmospheric sodioum, *Ap. J.*, **89**, 133–135.

BJORKMAN J. K. 1973. Meteors and meteorites in the ancient Near East, *Meteoritics*, **8**, 91–132.

BONE N. 1993. *Meteors*, Sky Publishing, Cambridge, MA, 176 pp.

BRONSHTEN V. A. 1983. *Physics of Meteoric Phenomena*, Kluwer, Dordrecht, Holland, 356 pp.

BURKE J. G. 1986. *Cosmic Debris – Meteorites in History*, University of California, Berkeley, CA, 445 pp.

CABANNES J., DUFAY J. & GAUZIT J. 1938a. Sodium in the upper atmosphere, *Ap. J.*, **88**, 164–172.

CABANNES J. & DUFAY J. 1938b. Sur la radiation jaune du ciel nocturne, *Compte Rendus*, **206**, 221–224.

CEVOLANI G. & FOSCHINI L. 1998. The effects of meteoroid stream enhanced activity on human space flight: an overview, *Planet. Space Sci.*, **46**, 1597–1604.

CHAPMAN S. 1939. Notes on atmospheric sodium, *Ap J.*, **90**, 309–316.

CHLADNI E. F. F. 1798a. Account of a remarkable fiery meteor seen in Gascony on the 24th of July 1790, *Phil. Mag.*, **2**, 225–231.

CHLADNI E. F. F. 1798b. Observations on a mass of iron found in Siberia by Prof. Pallas, *Phil. Mag.*, **2**, 1–8.

CHLADNI E. F. F. 1798c. Observations on fireballs and hard bodies which have fallen from the atmosphere, *Phil. Mag.*, **2**, 337–345.

DENNIS L. 1993. Buck Rogers technology – A step in the right direction, in *Meteor Burst Communications – Theory and Practice*, Edited by D. L. Schilling, Wiley, New York, pp. 1–7.

DESOURDIS J. 1993. Modeling and analysis of meteor burst communications, in *Meteor Burst Communications – Theory and Practice*, Edited by D. L. Schilling, Wiley, New York, pp. 59–342.

ESKA V., HÖFFNER J. & VON ZAHN U. 1998. Upper atmosphere potassium layer and its seasonal variations at 54°N, *J. Geophys. Res.*, **103**, 29,207–29,214.

ESKA V., VON ZAHN U. & PLANE J. M. C. 1999. The terrestrial potassium layer (75–110 km) between 71°S and 54°N: Observations and modeling, *J. Geophys. Res.*, **104**, 17,173–17,186.

FEDYNSKY, V. 1959. *Meteors*, Akademia Nauk SSSR, Moscow, 123 pp.

FIOCCO L. & SMULLIN L. D. 1963. Detection of scattering layers in the upper atmosphere (60–140 km) by optical radar, *Nature*, **199**, 1275–1276.

FIOCCO L. & COLOMBO G. 1964. Optical radar results and ionospheric Sporadic E, *J. Geophys. Res.*, **69**, 1795–1803.

FIOCCO L. 1965. Optical radar results and meteoric fragmentation, *J. Geophys. Res.*, **70**, 2213–2215.

FOSCHINI L. 1998. Electromagnetic interference from plasmas generated in meteoroid impact, *Europhys. Lett.*, **43**, 226–229.

FUJIWARA Y., UEDA M., SHIBA Y., SUGIMOTO M., KINOSHITA M. & SHIMODA C. 1998. Meteor luminosity at 160 km altitude from TV observations of bright Leonid meteors, *Geophys. Res. Lett.*, **25**, 285–288.

GARDNER C. S. 1989. Sodium resonance fluorescence lidar applications in atmospheric science and astronomy, *Proc. IEEE*, **77**, 408–418.

GLEGHORN G., ASAY J., ATKINSON D., FLURY W., JOHNSON N., KESSLER D., KNOWLES S., REX D. TODA S. VENIAMINOV S. & WARREN R. 1995. *Orbital Debris – A Technical Assessment*, National Academy Press, Washington, DC, 210 pp.

GREENHOW J. S. 1963. Limitations of Radar techniques for the study of meteors, in *Symposium on the Astronomy and Physics of Meteors*, Edited by F. L. Whipple, Smithsonian Institution, Cambridge, MA, pp. 5–17.

HARTUNG J. B. 1993. Corvid meteoroids are ejecta from the Giordano Bruno impact, *J. Geophys. Res.*, **98**, 9141–9144

HUGHES D. W. 1978. Meteors, in *Cosmic Dust*, Edited by J. A. M. McDonnell, Wiley, New York, pp. 123–185.

HUGHES D. W. 1995. The world's most famous meteor shower pictures, *Earth, Moon, Planets*, **68**, 311–322.

ISTOMIN V. G. & POKHUNKOV A. Z. 1963. Mass spectrometer measurements of atmospheric composition in the USSR, *Space Res.*, **3**, 117–131.

JOHNSON C. Y. & MEADOWS E. B. 1955. First investigation of ambient positive-ion composition to 219 km by rocket-borne spectrometer, *J. Geophys. Res.*, **60**, 193–203.

KRINOV E. L. 1960. *Principles of Meteoritics*, Pergamon, London, 535 pp.

KRONK G. W. 1988. *Meteor Showers – A Descriptive Catalog*, Enslow Publishers, Hillside, NJ, 291 pp.

LEACH M. & FRIED J., EDITORS 1984. *Funk and Wagnalls' Standard Dictionary of Folklore, Mythology, and Legend*, HarperCollins, San Francisco, London.

LINBLAD B. A. 1961. The relation between visual magnitudes of meteors and the duration of radar echoes, in *Symposium on the Astronomy and Physics of Meteors*, Edited by F. L. Whipple, Smithsonian Institution, Cambridge, MA, pp. 27–39.

LOVE S. G. & BROWNLEE D. E. 1993. A direct measurement of the terrestrial mass accretion rate of cosmic dust, *Science*, **262**, 550–553.

LOVELL A. C. B. 1954. *Meteor Astronomy*, Oxford University Press, London, 463 pp.

MCKINLEY D. W. R. 1961. *Meteor Science and Engineering* McGraw-Hill, New York, 309 pp.

MCSWEEN H. Y. 1994. What have we learnt about Mars from SNC meteorites, *Meteoritics*, **29**, 757–779

MILLMAN P. M. & MCKINLEY D. W. R. 1956. Meteor echo durations and visual magnitudes, *Can. J. Phys.*, **34**, 50–61.

MILLMAN P. M. & MCKINLEY D. W. R. 1963. Meteors, in *The Moon, Meteorites, Comets*, Edited by B. M. Middlehurst and G. P. Kuiper, University of Chicago Press, pp. 674–773.

NAGAOKA H. 1929. Possibility of disturnace of radio transmission by meteor showers, *Proc. Imp. Acad. Tokyo*, **5**, 233.

NARCISI R. S. & BAILEY A. D. 1965. Mass spectrometric measurements of positive ions at altitudes from 64 to 112 kilometers, *J. Geophys. Res.*, **70**, 3687–3700.

NARCISI R. S. 1968. Processes associated with metal-ion layers in the E-region of the ionosphere, *Space Res.*, **8**, 360–369.

OLIVIER C. P. 1925. *Meteors*, Williams and Wilkins, London, 276 pp.

ÖPIK, E. J. 1958. *Physics of Meteor Flight in the Atmosphere*, Interscience Publishers, New York, 174 pp.

PELLINEN-WANNBERG A., WESTMAN A., WANNBERG G. & KAILA K. 1998. Meteor fluxes and visual magnitudes from EISCAT radar event rates: a comparison with cross-section based magnitude estimates and optical data, *Ann. Geophysicae*, **16**, 1475–1485.

PORTER J. G. 1952. *Comets and Meteor Streams*, Wiley, New York, 123 pp.

RIETMEIJER F. J. M. 2000. Interrelationships among meteoric metals, meteors, interplanetary dust, micrometeorites, and meteorites, *Meteoritics Planet. Sci.* **35**, 1025–1041.

SKELLET A. M. 1931. The effect of meteors on radio transmission through the Kennelly–Heaviside layer, *Phys. Rev.*, **37**, 1668.

SKELLET A. M. 1935. The ionizing effects of meteors, *Proc. Inst. Radio Eng.*, **23**, 132–149.

SLIPHER V. M. 1929. Emission in the spectrum of the light of the night sky, *Publ. Astron. Soc. Pacific*, **41**, 262–263.

VON ZAHN U., GERDING M., HÖFFNER J., MCNEIL W. J. & MURAD E. 1999. Fe, Ca, and K atom densities in the trails of Leonids and other meteors: Strong evidence for differential ablation, *Meteoritics Planetary Sci.*, **34**, 1017–1027.

WEITZEN J. A. 1993. Meteor scatter communication: A new understanding, in *Meteor Burst Communication – Theory and Practice*, Edited by D. L. Schilling, Wiley, New York, pp. 9–58.

2
THE EVOLUTION OF METEOROID STREAMS

By IWAN P. WILLIAMS[†]

Astronomy Unit, Queen Mary, University of London, London, E1 4NS UK

Since the orbital speed of any body in the Solar System is much larger than any potential ejection speed of small particles off this body, the initial orbit of any meteoroid is fairly similar to that of the parent. However, with the passage of time the effects of solar radiation and gravitational perturbations from the planets will cause the orbits of the meteoroids to evolve differently from the parent. Initially, this may cause a meteor shower to be formed, after all the parent bodies do not generally collide with the Earth, but eventually it will lead to the dissipation of the stream. In this chapter, we will mainly consider the dynamics of the formation process and the subsequent effects of evolution, in particular focusing on what the detailed study of meteor showers an tell us about the formation dynamics and hence about the physics of the process and the nature of the parent body.

2.1. Introduction

Meteors must have been witnessed by the human race since antiquity. As mentioned in the introduction, early appearances were associated with myths and superstition. Whatever the reason for doing so, records of the appearance of meteors, described in terms like *many falling stars* date back for at least two millennia (see Hasegawa (1993)), but serious scientific analysis only dates back for a tenth of this time-span, or about two hundred years. One important reason for this must be based on the view of the Christian Church at that time that the heavens were perfect and so bits of solid material drifting about in interplanetary space could not exist. Hence, meteors were regarded as atmospheric phenomena akin for example to lightning. The first indication that this interpretation of meteors was not correct had come when Benzenberg & Brandes (1800) observed the same meteors simultaneously from two different locations and through parallax determined their height to be about 90 km, well above all usual visible atmospheric phenomenon.(This, as it turned out was a remarkably accurate determination of the typical height of meteors.)

The spectacular Leonid displays of 1799 and 1833 had prompted observations of meteors which, in turn, lead to establishing facts and formulating theories. Both Olmstead (1834) and Twining (1834) pointed out that shower meteors radiated from a fixed point, since called the *radiant*. A few years later, Herrick (1837, 1838) showed that the annual showers were periodic on a sidereal rather than a tropical year. These observations naturally led to the correct idea, namely that meteoroid streams exist in interplanetary space and that a meteor shower is observed whenever the Earth passed through such a stream. Observations of the Leonid storms also lead Adams (1867), Le Verrier (1867) and Schiaparelli (1867) to conclude that the orbits of the Leonid meteoroids were very similar to that of comet 55P/Tempel–Tuttle and that 33 years, being the interval between major displays, was very close to the orbital period of this comet. Hence, spectacular streams

[†] Email: I.P.Williams@qmul.ac.uk

are seen only when the parent comet is close to the Earth, in other words that comets and meteoroid streams were very closely related.

The behaviour of comet $3D/Biela$ around this time (1845–1872) also played a part in establishing a connection between meteor showers and comets. Prior to 1845, the comet was apparently a perfectly normal and rather unremarkable comet, named after an Austrian Army Captain who recovered it in 1826. It was subsequently identified as the same comet as had been seen in 1772 and 1805 and its period was derived as about 6.7 years. It was recovered again by Herschel in 1832. The comet was very poorly placed in relation to the Sun in 1839 and, in consequence, it was not recovered. In 1846, when it was recovered, a faint companion was also recognized one or two arcseconds to the North of the main comet and co-moving with it. At the 1852 return, both components were again present but fainter and further apart. This was the last time that comet 3D/Biela or its companion were seen. However, a very strong display of Andromedid meteors were seen in 1872, with weaker displays also in 1885, 1892 and 1899, dates when the Earth was close to the node of the orbit of comet 3D/ Biela. This established a generic as well as a dynamical link between comets and meteor showers. Unfortunately, it also lead to two misconceptions. The first was concerned with the nature of cometary nuclei. A meteor stream consists of myriads of small dust grains when a comet disintegrates we get a meteor stream, hence comets consist of a collection of dust grains, the 'bank' model for cometary nuclei as described for example by Proctor & Crommelin (1937) and defended by Lyttleton (1951). The second misconception partially follows from the first, namely that meteors were released through the disintegration of a cometary nucleus whereas in fact the release of meteoroids is part of the normal evolution of comets.

It is now generally accepted that meteoroid streams do form through the ejection of meteoroids from comets. It may also be possible that meteoroids can be ejected off the surface of asteroids, collisions being the most obvious driving mechanism. A review of this topic was given for example by Štohl & Porubčan (1993). Comet–meteoroid stream pairs or asteroid–meteoroid stream pairs, based on the similarity between the pair of orbits, have been identified for virtually all recognizable streams. To aid in this identification, objective measures of the differences between the orbits have been developed. Two such measures in common use are by Southworth & Hawkins (1963) and Drummond (1981). A more physical criterion, based on the two conserved quantities of angular momentum and orbital energy rather than the geometrical orbital elements, has also recently been proposed by Jopek (1993). A list of meteoroid stream–comet pairings was produced by Cook (1973) and most of the gaps in Cook's list have been filled in over the last quarter century. As observational data have improved with more meteor orbits being determined and catalogued (for example, Lindblad & Steel (1994), Betlem et al. (1997)), so the agreement between the mean orbit of the meteoroid stream and that of its parent body has improved, a good indication that the pairings are correct.

A new physical understanding of the relationship between meteoroids and comets, replacing the idea that the Earth simply passed through the debris from a disintegrated comet, became possible about half a century ago, when Whipple (1950) proposed a model for comets where a single cometary nucleus existed. The nucleus was composed of an icy matrix within which dust grains of varying sizes were embedded. As the comet approached the Sun, the ice would sublime, releasing the grains and accelerating them away from the nucleus through drag from the outflowing gas. The very small grains would also react to radiation pressure and flow nearly radially outwards to form the well known dust tail, while larger grains would remain bound and not have a large velocity relative to the nucleus and so move on fairly similar orbits to the comet – hence forming a meteor stream. Whipple (1951) considered the physics of the situation and obtained a formula for the ejection speed of the meteoroids relative to the nucleus in terms of nuclear radius and heliocentric distance, as well as the dimensions of the meteoroids. Generally, smaller meteoroids had a larger ejection velocity. A number of authors (e.g. Gustafson (1989a), Harris & Hughes (1995)) have suggested minor modifications to the details, such as non-spherical meteoroids and columated gas outflow, which results in a higher ejection velocity than that given by Whipple. All result in speeds between a few tens and a few hundred metres per second. For an asteroid, the obvious cause of the ejection of meteoroids is a collision between between it and a second asteroid. In this the main agent carrying the dust outwards arises through the vaporization of some of the parts of the surface directly in contact during the collision, so that a gas outflow is again responsible for dragging the grains, as in the cometary case. The actual value of the ejection velocity is thus very similar in both the cometary and the asteroidal case. Two main differences, however, exist. In the asteroidal case, collisions can occur at any point on the orbit but with aphelion, being nearer the asteroidal belt, perhaps more likely, whereas for comets activity is higher near perihelion and so dust ejection is much more likely near that locality. Second, cometary ejection takes place at every perihelion passage so that new meteoroids are released into the stream on every orbit while the asteroidal process is a 'one-off'event. Within the context of this discussion, the mechanism for ejecting meteoroids from their parent does not matter; we are concerned more with the dynamics of the consequential motion, based on the changes generated in the orbital energy and angular momentum.

If a meteoroid is to become a meteor, it must intersect the Earth's atmosphere, ablate and produce a visible trail. However, it is equally evident that the parent bodies of most streams have not collided with the Earth. Hence, the orbit of the parent can not be identical to those of the meteoroids. In this chapter, we discuss the processes that change the orbit from that of the original parent to the observed meteoroid orbit. This can be thought of as occurring in three separate stages. First, there is a process by which the meteoroid is lost from the parent body, which results in the meteoroids having a velocity (usually small) relative to the parent. Second, as the dimensions of the meteoroids are very different from those of the parent, additional forces such as radiation pressure may now be important, so that the physics governing the orbit is different for the meteoroid. Finally, more long-term effects such as planetary perturbations and drag cause differential orbital evolution. We shall discuss these in turn and conclude with a discussion of our current state of knowledge and understanding.

2.2. The Effect of the Initial Ejection Velocity

Any velocity given to a newly ejected meteoroid relative to its parent body implies that the meteoroid will move on a different orbit to that of the parent. It is convenient to think of the new orbit as having changed in two ways relative to the parent orbit: a change to the orbital plane (in effect a change in the inclination i and the longitude of the ascending node Ω) and a change in the dimensions of the orbit within this new plane (in effect a change in semi-major axis a, eccentricity e and argument of perihelion ω). Other effect, to be discussed later, will cause subsequent further changes to these orbital parameters.

2.2.1. *Changes in the Orbital Plane*

Let us first consider the change to the orbital plane of the meteoroid relative to that of the parent body. To generate any change, it is necessary to change the angular momentum per unit mass. In order to consider such a change, it is sensible to use a non-moving reference frame, and we use the Cartesian heliocentric–ecliptic frame with the Sun at the origin, the x-axis towards the first point of Aeres and the z-axis orthogonal to the ecliptic. In this frame the angular momentum per unit mass, $\mathbf{h}$, is given by

$$\mathbf{h} = (h_x, h_y, h_z) = \mathbf{r} \times \mathbf{V}$$

$$= (y\dot{z} - z\dot{y}, z\dot{x} - x\dot{z}, x\dot{y} - y\dot{x}). \tag{2.2.1}$$

Also, in terms of the orbital parameters,

$$\pm h_x = h \sin i \sin \Omega, \tag{2.2.2}$$

$$\mp h_y = h \sin i \cos \Omega, \tag{2.2.3}$$

and

$$h_z = h \cos i, \tag{2.2.4}$$

where i is the inclination of the orbit, Ω the longitude of the ascending node and h the magnitude of $\mathbf{h}$.

Hence

$$\tan \Omega = -\frac{h_x}{h_y}. \tag{2.2.5}$$

From this equation, it is easy to obtain

$$\sec^2 \Omega \Delta \Omega = -\left(\frac{\Delta h_x}{h_y} - \frac{h_x \Delta h_y}{h_y^2}\right) \tag{2.2.6}$$

$$= \tan \Omega \left(\frac{\Delta h_x}{h_x} - \frac{\Delta h_y}{h_y}\right), \tag{2.2.7}$$

where Δ represents the change in the relevant quantity between the meteoroid orbit and the parent orbit that was caused through the ejection process, that is, the initial velocity difference. Thus

$$\Delta \Omega = \sin \Omega \cos \Omega \left(\frac{\Delta h_x}{h_x} - \frac{\Delta h_y}{h_y}\right) \tag{2.2.8}$$

$$= -\frac{1}{h^2 \sin^2 i}(h_y \Delta h_x - h_x \Delta h_y), \tag{2.2.9}$$

on making use of equations (2.2.2) and (2.2.3).

But

$$\Delta h_x = y\Delta \dot{z} - z\Delta \dot{y}, \tag{2.2.10}$$

$$\Delta h_y = z\Delta \dot{x} - x\Delta \dot{z}. \tag{2.2.11}$$

since only the velocity, but not the position, is changed through the ejection process. Substituting Δh_x and Δh_y into equation (2.2.9), we have

$$\Delta \Omega = \frac{z}{h^2 \sin^2 i}(\mathbf{h} \cdot \Delta \mathbf{V}) = \frac{r \sin(\omega + f)}{h \sin i} v \sin \phi, \tag{2.2.12}$$

Here, f is the true anomaly of the point at which the ejection of the meteoroid occurred, r is the corresponding heliocentric distance, $\Delta \Omega$ represents the difference in the ascending node between the orbit of the meteoroid and that of the parent, v is the magnitude of the ejection velocity, ϕ is the angle between the direction of ejection velocity and the orbital plane of the parent body so that $v \sin \phi$ is the component of the ejection velocity perpendicular to the orbital plane of the parent.

Equation (2.2.12) indicates that, as one might expect, the ejection velocity (that is, both magnitude and direction) and the location of the ejection point on the orbit of the parent all affect the the longitude of the node of the new meteoroid orbit.

We note the obvious, namely that if $v \sin \phi$ is zero, there is no change in $\Delta \Omega$. More interestingly, we also note that at $f = -\omega$ and $\pi - \omega$, that is, at the two nodes of the cometsry orbit, $\Delta \Omega$ is also zero. Hence, if meteoroids are ejected at the nodes, irrespective of how large or small $v \sin \phi$ is, there is no change in Ω.

From equation (2.2.12) we can get a rough guide to the likely value of $\Delta \Omega$ fairly simply. It is generally likely that most of the meteoroids are ejected from the region close to the point of maximum activity of the comet, which is around perihelion. In this case, the heliocentric distance and orbital speed are nearly orthogonal to each other so that their product gives the the angular momentum per unit mass, h. In this event, apart from the sine of three angles, f, ϕ and i, $\Delta \Omega$ is simply the ratio of the ejection speed to the orbital speed, perhaps about 0.2%. Thus, remembering that in equation (2.2.12), $\Delta \Omega$ is in radians, its value in degrees is about 0.1, so that a meteor outburst, which is associated with a young stream where planetary perturbations have not had time to affect the stream, will generally last for a time of the order of 0.1 days, or a few hours. This is in excellent agreement with observations. We should, however, remember that for specific streams, the terms depending on the sine of either ω or i can be important, for example, for the Leonid meteor shower, $\sin \omega$ is only 0.13, while $\sin i$ is 0.3. Thus, in the Leonids, fortunately, the two terms almost cancel, leaving the original conclusion about the duration of the storm still valid.

The change in inclination Δi can also be obtained from the equation (2.2.4), giving

$$\tan i \Delta i = \frac{\Delta h}{h} - \frac{\Delta h_z}{h_z}, \tag{2.2.13}$$

and

$$h^2 = h_x{}^2 + h_y{}^2 + h_z{}^2 \tag{2.2.14}$$

For the general case, the algebra gets unwieldy, but it is clear that for any specific case Δi can be evaluated. In general, Δi is also small.

2.2.2. Changes to the Orbit

In addition to changing the orbital plane, the ejection velocity also changes the remaining orbital parameters, namely the semi-major axis a, the eccentricity e and the argument of perihelion ω. To a reasonable approximation, meteoroids move on Keplerian ellipses about the Sun, so that standard theory gives us the orbital energy per unit mass, U, as

$$\mathsf{U} = -\frac{GM_\odot}{2a}, \tag{2.2.15}$$

and that

$$P^2 = a^3, \tag{2.2.16}$$

where P is the orbital period, measured in years provided a is measured in astronomical units. Hence, we can obtain

$$\frac{\Delta \mathsf{U}}{\mathsf{U}} = -\frac{\Delta a}{a} = -\frac{2 \Delta P}{3P}. \tag{2.2.17}$$

If, as above, we denote the ejection velocity by $\mathbf{v}$, then the component in the orbital plane is $v \cos \phi$. It is convenient to consider this in two components, along the direction of the parent body's motion and perpendicular to this as $v \cos \theta \cos \phi$ and $v \sin \theta \cos \phi$, so that θ is the angle between the tangent to the parent's orbit and the direction of the ejection velocity component in the orbital plane. In a heliocentric frame, with V denoting the speed of the parent, twice the kinetic energy of the meteoroid is given by

$$(v \sin \phi)^2 + (v \cos \phi \sin \theta)^2 + (V + v \cos \phi \cos \theta)^2. \qquad (2.2.18)$$

Since the location of the parent body and the meteoroid is the same at this point, there is no change in potential energy per unit mass, hence

$$2\Delta \mathsf{U} = v^2 + 2vV \cos \theta \cos \phi \qquad (2.2.19)$$

represents the change in orbital energy per unit mass between the parent body and the meteoroid.

As mentioned already v is likely to be much less than V and so for most streams,

$$\Delta \mathsf{U} = vV \cos \theta \cos \phi. \qquad (2.2.20)$$

Inserting this into equation (2.2.15) and remembering that

$$V^2 = GM_\odot(2/r - 1/a) \qquad (2.2.21)$$

we obtain

$$\frac{\Delta a}{a} = 2\frac{v}{V} \cos \theta \cos \phi(\frac{2a}{r} - 1) \qquad (2.2.22)$$

In order to obtain some feeling of what the outcome is, if as before we take v/V as 0.2%, a as 10 AU and $\cos \theta$ and $\cos \phi$ both as 0.5, then Δa is about 0.2 AU. Hence the meteoroid stream will indeed follow an orbit that is hardly distinguishable from that of the original parent body. Again from equation (2.2.17), we can obtain $\Delta P/P$ as about 0.03, implying that after about 30 orbits, meteoroids will have deviated from the parent by a complete orbit, in other words meteors will be seen every year.

The angular momentum per unit mass, h, is also given by standard orbital theory as

$$h^2 = GM_\odot p,$$

where p is the semi-parameter of the orbit, that is $p = a(1 - e^2)$.
This yields

$$\frac{\Delta h}{h} = \frac{\Delta p}{2p} = \frac{\Delta a}{2a} - \frac{e\Delta e}{(1 - e^2)}. \qquad (2.2.23)$$

Unfortunately, the tangent to the orbit is not, in general, orthogonal to the radius vector, the angle between the radius vector and the velocity direction being ψ say, where

$$\sin \psi = \frac{1 + e \cos f}{(1 + e^2 + 2e \cos f)^{0.5}} \qquad (2.2.24)$$

This means that the direction of motion of the parent, which is the natural direction to select as one of our reference axis and used from equation (2.2.16) onwards, is not orthogonal to the solar direction and Δh, using this reference frame, is given by

$$\Delta h = rv(\cos \theta \cos \phi \sin \psi + \cos \phi \sin \theta \cos \psi) = rv \cos \phi \sin(\theta + \psi) \qquad (2.2.25)$$

For most realistic situations, as we have stated several times already, ejection is more likely to occur close to perihelion, that is at small values of f. From equation (2.2.24) we see that in this event $\sin \psi$ is close to unity so that ψ is very close to $\pi/2$.

Inserting our previously adopted typical values into equations (2.2.23) and (2.2.25) shows that Δe is also small again showing that the orbit of the meteoroid is initially very similar to that of the comet.

Wo also know that the point of ejection is on both the orbit of the parent body and the orbit of the ejected meteoroid. This gives

$$e \cos(f + \Delta\omega) - e \cos f = \frac{(2e + e^2 \cos f + \cos f)\,\Delta p}{2e}\frac{\Delta p}{p} - \frac{1 - e^2}{2e} \cos f \frac{\Delta a}{a} \quad (2.2.26)$$

from which $\Delta\omega$ can be determined for a given situation. In general, $\Delta\omega$ is also small.

In fact, the nodal distance r_N, is of more significance for determining the observability of a meteor shower than any of the fundamental orbital parameters as this has to be equal to the heliocentric distance of the Earth in order that a meteor shower is observed on Earth. The nodal distances are derived from the standard equation for an ellipse with the true anomaly being taken as $-\omega$ or $\pi - \omega$, that is

$$(1 - e \cos\omega)r_N = p \quad (2.2.27)$$

and

$$(1 + e \cos\omega)r_N = p. \quad (2.2.28)$$

Hence, we can obtain, remembering that $p = a(1 - e^2)$,

$$(1 - e \cos\omega)\frac{\Delta r_N}{r_N} = \frac{1 - e^2}{2e} \cos\omega \frac{\Delta a}{a} - \frac{(e^2 \cos\omega + \cos\omega - 2e)}{2e}\frac{\Delta p}{p} - e \sin\omega\Delta\omega \quad (2.2.29)$$

for the first node with a similar equation for the other node.

Hence, from considerations of the ejection velocity alone, all the orbital parameters of the new orbit are determinable. They are, as we expected, not very different from the orbital parameters of the parent body. The only two quantities that are important are the Period P and the nodal distance r_N, for small changes in either of these are important. In the first case this is because they are additive, and in the second because any small miss of the Earth by the meteoroid implies that no meteor is seen.

This then defines the situation immediately after ejection. Unfortunately for the study of meteor showers, this situation changes due to other effects and we now consider these.

2.3. Additional Forces

So far, we have consider only the change to the Keplerian orbit that comes about due to the meteoroid being given some velocity relative to the parent body. Meteoroids are in general small so that Solar radiation has some effect on their motion.

The most obvious effect of radiation is to produce a radially outwards force on the meteoroids, usually called the *radiation pressure* force. Since it linearly depends on the amount of radiation received from the Sun, it is also, as gravity, a force that is inversely proportional to the square of the heliocentric distance. The simplest way of dealing with this is to regard the gravitational constant G as being replaced by $G(1 - \beta)$, where β is the ration of the gravitational force to the radiation pressure. For a spherical meteoroid of radius b and bulk density σ, β is a non-dimensional constant whose numerical value depends on the solar luminosity, solar mass and the gravitational constant G. With b measured in centimetres and σ also in *cgs* units,

$$\beta = 5.75 \times 10^{-5}/(b\sigma). \quad (2.3.1)$$

One consequence of including radiation pressure is self evident, if $\beta \geq 1$, then the net force is outwards and the meteoroid will be quickly lost from the Solar System. Assuming that σ is around 0.5 g cm^{-3}, we see from the above equation that sub-micron sized meteoroids will be lost.

However, somewhat larger meteoroids can be lost from the Solar System due to the effects of radiation pressure. In Keplerian motion, the speed v at heliocentric distance r, we have

$$v^2 = GM_\odot \left(\frac{2}{r} - \frac{1}{a} \right) \qquad (2.3.2)$$

while for motion under gravity and radiation,

$$v^2 = G(1 - \beta)M_\odot \left(\frac{2}{r} - \frac{1}{a_1} \right). \qquad (2.3.3)$$

But immediately after ejection, v and r are the same, hence, as was first shown by Kresák (1976) the semi-major axis a_1 is given by

$$a_1 = ar(1 - \beta)(r - 2a\beta)^{-1}, \qquad (2.3.4)$$

while conservation of angular momentum allows the new eccentricity also to be determined from

$$e_1{}^2 = 1 - (1 - e^2)(r - 2a\beta)(1 - \beta)^{-2}r^{-1}. \qquad (2.3.5)$$

These equations show that when $\beta \geq r/2a$, a becomes infinite and e becomes unity; in other words the meteoroid has effectively escaped from the Solar System. The dimensions of the meteoroids that are lost through this mechanism depend on where they were ejected. At perihelion, r is at its minimum value, thus also giving the smallest value for β, or the largest meteoroids that are lost. For example for the Leonids, meteoroids with $\beta \geq 0.045$ ejected at perihelion will be lost. This corresponds to a meteoroid radius of about 25 microns.

From the point of view of the formation and evolution of meteoroid streams, any meteoroids that are lost are unimportant since they can play no further part. However, those meteoroids that remain are also affected by radiation pressure and their orbits also satisfy equations (2.3.4) and (2.3.5). These imply that the semi-major axis is increased and in consequence also the orbital period, with the meteoroid orbital period P_1 being given by

$$P_1 = P(1 - \beta)\left(1 - \frac{2a\beta}{r} \right)^{-3/2}. \qquad (2.3.6)$$

The right hand side of this equation can be expanded for small β to give

$$P_1 = P\left[1 + \beta\left(\frac{3a}{r} - 1 \right) \right] \qquad (2.3.7)$$

The effects considered so far come into play as soon as the meteoroid is ejected; that is, they define the initial orbit of the meteoroid. This orbit will not be significantly different from that of the parent body, but even small changes in the nodal distance and period will have a big effect on the observability of any consequent meteor shower. This initial orbit will be subject to a number of evolutionary effects and we now consider these.

2.4. The Evolution of the Meteoroid Orbit

There are a number of effects that causes the orbit of a meteoroid to evolve from its initial state, the two main ones that are generally included in any modelling being the Poynting–Robertson drag and planetary perturbations, though some have claimed that the Yarkovsky effect can also be important. The Yarkovsky effect is a force on a body caused because of anisotropic re-emission of radiation following from more radiation falling on the sunward face and rotation causing emission to be in a slightly different direction. According to Öpik (1951) this was first proposed by I.O. Yarkovsky in 1900 in a paper that was apparently lost. It has since been investigated by others including Burns *et al.* (1979). The effect of this force on asteroidal fragments has been investigated by several authors, for example Farinella *et al.* (1998), Bottke *et al.* (2000). For very small meteoroids, the rotation rate is high and randomly varying so that the effect becomes less important. Farinella *et al.* (1998) and Vokrouhlicky & Farinella (2000) state that the equations break down at sizes below 10 cm and ignored such small meteoroids in their discussion. In general, this effect has been ignored in discussion of the meteoroid streams.

2.4.1. *The Poynting–Robertson Drag*

For any moving body orbiting the Sun, radiation is absorbed in a reference frame that is stationary relative to the Sun. The resulting emission of radiation is, however, in a frame stationary on the body, and hence moving in relation to the Sun. Hence, momentum (relative to the Sun) is carried away by this radiation and this loss of momentum can be regarded as a drag since its effect is velocity dependent. The existence of this effect was first pointed out by Poynting (1903) and quantified for motion in the Solar System within a relativistic framework by Robertson (1937). Mathematical expressions for the rate of change of the orbital parameters have been derived by Wyatt & Whipple (1950) in particular,

$$\frac{da}{dt} = -\frac{\gamma(2+3e^2)}{a(1-e^2)^{3/2}} \tag{2.4.1}$$

and

$$\frac{de}{dt} = \frac{-2.5\gamma e}{2(1-e^2)^{1/2}a^2}. \tag{2.4.2}$$

Here,

$$\gamma = \frac{GM_\odot\beta}{c},$$

where c denotes the speed of light and β is as defined in equation (2.3.1).

Unfortunately, analytical solutions for a and e in terms of time do not exists. However, we can obtain a relationship between a and e. Dividing the two equations above gives

$$\frac{da}{de} = \frac{2a(2+3e^2)}{5e(1-e^2)}, \tag{2.4.3}$$

a differential equation where the variables are separable and which can be integrated to give

$$p = a(1-e^2) = Ce^{4/5}, \tag{2.4.4}$$

where C is a constant of integration and p is again the semi-parameter.

This equation was used for example by Arter & Williams (1997a) to show that a group of small meteoroids on smaller and more circular orbits than the April Lyrids were indeed related to this stream. We can however deduce a time scale for the variation of orbits in the inner solar system of the order of a^2/γ, or of the order of $b\sigma 10^7$ years. Major changes due to the Poynting–Robertson effect thus probably require a longer time than the age of most streams. However, since the only relevant criterion for a meteoroid is whether or not it hits the Earth, then changes may not have to be that significant so that an important change may take a factor of order 1000 less than implied above.

2.5. Planetary Perturbations

Since each planet generates a gravitational field, the motions of meteoroids are affected by the existence of planets. Further, since the field is proportional to the inverse square of the separation distance between the planet and meteoroid, planets that pass close to a meteoroid stream, for example the Earth, can have a bigger influence than consideration of their mass alone would suggest. Of course, planetary perturbations are in principle easy to understand. If the positions of all the planets and the meteoroids are known at a given instant of time, then the force on the meteoroid is calculable and hence the equations of motion of the meteoroid are determined at this instant in time.

The problem can be split into two parts, determining the positions of the planets at the given instant and determining the responses of the meteoroids to the gravitational field of these planets. Neither of the two problems has a solution that can be represented in a closed algebraic form. The first of these problems has in practice been approached in one of three ways. The positions of the planets can be regarded as known at fixed times from published ephemerides and the positions at the required times are then obtained by interpolation. Some averaging procedures over one orbit can be performed so that only a secular component for the perturbations is obtained and used to evaluate the effect on the meteoroids. Finally, the motions of the planets are integrated along with those of the meteoroids. The first method was in essence the same as the third, it is just that the computations of planetary positions have been carried out by somebody else. To be efficient it needs a large memory and in practice it is found to be generally less efficient than the third method.

2.5.1. *Secular Perturbation Methods*

The problem with all Solar System dynamical calculations is that the true anomaly, f, is not an algebraic function of time for motion on elliptical orbits. In fact, the eccentric anomaly, E, and time are related by Kepler's equation,

$$E - e \sin E = n(t - \tau), \tag{2.5.1}$$

where as before e is the eccentricity of the orbit, t the time and τ is the time of perihelion passage. The mean motion, n, is given by $n = (2\pi)/P$, where P is the orbital period. The eccentric anomaly and true anomaly are related by

$$\cos f = \frac{\cos E - e}{1 - e \cos E}, \tag{2.5.2}$$

and the heliocentric distance r is given by

$$r = a(1 - e \cos E) = p(1 - e \cos f)^{-1}, \tag{2.5.3}$$

where again a is the semi-major axis and p the semi-parameter of the ellipse. In order to evaluate the force on a meteoroid, we need the inverse of the distance from it to the relevant planets at any given time. Hence, a solution to equation (2.5.1) is needed. This can be obtained by some iterative scheme such as the Newton–Raphson method. Alternatively, in a semi-analytical approach, power series expansions are obtained for r^{-1}. Secular perturbation theories integrate over one orbit so that periodic terms vanish. For low values of the eccentricity, obtaining power series that are accurate after the inclusion of a few terms is easy but the problem becomes harder for high eccentricity. Unfortunately, for meteoroid streams, high values are the norm. This presented a severe difficulty as far as following the evolution of meteoroid streams was concerned until about half a century ago when Brouwer (1947) produced a mathematical algorithm that could be applied even for orbits of high eccentricity. This was used by Whipple & Hamid

(1950) to follow the evolution of the mean Taurid stream over an interval of 4700 years. By doing this, they demonstrating a similarity with the orbit of comet Encke in the past and claimed that the stream and comet were thus related.

At this time, computing resources and power were both scarce and so secular perturbation methods were the prime method for investigating the evolution of meteoroid streams under the effects of planetary perturbations. Plavec (1950) used the Gauss–Hill method to study the evolution of the Geminid stream, correctly concluding that the stream was sweeping past the Earth's orbit, being first visible around 1700 and ceasing to form meteor showers around the year 2100. The most popular secular perturbation method of the period was the Halphen–Goryachev method described for example in Hagihara (1972). This was used by Galibina (1972) and Galibina & Terentjeva (1980) to determine the effect of gravitational perturbations on the stability of a number of meteoroid streams (i.e. continued appearance of showers) over a time interval of tens of thousands of years. They reached the rather unsurprising conclusion that streams passing very close to Jupiter became unstable, while the effects of Jupiter could also be discerned in a 3500 year periodicity in most of the meteoroid orbital elements.

Rather than considering the mean meteoroid stream orbit, Zausaev (1972) took three actual meteoroid orbits from the Quadrantid stream and three from the δ Aquarid stream and followed the motion back for 4000 years, concluding that within this time interval there was no similarity between the streams. (Today this may appear to be a rather curious investigation, but it has to be seen within the context of the attempt to identify the parent body of all the major streams, and of the Quadrantids in particular.)

A significant volume of work using the Halphen–Goryachev method was carried out by Babadzhanov and Obrubov, who published mostly in Russian, but some works can also be found in English, examples being Babadzhanov & Obrubov (1980) and Babadzhanov & Obrubov (1983). In these works they investigated the evolution of both the Geminid and the Quadrantid streams and concluded that though both streams had existed as streams for thousands of years, they had only started intersecting the Earth about 100 years ago. This would appear to be in good agreement with observations, for King (1926) claimed that the Geminid shower was first seen in 1862, while Wartmann (1841) similarly claimed that the Quadrantids had first been observed in 1835. See also Fisher (1930) for a history of the Quadrantids up to 1927. We shall return to this point regarding the rapid evolution of this stream later when discussing direct integration methods.

The major drawback of the secular perturbation methods is that it deals with orbits rather than meteoroids. Since perturbations are averaged over an orbit, the true anomaly of a meteoroid plays no part in integration, the same answer is obtained for a meteoroid at its node or at its aphelion point when the Earth is at the node, a clearly absurd state of affairs for the meteoroids at the node, but not otherwise. Hence, the method is good for the evolution of the mean orbit and the majority of meteoroids, but probably not so good for the interesting meteoroids that form meteor storms or outbursts for example. Its great advantage at the time when it was primarily used was its speed, as computing power simply was not available to follow the evolution of individual meteoroids.

2.5.2. *Direct Integration Methods*

With the improvement in both the speed and memory of computers, direct numerical integration methods gained in popularity. Even without the aid of electronic computers, such methods had been used to follow the orbital evolution of a single body, usually a comet. Cowell developed a method that was used extensively to follow the evolution of comet $1P/Halley$ prior to it appearance in 1908, for example by Cowell & Crommelin (1907).

In 1970, Sherbaum (1970) adapted Cowell's method so that it could run on an electronic computer. This program was used by Levin *et al.* (1972) to investigate the effect of Jupiter on the structure of meteoroid streams, concluding that the perturbations increased the width of the stream. (Again, in the light of our knowledge today, this conclusion appears rather unremarkable, but it has to be viewed in the light of knowledge at the time, when some had argued that meteoroid streams accumulated into comets rather than the reverse. The arguments for this are given for example in Alfvén & Arrhenius (1976).) Kazimirchak-Polonskaya *et al.* (1972) also used Cowell's method to investigate the evolution of two streams, the α *Capricornids* and the α *Virginids*. The initial positions and velocities of a small number of meteoroids (10 α Virginids and 5 α Capricornids) were assumed and their subsequent motion integrated over an interval of 100 years.

The above were not the first investigations to use direct numerical integration methods; that honour probably belongs to Hamid & Youssef (1963), where the motion of six actual Quadrantid meteoroids was investigated. At that time, the main constraining factor was computing capacity. As this increased, so did the number of model meteoroids included and the length of time interval over which the evolution was followed. In 1979 Williams *et al.* (1979) used ten test meteoroids, but by the beginning of the 1980s, Hughes *et al.* (1981) had increased this to over 200, and two years later, Fox *et al.* (1983) had increased the number of meteoroids to 500 000. Since that time, the direct approach has become the main tool to investigate meteoroid stream evolution, with more test meteoroids being used and integrated over longer time intervals. Examples of numerical modelling for meteoroid streams are Hunt *et al.* (1985), Froeschlé & Scholl (1986), Jones & McIntosh (1986), Gustafson (1989b), Williams & Wu (1993a), Wu & Williams (1995a), Brown & Jones (1996), Steel & Asher (1996), Arter & Williams (1997b) and Jenniskens & van Leeuwen (1997). We will discuss the scientific results obtained from these and other investigations in a later section. For now, we concentrate on the methodology. All such numerical methods involve the numerical integration of second order differential equations. We now discuss such methods.

2.6. Numerical Integration Methods

The problems of integration within the context of the motion of meteoroids are all of the *initial value* kind, that is, the position R and velocity $\dot{R}$ of each body in the system is assumed to be known at some given time, t_0 say. The rates of change of each variable only depend on the position and velocity, and so they can also be calculated at this instant. Hence, in its simplest forms (Euler's method) we can evaluate the new value of each variable from

$$\mathbf{R}_1 = \mathbf{R}_0 + \frac{d\mathbf{R}}{dt}\Delta t \tag{2.6.1}$$

This gives the values of all the parameters, an interval of time Δt later and the whole process can be repeated.

Doing exactly what is outlined above is very inefficient, requiring Δt to be very small to retain any accuracy. Thus, more efficient methods have been developed. These fall into three broad categories and are briefly discussed in turn.

2.6.1. *Single Step Methods*

These methods are refined ways of doing exactly what was described above, namely progressing by a single step from the situation at t to the situation at $t + \delta t$. By far the best known method within the third category is the Runge–Kutta method. This works on the principle that, having obtained a rough solution for $\mathbf{R}_1$ for all meteoroids from equation (2.6.01) above, the estimates of $d\mathbf{R}/dt$ at other points within the time interval can be obtained and some average of these used to obtain the best estimate of $\mathbf{R}_1$. A very popular method is the Runge–Kutta 4th Order, and a description of this method can be found in any standard book on numerical analysis. This method was used in early work such as Williams *et al.* (1979) and Jones & McIntosh (1986).

However, it is possible to generate much higher order Runge–Kutta methods, that is, methods where the error generated is proportional to a much higher order in Δt, allowing far less computation. Also, variants such as the Runge–Kutta–Nystrom were developed, where by comparing the results for two orders such as the nth and the $(n + 1)^{\text{th}}$, the most efficient step size Δt is automatically selected. Arter & Williams (1997b) used this method with $n = 7$. An even higher order method was developed by Dormand El-Mikkaway & Prince (1987) and was used by Williams & Collander-Brown (1998), and Williams & Wu (1993a). In meteoroid stream problems, being able to vary the step size is not such a great advantage since the length is determined by the fastest changing set of parameters. In these problems, the relevant parameters are those of the Earth.

2.6.2. *Predictor–Corrector Methods*

The basic difference between these methods and the single step methods is that information from more than one time point in the past is used in order to make a prediction about the value at the next time point in the future as opposed to the single step methods using only one. This predicted value is then incorporated into a different formula in order to correct the initial prediction. The essential numerical analysis for this method was developed during the nineteenth century, when the actual calculations had to be carried out by hand. These mathematical formulations were subsequently modified in various ways to make use of the developing computer technology.

A very successful program was developed by Everhart (1969), which combined accurate determination of the evolution of a set of orbits by the above methods and a Monte-Carlo random walk to investigate the orbital evolution of thousands of comets, but which could equally be applied to meteoroid streams during the nineteenth century when the actual calculations had to be carried out by hand. A variation was developed specifically to deal with second order differential equations, called the Gauss–Jackson method. This was used by Fox *et al.* (1982) to investigate the nodal retrogression rate of the Geminids and by Fox & Williams (1984) to investigate fireballs seen in December.

An integration scheme was also developed by Bulirsch & Stoer (1966) and this was used by Froeschlé & Scholl (1986) to show that the Quadrantid stream, experiencing close encounters with Jupiter, was behaving chaotically.

All these methods are very efficient but have the drawback that data are required at more than one initial point. Hence, any modelling must start by using a method from the first subsection. In a general way, a second disadvantage is the inability to vary the step length continuously. Within the context of investigating meteoroid streams, this is not a big drawback since the Earth has to be included in the modelling and so the step length is likely to be determined by the Earth's orbit.

2.6.3. *Taylor Series Methods*

Methods based on obtaining a Taylor series give a very elegant solution. Consider the differential equation

$$\frac{dr}{dt} = f(r, t). \tag{2.6.2}$$

Then,

$$rt + \Delta t = r(t) + f(r, t)\Delta t + f'(r, t)(\Delta t^2)/2 + \cdots \tag{2.6.3}$$

and, if f can be differentiated analytically several times, then a very accurate determination of r at a later time Δt is obtained.

In the problems that we are considering, f can indeed be differentiated but the expressions are unwieldy. For this reason the Taylor series methods have not been used much. Perhaps the timing was unfortunate in that the need to integrate the equations became apparent before such algebraic packages as *MAPLE* or *Mathematica* became widely available. We should note that the Taylor series method was used in project LONGSTOP, the long-term integration of the outer planets (see Roy *et al.* (1988)).

By using any of the above-mentioned numerical techniques, the evolution of a given stream over a given time interval can easily be modelled and this has already been done for most major streams and we will discuss some of the results in the next section. There are, however, a number of problems with such numerical simulations; in particular, a numerical simulation can never include the correct number of test particles. Hughes & McBride (1989) have estimated the masses of some streams and it is clear that a meteoroid stream must consist of at least 10^{16} meteoroids, 10^{17} being more realistic. Hence, even in a simulation with say a million test meteoroids, each test meteoroid represents 10^{10} actual meteoroids and an observed flux of say $100\,000$ per hour, corresponding to a storm like the Leonids at its peak, is thus not even represented by one test meteoroid. Of course we can improve on this situation by not spreading our test particles uniformly in space and time, but in the end, storms like that of the Leonids are still represented by only a small number of test meteoroids.

2.7. The Current Situation

The first use that was made of the evolutionary methods described above was to investigate a potential link between a meteoroid stream and a comet, in this case between the Taurid stream and comet 2P/Encke by Whipple & Hamid (1950). The logic is simple; if the suspected parent and the stream are not on similar orbits now, this may be because their orbital evolution has been different so that integrating back will reveal a similarity. A particularly interesting case, still not fully resolved, concerns the Quadrantid stream. This stream causes one of the more active and regular showers to appear every January, though the first recorded sighting was not until 1798 (Hasegawa (1993)), suggesting that this stream is subject to a rather rapid evolution, while Jenniskens *et al.* (1997) have concluded that the observed stream is very young, only a few hundred years. Hence, the parent comet should be visible, but as evolution is rapid, it may be on a different orbit. The first suggestion regarding the pairing was by Hasegawa (1979), suggesting the comet 1491 I. Williams & Wu (1993b) integrated the mean orbit of the Quadrantids back and found a reasonable similarity to those of the comet. However, McIntosh (1990) suggested that another comet, 96P/Machholtz was the parent. Gonczi *et al.* (1992), Jones & Jones (1993) and Babadzhanov & Obrubov (1992) all concurred with this second identification. Just to confuse the issue further, Williams & Collander-Brown (1998) suggested that asteroid 5496 could be the missing parent.

The problem with uniquely determining the parent of the Quadrantids is that the stream is very rapidly evolving. The reason for this is not perhaps hard to find. The orbital elements taken from Wu & Williams (1992a) are

$$a = 3.11, e = 0.681, i = 71°.4, \Omega = 282°.17 \text{ and } \omega = 169°.45.$$

The most obvious fact to emerge from these data is that the aphelion distance, Q, is 5.227 AU, only marginally greater than the semi-major axis. Further, the mean orbit is moderately close to being in a 2:1 mean motion resonance with Jupiter. In fact, Wu

& Williams (1992a) identified a number of sub-groupings within the Quadrantid stream and many of these are closer to mean motion resonances. Wu & Williams (1992b) found that six mean motion resonances could be playing an important part in the evolution of the Quadrantids. The first investigation to suggest that Quadrantid meteoroids could behave significantly different from each other due to the proximity of Jupiter was Hughes *et al.* (1981), where large differences between the nodal retrogression rate of test particles were found. At about the same time, Froeschlé & Scholl (1986) suggested that the Quadrantids were behaving chaotically, an idea developed further in Froeschlé & Scholl (1986). In another investigation, Wu & Williams (1995b) the effects of mean motion resonance were found in the Perseid stream, gaps akin to the Kirkwood gaps in the Asteroid belt being clearly present in the semi-major axes (and hence period) distribution. Yet a further example of the importance of mean motion resonances in the evolution of meteoroid streams came following the Leonid storm observed in 1998. Most pre-appearance predictions for November 1998 suggested that if a storm did occur it would be coincident with, or just after, the normal Leonid peak activity. In fact, an outburst became visible about 16 hours before the traditional peak and Asher *et al.* (1999) showed that these were meteoroids ejected back in 1333 and trapped in a resonance.

As already mentioned, until about the mid 1970s much of the work on orbital evolution, because of the constraints placed by the available computing facilities, were concerned with changes to the mean orbit of the meteoroid shower, in particular the progression or retrogression of the nodes. This quantity is of course easily observed. Plavec (1950) had predicted a retrogression rate of 1.63 degrees per century for the Geminid stream, a remarkably similar value to that of 1.6 obtained much later by Babadzhanov & Obrubov (1980) and 1.57 obtained by Fox *et al.* (1982). This shows a commendable consistency in the results obtained from different orbital integration methods, but, as was pointed out by Fox *et al.* (1982) it also reveals a worrying problem. Observations of the Geminids since 1893 show no change in the nodal position. Since excellent agreement between integrations and observations of the nodal retrogression rate has been obtained for other streams (for example Hughes *et al.* (1979)), the results suggested some peculiarity with the behaviour of the Geminid stream. Fox *et al.* (1982) investigated a number of possibilities to account for this behaviour; they included general relativistic effects and perturbations due to asteroids, to no avail. They suggested a possible explanation, namely that the actual cross section of the Geminids was elongated with the motion of this cross section across the ecliptic not being parallel to its long axis. In this way, while the centre of the cross-sectional area retrogressed, the centre of the path of the Earth through this did not. With increasing computing power, Fox *et al.* (1983) were able to produce a theoretical cross section based on the meteoroid ejection model proposed by Whipple (1951). This was indeed elongated in precisely the way required to explain the nodal retrogression problem. This work also paved the way for theoretical cross sections of other streams to be produced and this is now common practice (see for example, Jones (1985), Williams & Wu (1993a), Williams & Wu (1994), Brown & Jones (1996), Arter & Williams (1997b) and McNaught & Asher (1999)). From such cross sections, predictions regarding both the time of maximum activity in a shower and its maximum level of activity (which is some measure of the Zenithal hourly rate) can be obtained.

This ability to produce a theoretical cross section, especially the ability to do this at small spatial scales demonstrated by Asher (1999), proved to be very useful in discussing the Leonid outburst of 1999. It was this capability that allowed McNaught & Asher (1999) to make their remarkably accurate prediction of the storm in 1999.

Both in the scientific study of meteors and in the perception of the the general public, meteor storms, or outbursts, have always been important. It was the Leonid storms of

1799 and 1833 that got the subject started in the nineteenth century. Over the last decade or so, if we take the definition of an outburst given by Jenniskens (1995), namely *an event where the activity stands out significantly above the random variations of annual activity*, we have been fortunate in that a number of such outbursts have been seen. In consequence, a significant amount of new evidence has been gathered. Outbursts can occur for a variety of reasons, but within this discussion, we will concentrate only on those that are the result of the parent body being near to the Earth at the relevant time. In this situation we observe very young meteoroids, possibly young enough that to first order, planetary perturbations can be neglected. In this case any dispersion in the orbits is due primarily to the ejection velocity and thus, observations of the outburst allows a limit to be placed on this. This was recently investigated by Ma & Williams (2001) who found that the ejection velocity in both the Perseid and the Leonid outbursts were of the order of 70–80 m s^{-1}. In the past a very wide range of ejection velocities have been used by authors, from as low as 25 m s^{-1} by Asher (1999), 40 m s^{-1} by Crifo (1995), 100 m s^{-1} by Hughes (2000), a mean of 150 m s^{-1} by Wu & Williams (1996) and a high of 600 m s^{-1} by Harris *et al.* (1995).

2.8. Conclusions

A meteoroid stream has three distinct phases: generation from a parent body, evolution as a set of freely moving particles, and death as a shower when the meteoroids ablate in the Earth's atmosphere. In this chapter, we have concentrated only on the middle stage, following the physical processes and the consequential dynamics. Hopefully, through observations of the last phase, this work can then be used to reach conclusions regarding the first phase.

REFERENCES

ADAMS, J. C. 1867. On the orbit of the November meteors, *Mon. Not. R. Astr. Soc.*, **27**, 247–252.

ALFVÉN, H. & ARRHENIUS, G. 1976. *Evolution of the Solar System*, NASA-SP345, Washington D.C.

ARTER, T. R. & WILLIAMS, I. P. 1997a. The mean orbit of the April Lyrids, *Mon. Not. R. Astr. Soc.*, **289**, 721–728.

ARTER, T. R. & WILLIAMS, I. P. 1997b. Periodic behaviour of the April Lyrids, *Mon. Not. R. Astr. Soc.*, **286**, 163–172.

ASHER, D. J. 1999. The Leonid Meteor storms of 1833 and 1966, *Mon. Not. R. Astr. Soc.*, **307**, 919–929.

ASHER, D. J., BAILEY, M. E. & EMEL'YANENKO, V. V. 1999. Resonant meteoroids from comet Tempel–Tuttle in 1333: the cause of the unexpected Leonid outburst in 1998, *Mon. Not. R. Astr. Soc.*, **304**, L53–L57.

BABADZHANOV, P. B. & OBRUBOV, Y. Y. 1980. Evolution of orbits and intersection condition with the Earth of Geminid and Quadrantid meteor Streams, in *Solid Particles in the Solar System* EDS HALLIDAY, I. & McINTOSH, B. A., D. Reidel, Dordrecht, 157–162.

BABADZHANOV, P. B. & OBRUBOV, Y. Y. 1983. Some features of evolution of meteor streams, in *Highlights in Astronomy*, ED WEST, R. M., D. Reidel, Dordrecht, 411–419.

BABADZHANOV, P. B. & OBRUBOV, Y. Y. 1992. Comet Machholtz 1986VII and the Quadrantid meteor swarm – orbital evolution and relationship, *Solar Sys. Res.*, **26**, 288.

BENZENBERG, J. F. & BRANDES, H. W. 1800. Versuch die entfernung, die geschwindigkeit und die bahn der sternschnuppen zu bestimmen, *Annalen der Phys*, **6**, 224–232.

BETLEM, H., TER KUILE, C. R., DE LIGNE, M., VAN'T LEVEN, J., JOBSE, K., MISKOTTE, K. & JENNISKENS, P. 1997. Precission meteor orbits obtained by the Dutch Meteor Society – Photographic Meteor Survey (1981–1993), *Astron. Astrophys. Suppl. Ser.*, **128**, 179–185.

BOTTKE, W. F., RUBINCAM, D. P. & BURNS, J. A. 2000. Dynamical evolution of main belt meteoroids: Numerical simulations incorporating planetary perturbations and the Yarkovsky thermal force, *Icarus*, **145**, 301–331.

BROUWER, D. 1947. Secular variations of the elements of Encke's comet, *Astron. Jl.*, **52**, 190–198.

BROWN, P. & JONES, J. 1996. Dynamics of the Leonid Meteoroid Stream: a numerical approach, in *Physics, Chemistry and Dynamics of Interplanetary Dust*, EDS GUSTAFSON, B. A. S. & HANNER, M. S., ASP Conf. Ser, 113–116.

BULIRSCH, R. & STOER, J. 1966. *Numer. Math.*, **8**, 1.

BURNS, J. A., LAMY, P. L. & SOTER, S. 1979. Radiation forces on small particles in the Solar System, *Icarus*, **40**, 1–48.

COOK, A. F. 1973. A working list of meteor streams, in *Evolutionary and Physical Properties of Meteoroids*, EDS HEMENWAY, C. L., MILLMAN, P. M. & COOK, A. F., NASA SP-319, Washington DC, 183–191.

CRIFO, J. F. 1995. A general physiochemical model of the inner coma of active comet I. Implications of spatially distributed gas and dust production, *Astrophys. Jl.*, **445**, 470–488.

COWELL, P. H. & CROMMELIN, A. C. D. 1907. The perturbations of Halley's comet in the past: First paper 1301–1531, *Mon. Not. R. Astr. Soc.*, **68**, 173–179.

DORMAND, R. J., EL-MIKKAWAY, M. E. A. & PRINCE, P. J. 1987. High order embedded Runge–Kutta Nystrom formulae, *IMA Jl. Numer. Anal.*, **7**, 423–430.

DRUMMOND, J. D. 1981. A test of comet and meteor shower associations, *Icarus*, **45**, 545–553.

EVERHART, E. 1969. Close encounters of comets and planets, *Astron. Jl.*, **74**, 735–750.

FARINELLA, P., VOKROULICKY, D. & HARTMANN, W. K. 1998. Meteorite Delivery via Yarkovsky orbital drift, *Icarus*, **132**, 378–387.

FISHER, W. J. 1930. The Quadrantid Meteors; history to 1927, *Circ. Harvard College Obs.*, No. **346**.

FOX, K. & WILLIAMS, I. P. 1984. A possible origin for some December fireballs, *Mon. Not. R. Astr. Soc.*, **217**, 407–411.

FOX, K., WILLIAMS, I. P. & HUGHES, D. W. 1982. The evolution of the orbit of the Geminid meteor stream, *Mon. Not. R. Astr. Soc.*, **199**, 313–324.

FOX, K., WILLIAMS, I. P. & HUGHES, D. W. 1983. The rate profile of the Geminid meteor shower, *Mon. Not. R. Astr. Soc.*, **205**, 1155–1169.

FROESCHLÉ, C. & SCHOLL, H. 1982. A systematic exploration of three dimensional asteroidal motion at the 2:1 resonance, *Astron. Astrophys.*, **111**, 346–356.

FROESCHLÉ, C. & SCHOLL, H. 1986. Gravitational splitting of Quadrantid-like meteor streams in resonance with Jupiter, *Astron. Astrophys.*, **158**, 259–265.

GALIBINA, I. V. 1972. Secular perturbations on the minor bodies of the solar system, in *The Motion, Evolution of Orbits, and Origin of Comets*, EDS CHEBOTAREV, G. A. KAZIMIRCHAK-POLONSKAYA, H. I. & MARSDEN, B. G., D. Reidel, Dordrecht, 440.

GALIBINA, I. V. & TERENTJEVA, A. K. 1980. Evolution of meteors over milenia, in *Solid Particles in the Solar System*, EDS HALLIDAY, I. & McINTOSH, B. A., D. Reidel, Dordrecht, 145–148.

GONCZI, R., RICKMAN, H. & FROESCHLÉ, C. 1992. The connection between comet P/Machholtz and the Quadrantid meteors, *Mon. Not. R. Astr. Soc.*, **254**, 627–634.

GUSTAFSON, B. Å. S. 1989a. Comet ejection and dynamics of nonspherical dust particles and meteoroids, *Astrophys. Jl.*, **337**, 945–949.

GUSTAFSON, B. Å. S. 1989b. Geminid meteoroids traced to cometary activity on Phaethon, *Astron. Astrophys*, **225**, 533–540.

HAGIHARA, Y. 1972. *Celestial Mechanics*, MIT, Cambridge.

HAMID, S. E. & YOUSSEF, M. N. 1963. A short note on the origin and age of the Quadrantids, *Smithson. Cont. Astrophys.*, **7**, 309–311.

HARRIS, N. W. & HUGHES, D. W. 1995. Perseid meteors: The relationship between mass and orbital semi-major axis, *Mon. Not. R. Astr*, **273**, 992–998.

HARRIS, N. W., YAU, K. C. & HUGHES, D. W. 1995. The true extent of the nodal distribution of the Perseid meteoroid stream, *Mon. Not. R. Astr*, **273**, 999–1015.

HASEGAWA, I. 1979. Orbits of ancient and medieval comets, *Pub. Astron. Soc. Japan*, **31**, 257–270.

HASEGAWA, I. 1993. Historical records of meteor showers, in *Meteoroids and their Parent Bodies*, EDS ŠTOHL, J. & WILLIAMS, I. P., Slovak Academy of Sciences, Bratislava, 209–223.

HERRICK, E. C. 1837. On the shooting stars of August 9th and 10th 1837, and on the probability of the annual occurrence of a meteoric shower in August, *American Jl. Sci.*, **33**, 176–180.

HERRICK, E. C. 1838. Further proof of an annual Meteoric Shower in August, with remarks on Shooting Stars in general, *American Jl. Sci.*, **33**, 354–364.

HUGHES, D. W. 2000. On the velocity of large cometary dust particles, *Plan. Sp. Sci.*, **48**, 1–7.

HUGHES, D. W. & MCBRIDE, N. 1989. The mass of meteoroid streams, *Mon. Not. R. Astr. Soc.*, **240**, 73–79.

HUGHES, D. W., WILLIAMS, I. P. & MURRAY, C. D. 1979. The orbital evolution of the Quadrantid meteor stream between AD 1830 and 2030, *Mon. Not. R. Astr. Soc.*, **189**, 493–500.

HUGHES, D. W., WILLIAMS, I. P. & FOX, K. 1981. The mass segregarion and nodal retrogression of the Quadrantid meteor stream, *Mon. Not. R. Astr. Soc.*, **195**, 625–637.

HUNT, J., WILLIAMS, I. P. & FOX, K. 1985. Planetery perturbations on the Geminid meteor stream, *Mon. Not. R. Astr. Soc.*, **217**, 533–538.

JENNISKENS, P. 1995. Meteor stream activity II Meteor outbursts, *Astron. Astrophys.*, **295**, 206–235.

JENNISKENS, P. & VAN LEEUWEN, G. D. 1997. The α-Monocerotid Meteor outburst: the cross section of a comet dust tail, *Planet. Space Sci.*, **45**, 1649–1652.

JENNISKENS, P., BETLEM, H. DE LIGNE, M., LANGBROEK, M. & VAN VLIET, M. 1997. Meteor stream activity V: the Quadrantids, a very young stream, *Astron. Astrophys.*, **327**, 1242–1252.

JONES, J. 1985. The structure of the Geminid meteor stream: I. The effect of planetar perturbations, *Mon. Not. R. Astr. Soc.*, **217**, 523–532.

JONES, J. & JONES, W. 1993. Comet Machholtz and the Quadrantid meteor stream, *Mon. Not. R. Astr. Soc.*, **261**, 605–611.

JONES, J. & MCINTOSH, B. A. 1986. On the structure of the Halley comet meteor stream, in *Exploration of Comet Halley*, ESA-SP 250, Paris, 233–243.

JOPEK, T. J. 1993. Remarks on the Meteor Orbital Similarity D-Criterion, *Icarus*, **106**, 603–607.

KAZIMIRCHAK-POLONSKAYA, E. I. BELYAEV, N. A. & TERENTEVA, A. K. 1972. Orbital evolution of the α Virginid and α Capriconid meteor streams, in *The Motion, Evolution of Orbits, and Origin of Comets*, EDS CHEBOTAREV, G. A., KAZIMIRCHAK-POLONSKAYA, H. I. & MARSDEN, B. G., D. Reidel, Dordrecht, 462–471.

KING, A. 1926. An ephemeris of the geminid radiant-point, *Mon. Not. R. Astr. Soc.*, **86**, 638–641.

KRESÁK, L. 1976. Orbital evolution of the dust streams released from comets, *Bull. Astron. Inst. Czech.*, **27**, 35–46.

LE VERRIER, U. J. J. 1867. Sur les etoiles filantes de 13 Novembre et du 10 Aout, *Comptes Rendus*, **64**, 94–99.

LEVIN, B. Y., SIMONENKO, A. N. & SHERBAUM, L. M. 1972. Deformation of a meteor stream caused by an approach to Jupiter, in *The Motion, Evolution of Orbits, and Origin of Comets* EDS CHEBOTAREV, G. A., KAZIMIRCHAK-POLONSKAYA, H. I. & MARSDEN, B. G., D. Reidel, Dordrecht, 454–461.

LINDBLAD, B. A. & STEEL, D. I. 1994. Meteoroid orbits available from the IAU meteor data center, in *Asteroids, Comets, Meteors 1993*, EDS MILANI, A., DIMARTINO, M. & CELLINO, A., Kluwer, Dordrecht, 497–501.

LYTTLETON, R. A. 1951. On the structure of comets and the formation of tails, *Mon. Not. R. Astr. Soc.*, **111**, 268–277.

MA, Y. & WILLIAMS, I. P. 2001. The ejection velocity of meteoroids from cometary nuclei deduced from the observations of meteor shower outbursts, *Mon. Not. R. Astr. Soc.*, **325**, 379–384.

MCINTOSH, B. A. 1990 Comet P/Machholtz and the Quadrantid meteor stream, *Icarus*, **86**, 299–304.

MCNAUGHT, R. H. & ASHER, D. J. 1999. Leonid dust trails and meteor storms, *WGN*, **27:2**, 85–102.

OLMSTEAD, D. 1834. Observations on the meteors of 13 Nov. 1833, *American Jl. Sci.*, **25**, 354–411.

ÖPIK, E. J. 1951. Collisional probabilities with the planets and the distribution of interplanetary matter, *Proc. Roy. Irish Acad.*, **A54**, 165–199.

PLAVEC, M. 1950. The Geminid meteor shower *Nature*, **165**, 362–363.

POYNTING, J. H. 1903, Radiation in the solar system: its effect on temperature and its pressure on small bodies, *Proc. R. Soc. London*, **72**, 265–267.

PROCTOR, M. & CROMMELIN, A. C. D. 1937. *Comets: their Nature, Origin and Place in the Science of Astronomy*, Technical Press, London.

ROBERTSON, H. P. 1937. Dynamical effects of radiation in the Solar System, *Mon. Not. R. Astr. Soc.*, **97**, 423–438.

ROY, A. E. WALKER, I. W., MacDONALD, A. J., WILLIAMS, I. P., FOX, K., MURRAY, C. D., MILANI, A., NOBILI, A. M., MESSAGE, P. J., SINCLAIR, A. T. & CARPINO, M. 1988. Project LONGSTOP, *Vistas in Astronomy*, **32**, 95–116.

SCHIAPARELLI, G. V. 1867. Sur la relation qui existe entre les cometes et les etoiles filantes, *Astronomische Nachrichten*, **68**, 331–332.

SHERBAUM, L. M. 1970. *Vestn. Kiev Un-ta Se Astron.*, **12**, 42–XX.

STEEL, D. S. & ASHER, D. J. 1996. When might 2P/Encke have produced meteor storms? in *Physics, Chemistry and Dynamics of Interplanetary Dust*, EDS GUSTAFSON, B. Å. S. & HANNER, M. S., *Pub. Astron. Soc. Pacific Conference Series*, 125–132.

ŠTOHL, J. & PORUBČAN, V. 1993. Meteor streams of asteroidal origin, in *Meteoroids and their Parent Bodies*, EDS ŠTOHL, J. & WILLIAMS, I. P., Slovak Academy of Sciences, Bratislava, 41–47.

SOUTHWORTH, R. B. & HAWKINS, G. S. 1963. Statistics of Meteor Streams, *Smith. Cont. Astrophys.*, **7**, 261–285.

TWINING, A. C. 1834. Investigations respecting the meteors of Nov. 13th, 1833, *American Jl. Sci.*, **26**, 320–352.

VOKROUHLLCKY, D. & FARINELLA, P. 2000. Efficient delivery of meteorites to the Earth from a wide range of asteroid parent bodies, *Nature*, **407**, 606–608.

WARTMANN, M. 1841. Observations de Genève, *Bull. Acad. R. Belg. Cl. Soc.*, **8**, 225–231.

WHIPPLE, F. L. 1950. A comet model 1: The acceleration of Comet Encke *Astrophys. Jl.*, **111**, 375–394.

WHIPPLE, F. L. 1951. A comet model II. Physical relations for comets and meteors, *Astrophys. Jl.*, **113**, 464–474.

WHIPPLE, F. L. & HAMID, S. E. 1950. On the origin of the Taurid meteors, *Astron. Jl*, **55**, 185–186.

WILLIAMS, I. P. & COLLANDER-BROWN, S. J. 1998. The parent of the quadrantid meteoroid stream, *Mon. Not. R. Astr. Soc.*, **294**, 127–138.

WILLIAMS, I. P., MURRAY, C. D. & HUGHES, D. W. 1979. The long term orbital evolution of the Quadrantid stream, *Mon. Not. R. Astr. Soc.*, **189**, 483–492.

WILLIAMS, I. P. & WU, Z. 1993a. The Geminid stream and asteroid 3200 Phaethon, *Mon. Not. R. Astr. Soc.*, **262**, 231–248.

WILLIAMS, I. P. & WU, Z. 1993b. The Quadrantid meteoroid stream and comet 1491 I, *Mon. Not. R. Astr. Soc.*, **264**, 659–664.

WILLIAMS, I. P. & WU, Z. 1994. The current Perseid meteor shower, *Mon. Not. R. Astr. Soc.*, **269**, 524–528.

WU, Z. & WILLIAMS, I. P. 1992a. On the Quadrantid meteoroid stream complex, *Mon. Not. R. Astr. Soc.*, **259**, 617–628.

WU, Z. & WILLIAMS, I. P. 1992b. The Quadrantid stream. Chaos or not? in *Chaos, Resonance and Collective Phenomena in the Solar System*, ED. FERRAZ-MELLO, S., Kluwer, Dordrecht 329–332.

WU, Z. & WILLIAMS, I. P. 1995a. P/Giacobini-Zinner and the Draconid meteor shower, *Plan. Sp. Sci.*, **43**, 723–731.

WU, Z. & WILLIAMS, I. P. 1995b. Gaps in the distribution of semi-major axes of the Perseid meteors, *Mon. Not. R. Astr. Soc.*, **276**, 1017–1023.

WU, Z. & WILLIAMS, I. P. 1996. Leonid meteor storms, *Mon. Not. R. Astr. Soc.*, **280**, 1210–1218.

Wyatt S. P. & Whipple F. L. 1950. The Poyntin-Robertson effect on Meteor orbits, *Astrophys. Jl.*, **111**, 134–141.

Zausaev A. F. 1972. The use of the Halphen–Goryachev method in the study of the evolution of the orbits of the Quadrantid and δ Aquarid meteor streams, in *The Motion, Evolution of Orbits, and Origin of Comets*, Eds Chebotarev, G. A., Kazimirchak-Polonskaya, H. I. & Marsden, B. G., D. Reidel, Dordrecht, 441.

Part 1
EXOATMOSPHERIC MEASUREMENTS
OF METEORIC DUST

3
SPACE DUST MEASUREMENTS

By EBERHARD GRÜN[1†], VALERI DIKAREV[1,2‡], HARALD KRÜGER[1¶], AND MARKUS LANDGRAF[3‖]

[1]Max-Planck-Institut für Kernphysik, Heidelberg, Germany
[2]Astronomical Institute of St. Petersburg University, Russia
[3]ESOC, Darmstadt, Germany

Space dust consists of particles typically 0.1 mm in diameter or less. They overlap with the sizes of the smallest radar meteor particles. Clouds of space dust can be observed by scattered sunlight, such as that scattered by comet tails, planetary rings or the faint zodiacal cloud. The size distribution of interplanetary dust from a few micrometers to millimeters was determined by the analysis of lunar microcraters and verified by near-Earth satellites. In situ measurements of space dust provide information on spatial and orbital distributions, and on physical and chemical properties of dust in interplanetary space. Spaceprobes measured interplanetary dust at distances of 0.3 to 18 AU from the Sun. Models of the interplanetary dust cloud have been developed on the basis of zodiacal light, thermal infrared observations, in situ measurements, microcrater statistics, and radar meteor observations. These models provide spatial density, dust flux, and line-of-sight brightness of interplanetary dust. Dust rings have been observed orbiting all giant planets. In situ studies of dust in the jovian system found various sources of dust: all the larger satellites and even the volcanos on Jupiter's moon Io. The dust detectors on board the Ulysses and Galileo spaceprobes identified submicron-sized interstellar dust sweeping through the solar system.

3.1. Introduction

Dust particle sizes range from about 0.1 mm, comparable to the smallest radar meteor particles, down to micron- and even nanometer-sized particles. Dust particles are the small brothers and sisters of meteor particles. They are too small to generate significant electromagnetic radiation or produce an ion trail upon entry in the Earth's atmosphere. However, there are several other methods available to analyze space dust (Fig. 3.1). The methods are distinguished by the size or mass range of particles that can be studied.

The earliest method was ground-based zodiacal light observations. In 1683 Cassini presented the correct explanation of this phenomenon: it is sunlight scattered by dust particles orbiting the Sun (Fechtig *et al.* 2001). With the onset of space flight, in situ detection by space instrumentation provided new information on small interplanetary dust particles. Among the first reliable instruments were simple one-shot detectors that recognized the penetration of a thin wall when a dust particle punctured it, whereas, modern impact ionization detectors allow us not only to detect but also to analyze the composition of micrometeoroids. There are more methods to study space dust near the Earth: interplanetary dust particles are collected in the stratosphere or micrometeoroids can be extracted from the Antarctic ice where contamination by terrestrial dust is low (cf. Jessberger *et al.* 2001, and Chapters 4 and 9 of this book). Microcraters on lunar rocks or

[†] Email: eberhard.gruen@mpi-hd.mpg.de
[‡] Email: dikarev@galileo.mpi-hd.mpg.de
[¶] Email: krueger@galileo.mpi-hd.mpg.de
[‖] Email: Markus.Landgraf@esa.int

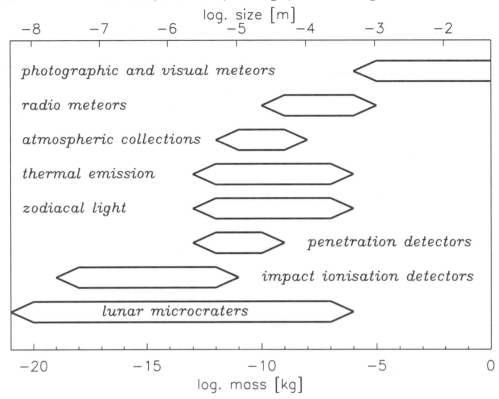

FIGURE 3.1. Comparison of meteoroid sizes and masses covered by different observational methods. Most methods have no real upper limit but for practical reasons meteoroids above a certain size will not significantly contribute to the total observational data set.

on satellite surfaces that were returned to Earth (e.g. LDEF) provide knowledge of the overall dust flux over a wide size range from sub-micron to millimeter sizes. Modern space-based infrared observatories allow observation of thermal emission from interplanetary dust in the outer solar system.

An early motivation for the study of dust in space was the risk imposed by impacts of natural meteoroids onto man-made satellites. A large number of simple dust detectors were launched into near-Earth space. After it was recognized that the natural meteoroid flux is low and that it can be countered by simple means space engineers' interest decreased. Subsequently, the motivation for the study of dust in space with more sophisticated instruments shifted to astrophysical questions about the physical and chemical properties of the grains, their contemporary sources and sinks, and their significance for the formation of the planetary system as a whole. This is documented by dust measurements (Grün et al. 2001) in the inner planetary system (Helios), at comets (Giotto, VeGa, Stardust), in the vicinity of Mars, Jupiter, and Saturn (Nozomi, Galileo, Cassini) and above the poles of the Sun (Ulysses). The realization that manned activity in near-Earth space leaves an increasing amount of space debris led to the deployment of an increasing number of dust impact detectors in low Earth orbit (cf. Chapter 10). However, within the Earth debris belts these instruments can not reliably characterize the natural meteoroid populations.

The dust detectors of the Galileo, Ulysses and Cassini missions have a large (0.1 m^2) sensitive area and enable, thereby, the study of effects that cannot be reliably identified by smaller detectors because of too few of recorded dust impacts, such as the passage of interstellar grains through the planetary system. An impact ionization detector records the ions and electrons that are released upon the impact of a fast ($\gg 1$ km s^{-1}) dust particle onto a solid surface. The combination of an impact ionization detector with a time-of-flight mass spectrometer provides the capability to undertake chemical analysis of the dust grains that are detected. Such instruments have been used on Halley missions and the latest one flies on the Stardust mission to comet Wild 2. In the most versatile dust detector, the Cosmic Dust Analyzer on the Cassini mission, a large impact area (0.1 m^2) is combined with a mass spectrometer. In addition, this instrument measures the electrical charge of the dust particles, which provides a refined measurement of the dust speed and trajectory.

The dynamics of dust particles differs from that of larger particles because other forces than just the gravitational pull of the Sun become increasingly important. For particles with masses $> 10^{-11}$ kg, solar gravity is by far the most dominant force. As a consequence, they move on Keplerian orbits, which are conic sections with the Sun in one focus – other forces are only small disturbances. Certainly, all observations of large meteoroids (e.g. meteor particles) are compatible with such orbits. However, micron-sized particles also feel the repulsive force of solar radiation pressure and the electromagnetic interactions with the interplanetary magnetic field.

The pressure exerted on dust in interplanetary space by solar radiation decreases with the inverse square of the distance to the Sun (Burns *et al.* 1979); this is the same dependence as that of gravitational force. Therefore, the ratio of the forces of gravity and radiation pressure is a constant for each particle everywhere in the solar system and is only dependent on the particle's size and material properties. This ratio is generally termed β. It is inversely proportional to the size of the particle for particles larger than the effective wavelength of sunlight (~ 0.5 µm). As a consequence, β increases for smaller sizes and reaches maximum values between 0.1 and 1 microns. The maximum value is about 0.5 for dielectric (transparent) materials and reaches values of 3 to 10 for strongly light-absorbing particles.

There are important consequences for the dynamics of small particles because of the radiation pressure. Small particles that are generated from large bodies, e.g. by emission from comets or by impact ejection from meteoroids or asteroids, carry the specific kinetic energy of their parents. However, because of radiation pressure they feel a reduced attraction to the Sun. As a consequence, they move on different orbits than do their parents. For example, a dust particle with radiation pressure constant $\beta > 0.5$ released from a large parent object on a circular orbit will leave the solar system on a hyperbolic orbit.

Besides the direct effect of radiation pressure on the trajectories of small dust grains there is also the more subtle Poynting–Robertson effect. It is caused by a small component of the radiation pressure that acts opposite to the dust particle's motion. This drag force leads to a loss of angular momentum and orbital energy of the particle. The effect is strongest when the particle speed is highest, i.e. close to the Sun and at its perihelion. Therefore, particle orbits get slowly circularized while they spiral towards the Sun. For a centimeter-sized particle on a circular orbit at the Earth's distance from the Sun the time to spiral to the Sun is 7×10^6 years.

Any meteoroid in interplanetary space will be electrically charged (Horányi 1996). Irradiation by solar UV light frees photoelectrons which leave the grain. Electrons and ions are collected from the ambient solar wind plasma. Energetic ions and electrons cause the emission of secondary electrons. In interplanetary space the dust particle reaches a surface potential of about $+5$ V due to the dominant photoeffect. Electrically charged dust particles then interact with the interplanetary magnetic field that is carried by the outward streaming (away from the Sun) solar wind. This electromagnetic effect increases for smaller grain sizes and with larger heliocentric distances. For example, for a 10 nm-sized particle at Jupiter's distance from the Sun the Lorentz force exceeds solar gravity by more than a factor of a thousand. Therefore, the trajectories of these particles are totally dominated by the interaction with the interplanetary field. Zook *et al.* (1996) demonstrated that the particles that constitute the Jupiter dust streams are of this size and show this behavior.

The overall polarity of the solar magnetic field changes with the solar cycle of 11 years. For one solar cycle positive magnetic polarity prevails in the northern solar hemisphere and negative polarity in the southern hemisphere. As a consequence, interstellar particles that enter the heliosphere are either deflected towards the solar equatorial plane (which roughly corresponds to the ecliptic plane) or away from it, depending on the overall polarity of the magnetic field (Morfill *et al.* 1986). Therefore, small interstellar particles are either prevented from reaching the inner solar system (during one solar cycle) or are concentrated in the ecliptic plane (in the other solar cycle). However, one has to recognize that interstellar grains need about 20 years to traverse the distance from the heliospheric boundary (expected to be at about 100 AU from the Sun) to the inner solar system and, hence, encounter two opposing solar cycles. Focusing magnetic configuration occurred during the solar cycles from 1956 to 1967 and from 1978 to 1989. In the period from 1989 to 2000 the overall magnetic field had an unfavorable configuration; therefore, after some time lag only large (micrometer-sized) interstellar particles should reach the inner solar system.

An important effect of meteoroids in interplanetary space is mutual collisions. It has been shown (Whipple 1967; Grün *et al.* 1985) that the meteoritic complex in the inner solar system is self-destructive on a time scale of the order of 10^5 years. Grün *et al.* (1985) have shown that meteoroids below 10^{-6} g inside 1 AU are generated by collisions of larger meteoroids. The lifetimes of larger meteoroids are controlled by collisions rather than by the Poynting–Robertson effect. A direct consequence of collisions in space is the generation of fragments, part of which become β-meteoroids. β-meteoroids, therefore, provide the major loss mechanism for meteoritic matter in the inner solar system.

A consequence of all these different effects is that dust is short-lived and it must have contemporary sources, such as comets, asteroids, or larger meteoroids. Therefore, by studying dust grains in space one also obtains information about their parents. Future attempts to study space dust will include instruments that combine dust trajectory analysis for identification of the dust sources with compositional analysis of dust grains themselves.

The organization of this chapter is as follows: in section 2 some details of measurements of space dust are given. The following sections discuss results from dust measurements in space, beginning with circumplanetary dust (section 3), interplanetary dust (section 4), and interstellar dust (section 5) traversing the planetary system. A summary concludes this chapter.

3.2. Analysis of Space Dust

In this section we describe the main methods to characterize space dust. First, we discuss observations of the zodiacal light and the thermal emission in order to determine the large-scale structure of the interplanetary dust cloud. The cratering records on lunar samples and on man-made satellites are shown to provide the size distribution of inter-planetary dust over a wide size range. In situ analyses of individual interplanetary grains, i.e. trajectories, sizes, and particle properties, are provided by modern dust detectors on satellites and space probes. And, finally, we discuss simulation techniques that enable the development and calibration of versatile space dust detectors.

The earliest method to detect dust in interplanetary space was observation of the zodiacal light. The brightness of zodiacal light arrives from sunlight scattered by a huge number of particles in the direction of observation. The scattering angle, that is the Sun–particle–observer angle, varies systematically along the line of sight. This angle is largest closest to the observer and can approach 180°. For one particle the scattered light intensity is strongly dependent on the scattering angle. For particles larger than the wavelength of the scattered light this scattering function is strongly peaked in the forward direction (scattering angle = 180°). For particles much smaller than the wavelength the scattering function is more uniform. Variable particle structure and composition affect the scattering function as well. Therefore, the observed zodiacal brightness is a mean value, averaged over all sizes, compositions and structures of particles along the line of sight. The increased zodiacal light brightness towards the Sun is in part the effect of the enhancement in the scattering function.

At visible wavelengths the spectrum of the zodiacal light follows closely the spectrum of the Sun. A slight reddening (i.e. the ratio of red and blue intensities is larger for zodiacal light than for the Sun) indicates that the majority of particles is larger than the mean visible wavelength of 0.54 µm. In fact, most of the zodiacal light is scattered by 10- to 100-µm-sized particles. Therefore, the dust seen by zodiacal light is only a subset of the interplanetary dust cloud. Submicron- and micron-sized particles as well as millimeter-sized and larger particles are not well represented by the zodiacal light.

It was quite obvious that zodiacal light observations have many advantages when they are done from an Earth satellite or even from a deep spaceprobe. No detrimental weather effects, no dust and light pollution in the atmosphere, no airglow disturb the measurements. There is the possibility of observing in a wide range of directions and from different positions in space and thereby, obtaining the large-scale structure of the zodiacal dust cloud. Leinert *et al.* (1981) succeeded with the photometers on board Helios in measuring zodiacal light between 0.3 and 1 AU. The radial brightness profile was determined and an inclination of 2° of the symmetry plane of the zodiacal cloud with respect to the ecliptic plane was found. Outside the Earth's orbit zodiacal light was observed by photometers on board Pioneers 10 and 11 (Weinberg *et al.* 1974). The zodiacal light brightness was found to be higher than that of the background out to a distance from the Sun of 3.3 AU (Hanner & Weinberg 1973).

Only less than 10% of the incident sunlight is scattered by interplanetary dust and contributes to the zodiacal light. The rest of the absorbed energy (> 90%) is re-emitted as thermal infrared radiation mostly in the 10 to 50 micron wavelength range. Because of that effect zodiacal infrared emission is a much more prominent astronomical phenomenon than zodiacal light – much to the dismay of astronomers who are interested in more distant objects that are blended by the foreground zodiacal emission. Once the technology for space-based infrared observations was developed in 1984, the IRAS satellite and later the COBE satellite obtained unprecedented information on the overall structure of the zodiacal cloud. Like the zodiacal light brightness the intensity of the thermal emission is an integral over the emission from all particles along the line of sight. Since the distance to the Sun varies along the line of sight each observation samples particles at different temperatures. Additionally, the thermal emission from a particle depends on its size and material properties. Typically, a particle of a given size can effectively emit only radiation at wavelengths greater than its diameter. Therefore, infrared radiation from 5 to 500 microns refers only to particles in this size range or larger. Beside the large-scale structure of the zodiacal cloud, e.g. its latitudinal width and symmetry plane,

TABLE 3.1. Characteristics of in situ dust instruments on planetary and interplanetary missions. Distance ranges are those in which dust measurements were obtained. The missions to comet P/Halley (VeGa and Giotto) carried several dust instruments. The mass threshold refers to the most sensitive instrument at 20 km s^{-1} impact speed

Mission	Launch year	Distance range (AU)	Detection technique	Sensitive area (cm^2)	Mass threshold (g)	Dust science
Pioneer 8	1967	1–1.1	Impact ionization, microphone	94	2×10^{-13}	Interplanetary dust at 1 AU
Pioneer 9	1968	0.7–1	Impact ionization, microphone	74	2×10^{-13}	Inner solar system dust
Pioneer 10	1972	1–18	Film penetration	2600[a]	2×10^{-9}	Outer solar system dust
HEOS 2	1972	1	Impact ionization	100	2×10^{-15}	Earth system and interplanetary dust
Pioneer 11	1973	1–10	Film penetration	2600[a]	10^{-8}	Outer solar system dust
Helios 1, 2	1974/76	0.3–1	Impact ionization	120	9×10^{-15}	Inner solar system dust
VeGa 1, 2	1984	0.8	Film penetration, impact ionization, microphone, PVDF	5–500	4×10^{-15}	Comet P/Halley dust
Giotto	1985	0.9	Film penetration, impact ionization, microphone	5–20000	4×10^{-15}	Comet P/Halley dust
Galileo	1989	0.7–5.4	Impact ionization	1000	4×10^{-15}	Jupiter system and interplanetary dust
Hiten	1990	1	Impact ionization	100	2×10^{-15}	Earth system and interplanetary dust
Ulysses	1990	1–5.4	Impact ionization	1000	4×10^{-15}	3-dimensional dust distribution
Cassini	1997	0.7–10	Impact ionization, PVDF	1000	4×10^{-15}	Saturn system and interplanetary dust
Nozomi	1998	1–1.5	Impact ionization	140	2×10^{-15}	Mars system and interplanetary dust
Stardust	1999	1–2.8	Impact ionization	90	10^{-14}	Comet P/Wild 2 and interstellar dust
SPADUS	1999	1	PVDF	500	5×10^{-11}	Near-Earth dust environment

[a]Initial area, actual area decreased as cells were punctured.

broad asteroidal bands and an Earth-shepherding ring, as well as narrow comet trails were discovered.

The size distribution of interplanetary dust particles is represented in the lunar micro-crater record. Microcraters on lunar rocks have been found ranging from 0.02 microns to millimeters in diameter. Laboratory simulations of high velocity impacts on lunar-like materials have been performed in order to calibrate crater sizes with projectile sizes and impact speeds. At the typical impact speed of interplanetary meteoroids on the moon of about 20 km s^{-1} the crater diameter to projectile diameter varies from 2 for the smallest microcrater to about 10 for centimeter-sized projectiles.

The difficulty for deriving the impact rate from a crater count on the moon is the generally unknown exposure geometry (e.g. shielding by other rocks) and exposure time to the dust flux of any surface on a rock. Therefore, the crater size or meteoroid distribution has to be normalized with the help of an impact rate or meteoroid flux measurement obtained by other means. In situ detectors or recent analyses of impact plates that were exposed to the meteoroid flux in a controlled way provided this flux calibration.

In 1984 NASA released the Long Duration Exposure Facility (LDEF) into near-Earth space at about 450 km altitude in order to study the effects on materials during prolonged exposure to space environment. Of primary interest was the effect of meteoroids. Six years after launch LDEF was retrieved by the Space Shuttle and brought back to the ground. Near Earth the meteoroid flux is about a factor of two higher than that in deep space because of gravitational concentration by the Earth. Micron-sized natural meteoroids are outnumbered (by a factor of three) by man-made space debris. Craters produced by space debris particles are identified by chemical analyses of residues in the craters. Residues have been found from space materials and signs of human activities in space such as paint flakes, plastics, aluminum, titanium, and human excretions.

The main method of gaining information on individual dust grains are in situ measurements by dust impact detectors on board interplanetary spacecraft and Earth-orbiting satellites. In situ measurements of interplanetary dust have been performed in the heliocentric distance range from 0.3 AU out to 18 AU (Table 3.1).

Several types of impact detectors were used for interplanetary dust measurements: (1) microphones, which record the momentum transferred to the detector upon impact of a dust grain of 10^{-12} g or larger, (2) penetration detectors (where the micrometeoroid has to penetrate 25 to 50 μm thick metal films) with detection thresholds of 10^{-9} and 10^{-8} g, (3) polyvinylidene fluoride (PVDF) detectors, which provide a charge signal from the excavation of polarized material at a detection threshold of 10^{-12} g, and (4) impact ionization detectors, which detect the charge released upon impact with detection thresholds of 10^{-16} to 10^{-13} g. These detection thresholds refer to a typical impact speed of 20 km s^{-1}. At lower impact speeds the minimum detectable particle mass is larger and vice versa. In the following paragraphs we discuss some of these detectors in detail.

A dust particle that hits a solid target at a speed above a few kilometers per second produces an impact crater. During this process part of the projectile and target material are strongly compressed, heated, and vaporized. This vapor is partially ionized. By an electric field applied to the impact area positive and negative charges (ions and electrons) can be separated and recorded. This phenomenon has been successfully applied for the first time in the dust detectors flown on Pioneers 8 and 9 (Berg & Richardson 1969), which also included microphones for coincidence measurements. These detectors made the first important dust observations in interplanetary space. The dust detector flown on the HEOS-2 satellite (Dietzel *et al.* 1973) employed an impact ionization detector of about 100 cm^2 sensitive area. From measurements of the signal amplitude and the rise-time, the mass and speed of the impacting micron- and submicron-sized particle were derived. A significant enhancement of this simple dust impact detector was the ten times increase of the sensitive area (1000 cm^2) of the dust instruments on the Galileo and Ulysses missions Grün *et al.* (1992*a,b*) that take measurements in the outer solar system.

The next step was to combine an impact ionization detector with a mass spectrometer and thus to analyze the ions released. In the laboratory the first compositional measurements using a time-of-flight (TOF) mass spectrometer were reported by Auer & Sitte (1968) and by Hansen (1968). The mass resolutions of the spectra obtained in the laboratory were low: only elements up to mass 50 amu (atomic mass units) could be resolved in the best cases. The compositional analyzer for the Helios mission was described by Dietzel *et al.* (1973); however, this detector had a mass resolution of only $M/\Delta M \sim 10$ (Leinert & Grün 1990). A major step forward towards making impact ionization mass spectrometers an analytic tool was made by Kissel (1986) with the dust mass analyzers for the Halley missions Giotto and VeGa. An electrostatic reflector was included in the time-of-flight mass spectrometer to improve the mass resolution beyond $M/\Delta M = 100$. With these instruments even isotopic analyses of cometary dust could be performed. An updated version of this mass analyzer flies on the Stardust mission. This instrument has an enlarged sensitive area of *c.* 100 cm^2 compared with the 5 cm^2 of the Halley instruments.

The Cassini Cosmic Dust Analyzer, CDA (Srama & Grün 1996), is a combination of a Galileo-type dust detector with a linear impact ionization mass analyzer. CDA measures the mass, coarse composition, electric charge, speed, and flight direction of individual dust particles. It measures impacts from as low as one impact per month up to 10^4 impacts per second. The detection of dust particle impacts is accomplished by two different methods: (1) a Dust Analyzer using impact ionization. This Dust Analyzer measures the electric charge carried by dust particles, impact direction, impact speed, mass and chemical composition, and (2) a High Rate Detection system, using two separate PVDF sensors (Simpson & Tuzzolino 1985; Tuzzolino 1992, polyvinylidene fluoride) for the determination of high impact rates during Saturnian ring plane crossings.

Figure 3.2 shows a schematic cross section of the sensor with its charge-sensing electrodes, and grids. The grid system in the front provides the measurement of the dust charge and of components of the velocity vector. A charged dust particle entering the sensor will induce a charge which corresponds directly to its own charge. The output voltage of the amplifier will rise until the particle passes the second grid. As long as the particle is located between the second and third grid the output voltage remains more or less constant. As soon as the dust particle has passed the third grid, the voltage begins to fall until the fourth grid is passed. Due to the inclination of 9° for the inner two grids, the path length between the grids depends on position and angle of incidence, and allows a determination of the directionality of the incident particle in one plane. The choice of 9° is a compromise between angular resolution and tube length of the detector. The detection of particle charges as low as 10^{-15} C has been achieved.

Dust impacts onto the target produce an impact plasma, i.e. a cloud comprised of neutral atoms, ions and electrons. Electrical charges produced by impacts on the large gold plated impact ionization target are collected on the target and on the negatively biased ion collector. Mass and speed can be derived from the measured signal rise-times through empirical calibration. Positive ions produced by impacts onto the chemical analyzer target will be mass analyzed. The strong electric field between the small rhodium target and the grid separates the impact charges very quickly and accelerates the ions towards the

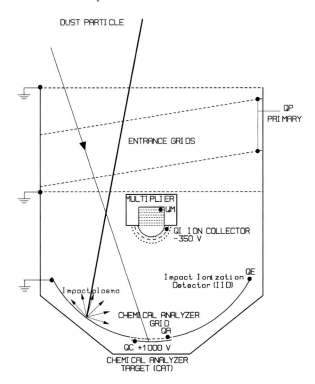

FIGURE 3.2. Schematic cross section of the Cosmic Dust Analyzer, CDA. The sensor consists of four charge-sensing entrance grids, the hemispherical target, and the ion collector with the multiplier in the center. The innermost and outermost of the four entrance grids are grounded, the two inclined grids are connected to a charge-sensitive amplifier (QP) which provides measurements of the induced dust particle charge and of the particle velocity. Dust particles (two cases are indicated) can impact either on the large gold plated impact ionization target (IID, diameter 0.41 m) or the small rhodium chemical analyzer target (CAT, diameter 0.16 m) in the center. An electric field of 350 volts separates electrons (collected by the targets) and ions (collected by the ion collector grid). Charge-sensitive amplifiers collect the charges at the two target electrodes (QE, QC) and the acceleration grid (QA). The acceleration grid is located 3 mm in front of the target and is electrically grounded, whereas CAT is at a potential of +1000 volts. The ion collector is located in the center of the detector. Amplifiers are connected to the ion grid (QI), and the multiplier (QM) and provide measurements of the total ion charge released and the time-of-flight spectrum over a 23 cm drift distance. The multiplier signal is sampled at a rate of 100 MHz.

multiplier. The curved shape of target and grid provides spatial focusing of the ions onto the multiplier. This time-of-flight mass spectrometer has a flight path length of 230 mm and gives information about the elemental composition of the micrometeoroids (Ratcliff *et al.* 1992). The signals at the output of the electron multiplier are analyzed over a large dynamic range (10^6).

The measurable particle mass ranges are 5×10^{-19} kg to 10^{-13} kg for 40 km s^{-1} impact speed and 10^{-16} kg to 10^{-10} kg for 5 km s^{-1} impact speed. The detection threshold for the impact speed is about 1 km s^{-1}. There is no upper speed limit for the detection of particles, but speed determination will be difficult beyond 80 km s^{-1}. Particle charges are measured from 10^{-15} to 5×10^{-13} C for both negative and positive charges. The trajectory of charged particles (above 10^{-14} C) can be measured with an accuracy of 2°

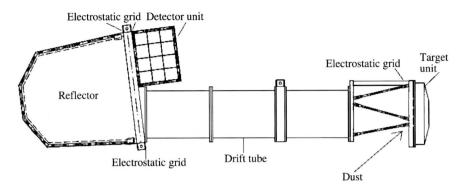

FIGURE 3.3. Cross section through the Cometary and Interstellar Dust Analyzer, CIDA. CIDA is a time-of-flight impact ionization mass analyzer of of approx. 0.009 m² target size. The target unit houses the positively biased (+1000 V) impact target and the grounded acceleration grid in front of it. The open structure in front of the target assembly allows dust particles to reach the target. The ion drift tube of 0.55 m length is on the axis with the target normal to it. The reflector unit is separated by a grid from both the drift tube and the ion detector. The electrostatic reflector deflects the ions onto the ion detector in such a way that ions of the same mass arrive at the detector at about the same time. The ion detector is a large-area open electron multiplier.

in one plane. The mass resolution of the time-of-flight mass spectrometer lies between 20 and 50.

An instrument that is optimized for the compositional analysis of dust is the CIDA instrument on board the Stardust spacecraft. Figure 3.3 shows a schematic view of the CIDA instrument. When a dust particle hits the solid silver target at a speed well above 1 km s⁻¹, solid ejecta, neutral and ionized molecules, electrons from the target and the projectile are emitted. Both positive and negative ions can be analyzed with a mass spectrometer by changing the voltage in the instrument. A charge-sensitive amplifier is hooked up to the biased part of the target. Accelerated by the electric field in front of the target, the ions travel into the drift tube of a time-of-flight mass spectrometer at the end of which an electrostatic reflector is located. This reflector deflects the ions onto an electron multiplier, and at the same time compensates for the flight time errors due to different initial starting energies the ions might have. Amplifiers connected to the multiplier allow the measurement of the ions' time-of-flight spectrum. For positive ions the bias at the target and the multiplier front stage are +1 kV and −1.3 kV, respectively. The layout of the device is determined by the size of the target and the desired time resolution. While the target size is limited by the size of the ion detector, the time resolution is mostly limited by the instrument's electronics. CIDA has a single stage ion reflector, followed by an open electron multiplier with a 30 mm diameter sensitive surface area. The mass resolution of $M/\Delta M = 250$, or time resolution $t/\Delta t = 500$, is achieved by a digitization frequency of 80 MHz. The maximum mass able to be detected is about 1000 atomic mass numbers. The useful target size is 12 cm in diameter. For an impact angle of 40° from the target normal, this is an area of 86.6 cm².

Dust particles' trajectories are determined by in situ dust detectors with narrow apertures and by the measurements of the electric charge signals that are induced when the charged grains fly through appropriately configured grid systems. Modern in situ dust detectors are capable of providing mass, speed, physical and chemical information on dust grains in space. A "dust telescope" is, therefore, a combination of detectors for dust particle trajectories and detectors for physical and chemical analysis of dust particles.

Future dust telescopes may consist of an array of parallel-mounted dust instruments, which share a common impact plane of at least one square meter in size.

A dust telescope can use two different, complementary strategies to establish the origin of small, and large dust grains. For large grains ($m > 10^{-11}$ g), the primary electrostatic charge is sufficiently high (10^{-15} C) to determine the impact velocity and direction, by use of a large area charge and trajectory sensing instrument. For small grains ($m < 10^{-11}$ g) trajectory determination can be achieved by instruments with a narrow field of view of only about 25° half-width. Distinction between interstellar and interplanetary grains of cometary or asteroidal origin is achieved by such methods. Once the origin of the different populations of dust grains has been determined their chemical characteristics can be established. The dust telescope will use techniques for the chemical analysis of dust which have previously successfully been applied to cometary dust. For the isotopic analysis of the detected grains, the mass spectrometer must have a mass resolution on the order of $M/\Delta M \sim 100$. A lower mass resolution of $M/\Delta M \sim 20$–50 is sufficient to obtain an inventory of heavy elements in the detected grains.

A pre-condition for the development of advanced space dust detectors is the availability of appropriate simulation techniques. In order to simulate impact and collision phenomena in space, it was necessary to have accelerators that provide projectiles in the speed range at and above 10 km s^{-1}. Since projectiles from military guns reached speeds of only a few km s^{-1}, new developments were necessary. The workhorse of accelerators for millimeter and larger projectiles is the light gas gun. This gun consists of a conventional powder gun that pushes a piston into a barrel containing hydrogen gas. This barrel is sealed by a steel diaphragm which breaks when the pressure exceeds several kilobars. In front of the diaphragm sits the projectile, which is accelerated by the expanding gas. Because of the low molecular mass of the compressed and heated gas its expansion speed is high, much higher than that of the exhaust from the explosive. By this method projectile speeds up to 12 km s^{-1} were reached.

For smaller (0.1-mm-sized) projectiles the Plasma Drag Gun has been developed by Igenbergs & Kuczera (1979). In this gun an electric discharge vaporizes a metal film, the ionized vapor (plasma) of which is compressed and accelerated by a high current flowing through a metal coil. Plasma speeds of several tens of kilometers per second are reached. In front of the coil is a thin plastic film to which several 100-micron-sized projectiles are loosely attached. When the plasma beam hits the film it vaporizes and the projectiles are accelerated by plasma drag. Glass projectiles can reach speeds of up to 20 km s^{-1}.

Micron-sized and smaller particles are accelerated by electrostatic accelerators that were developed by Friichtenicht (1962; see also Fechtig *et al.* 1978). These accelerators are similar to nuclear physics devices that accelerate ions by high electric voltages. Voltages of several megavolts are obtained by van-de-Graaff generators. In an electrostatic dust accelerator the ion source is replaced by a dust source in which dust is electrically charged. High charges on electrically conducting particles are obtained by bringing these particles in contact with the fine tip of a tungsten needle which is at a high electric potential. As the field strength for electron field emission is smaller than the field strength for ion emission by a factor of 10 to 100, positive potentials are used to charge-up dust particles in a dust accelerator. Spherical iron particles can be charged so that the surface electric field reaches values close to the ion field emission limit of 10^{10} V m^{-1}. In a 2 MV accelerator micron-sized iron particles reach speeds of 12 km s^{-1}, and 0.1-micron-sized particles reach speeds of 35 km s^{-1}. Speeds as high as 70 km s^{-1} have been observed for very small grains. Because of the charging process only conducting particles can be accelerated. Up to now particles consisting of iron, carbon, aluminum, metallic coated glass and conducting plastic spheres have been used in these accelerators. It is hoped that the range of projectile speeds and compositions can be extended in the future.

3.3. Circumplanetary Dust

Dust exists in the circumplanetary space around many planets of our solar system, and is most prominently seen in the ring systems surrounding the giant planets Jupiter, Saturn, Uranus and Neptune. The abundance of dust around the terrestrial planets is much smaller. The only terrestrial planets where tenuous dusty rings may exist are Mars and Pluto. Dust rings are valuable tracers of otherwise unrecognized moons.

Grain lifetimes in a circumplanetary environment are typically orders of magnitude smaller than the age of the solar system, so continuous dust supply must be maintained. Dust grains are generated by collisions between macroscopic bodies (e.g. in Saturn's main rings and the Neptunian ring), bombardment of moons and ring particles by interplanetary impactors (e.g. the Jovian rings or Saturn's E ring), as well as volcanic plumes (on Io). All of the processes that act to remove or alter dust in circumplanetary space are functions of particle size, and studies of the grain size distributions give information about the physical processes acting on the grains (Colwell 1996).

The forces dominating particle dynamics are strongly grain size dependent. Once released from their parent body, dust particles collect an electrical charge which makes them susceptible to electromagnetic forces (cf. Section 3.1). The most important forces in a circumplanetary environment are typically the gravity of the parent planet, the Lorentz force, solar radiation pressure, plasma and Poynting–Robertson drag. Circumplanetary dust is generally confined to the planetary environment, although streams of tiny electromagnetically interacting grains ejected from the Jovian system into interplanetary space are a (minor) source of interplanetary dust. Circumplanetary dust grains may eventually become meteors in a planetary atmosphere.

3.3.1. *Jupiter*

Dust grains have been detected throughout Jupiter's magnetosphere, and the Jovian system is perhaps the best studied circumplanetary dust environment in our solar system. At least, much more is presently known about Jupiter dust than about dust in the Earth environment. Apart from the Jovian ring system, dust streams originating from Io have been measured throughout the whole magnetosphere and even in interplanetary space. Circum-satellite dust clouds have been found surrounding the Galilean satellites Europa, Ganymede and Callisto, and a very faint dust ring exists in the region between the Galilean moons and much further away from the planet. These 'dusty' phenomena are reviewed in the next paragraphs.

Jupiter's ring system was discovered on a Voyager 1 image in 1979. It was the first dusty ring identified from space. Earlier hints that this faint ring might exist had already come from Pioneer 11 charged particle measurements and from impact events recorded by the Pioneer 10/11 dust detectors (Humes 1976). Galileo imaging confirmed the earlier Voyager results that the tenuous Jovian rings have at least three components: the main ring, the halo and the gossamer ring. The first two of these components have typical normal optical depths of a few times 10^{-6} while that of the gossamer ring is at least an order of magnitude smaller. All components contain large fractions of micrometer-sized dust (dust cross section $\sim 50\%$ of the total cross section in the main ring and 100% in the halo and gossamer rings) (Burns *et al.* 2001). Embedded in the ring system are the small moons Metis, Adrastea, Amalthea and Thebe, which are sources of ring dust via meteoroid impact erosion. Typical particle lifetimes in the main ring are only 10 years.

At the inner edge of the main ring, the Jovian ring expands vertically into a toroidal "halo". The vertical extension requires that electromagnetic forces act on the grains and implies that the halo grains are sub-micrometer in size. As their orbits decay through plasma drag out of the main ring their orbit inclination is increased by Lorentz resonances near the inner edge of the ring (Burns *et al.* 1985). Voyager images indicate that halo grains are smaller – on average – than grains in the main ring (Showalter *et al.* 1987).

Galileo and ground-based imaging revealed that the gossamer ring – lying exterior to the main ring – has two main components, each of which is fairly uniform: one lies just interior to Amalthea's orbit while the other one is located interior to Thebe's orbit (Ockert-Bell *et al.* 1999; De Pater *et al.* 1999). Very faint material could even be detected outside Thebe's orbit. The vertical thicknesses of the gossamer rings match quite well the maximum elevations of these satellites off Jupiter's equatorial plane and the rings have greater intensities along their top and bottom edges. This implies that the ring material originates from the satellites by meteoroid impact erosion (Burns *et al.* 1999). Particles derived from the satellites are probably driven towards Jupiter by Poynting–Robertson drag and their collisional lifetimes are 10^2 to 10^4 years. Grain sizes derived from Voyager (and confirmed by Galileo and ground-based) imaging are about 1.5 µm in radius (Showalter *et al.* 1985; Ockert-Bell *et al.* 1999; De Pater *et al.* 1999), whereas one single impact of a particle at least 6 µm in size detected by the Pioneer 11 dust detector (Humes *et al.* 1974) shows that larger grains also exist.

In 1992, when Ulysses flew by Jupiter, the dust detector instrument on board the spacecraft (Section 3.2) recorded several periodic bursts of submicrometer dust particles with durations ranging from several hours to two days and occurring at approximately monthly intervals (28 ± 3 days, Grün *et al.* (1993)). These dust streams were observed in interplanetary space within 2 AU from Jupiter, and the particles arrived at Ulysses in collimated streams radiating from close to the line of sight to Jupiter, suggesting a Jovian origin. The Galileo dust detector, which is identical to Ulysses' dust detector, confirmed the dust streams while the spacecraft was approaching the planet and has continued to observe them in the Jovian system since December 1995 (Grün *et al.* 1996, 1998). Strong electromagnetic interaction of the charged dust grains was found with Galileo inside the Jovian magnetosphere: the impact rates fluctuated by more than two orders of magnitude with periods of 10 and 5 hours, which are Jupiter's rotation and half Jupiter's rotation period, respectively (Fig. 3.4). These fluctuations were correlated with the position of Galileo in Jupiter's magnetic field. Due to a tilt of 9.6° of Jupiter's magnetic axis with respect to the planet's rotation axis the magnetic equator sweeps over the spacecraft in either an up- or a downward direction every 5 hours. Thus, with the Galileo dust detector, the electromagnetic interaction of charged dust grains with a planetary magnetosphere could be demonstrated for the first time (Grün *et al.* 1998).

Immediately after it was recognized that the dust streams originated from within the Jovian system the question of the source of the particles arose. Horányi *et al.* (1993a, 1993b) suggested Io as the source whereas Hamilton & Burns (1993) presented an alternative model where particles originated in Jupiter's gossamer ring. The exclusion of comets (e.g. Shoemaker–Levy 9) as sources for the streams was relatively straightforward (Grün *et al.* 1994b). The final answer regarding the source came from a frequency analysis of Galileo dust data collected within Jupiter's magnetosphere: the impact rate showed a characteristic frequency of Io's orbital period (~ 42 hours) and amplitude modulations of Io's orbital period and Jupiter's rotation period, which clearly shows that Io acts as a single point source for dust particles (Graps *et al.* 2000).

Analysis of the Ulysses dust stream measurements from interplanetary space together with dynamical modelling implied particle speeds in excess of $200 \, \mathrm{km \, s^{-1}}$ and particle

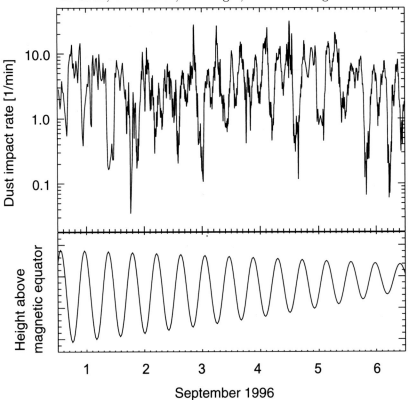

FIGURE 3.4. Top: Impact rate of Jovian dust stream particles measured in September 1996 (Galileo's G2 orbit). In the time period shown the spacecraft approached the planet from $60\,R_J$ to $10\,R_J$ jovicentric distance (Jupiter radius, $R_J = 71{,}492\,km$). Note the strong fluctuations with 5 and 10 hour periods. Bottom: The height of Galileo spacecraft above the Jovian magnetic equator. A dipole tilted at $9.6°$ with respect to Jupiter's rotation axis has been assumed.

radii in the range $5\,nm \leq s \leq 15\,nm$ (Zook *et al.* 1996). Modelling of the streams observed within the magnetosphere implied that the particle speeds must even exceed $300\,km\,s^{-1}$ (Horányi *et al.* 1997; Grün *et al.* 1998). Very recent measurements of the dust streams with two spacecraft, Galileo and Cassini, in December 2000 suggest particle speeds of $400\,km\,s^{-1}$.

Grains are accelerated to such high speeds by Jupiter's corotating electric field. After being released from Io's volcanic plumes, the grains eventually collect a positive charge of $+3\,V$ in Io's plasma torus, which makes them susceptible to electromagnetic interaction. Particles with radii between 9 and 180 nm get ejected from the Jovian system (Grün *et al.* 1998). Smaller grains remain tied to the Jovian magnetic field whereas larger grains move on gravitationally bound orbits. Because of the $9.6°$ tilt of Jupiter's magnetic field with respect to the planet's rotation axis, the particles also experience a significant out-of-plane component of the Lorentz acceleration: particles continuously released from Io move away from Jupiter in a warped dust sheet which has been nick-named "Jupiter's dusty ballerina skirt" (Horányi *et al.* 1993b). The dust measurements obtained with Galileo since 1996 imply that about $10\,g\,s^{-1}$ to $10\,kg\,s^{-1}$ of dust are continuously released from Io. This is less than 1% of the total plasma mass ejected from Io into the torus. These numbers also indicate that Io is a minor source of interplanetary dust compared with comets or main belt asteroids (Whipple 1987). In principle, the Galileo dust measurements can be used

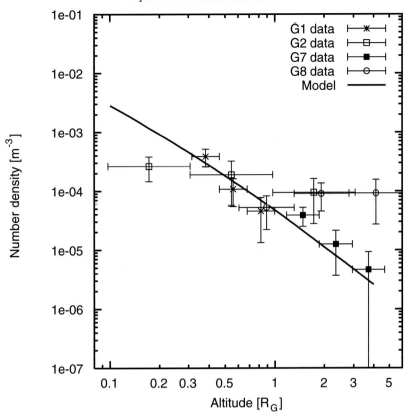

FIGURE 3.5. Number density of dust as a function of altitude above the surface of Ganymede. Vertical error bars reflect statistical errors due to the small number of impacts. The solid line is the theoretical distribution of impact ejecta expected for interplanetary impactors with a plausible set of model parameters (Krivov *et al.*, 1994, Krivov *et al.*, 1999, Krüger *et al.*, 1999d).

as a monitor of Io's volcanic plume activity; the results, however, are not yet conclusive (Krüger & Grün 2002).

During many of Galileo's close fly-bys at Europa, Ganymede and Callisto the impact rate showed a sharp peak within about half an hour centered on the closest approach to the satellite (Grün *et al.* 1998; Krüger *et al.* 1999*a*). This indicated that dust concentrations surround these moons. Analysis of the impact directions and impact speeds of the grains implied that the particles belonged to a steady-state dust cloud surrounding the satellites (Krüger *et al.* 1999*b*, 2000).

The measured radial density profiles of the dust clouds together with detailed modelling of the impact–ejection process implied that the particles had been kicked up by hypervelocity impacts of micrometeoroids onto the satellite's surface (Fig. 3.5). The projectiles were most likely to be interplanetary particles. The measured mass distribution of the grains was consistent with such an ejection mechanism with grain sizes being mostly in the range $0.5\,\mu m \leq r \leq 1.0\,\mu m$. Grain dynamics is dominated by gravitational forces, whereas non-gravitational forces, especially electromagnetic forces, were negligible. Most ejected grains follow ballistic trajectories and fall back to the surface within minutes after they have been released. Only a small fraction of the ejecta has sufficient energy to remain at high altitudes for several hours to a few days. The total amount of debris contained in such a steady-state dust cloud is roughly 10 tons. The optical thickness of the cloud is by far too low to be detectable with imaging techniques.

The Galileo measurements are the first successful in situ detection of satellite ejecta in the vicinity of a source body. All celestial bodies (planets, satellites, asteroids, large meteoroids) without gaseous atmospheres should be surrounded by an ejecta dust cloud. The Galileo measurements can be considered as a unique natural impact experiment which complements laboratory experiments in an astrophysically relevant environment.

Apart from the Io dust streams and circum-satellite ejecta clouds the Galileo dust detector has recorded a population of mostly micrometer-sized dust grains throughout Jupiter's magnetosphere (Grün et al. 1998). Most of these impacts occurred in the region between the Galilean satellites and the majority of them are grains which have escaped from the circum-satellite dust clouds (Krivov et al. 2002). These grains form a tenuous dust ring with a number density of about 4×10^2 km^{-3} at Europa's orbit. Grains derived from the satellite-ejecta clouds are mostly on prograde orbits. A population of grains on retrograde orbits about Jupiter, however, must exist as well (Thiessenhusen et al. 2000). These are most likely to be interplanetary or interstellar grains captured by the Jovian magnetosphere (Colwell et al. 1998). Earlier indications for the existence of this very faint dust ring have already been found in the Pioneer 10/11 (Humes et al. 1974) and the Ulysses dust data. Recent Galileo measurements out to a jovicentric distance of about 280 R$_J$ indicate that grains originating from the outer Jovian retrograde satellites can also be recognized in the dust data (Krivov et al. 2002).

3.3.2. Saturn

Dust has been found throughout Saturn's ring system. The most prominent rings, designated A, B and C, mainly consist of macroscopic bodies (dust fractions below $\sim 3\%$). These rings are optically thick and the time of flight between ring particles is short so that no significant orbital evolution of the dust particles can take place. Water ice is the dominant constituent of Saturn's main rings and inner moons.

The so-called "spokes" in Saturn's B ring are the primary sites of dust within the main rings. They are wedge-shaped radial markings rotating with the B ring, predominantly on the morning ansa where the ring material first emerges from Saturn's shadow (Smith et al. 1981). The spokes are composed of sub-micron grains lifted off the larger B ring particles by radially moving plasma created by meteoroid impacts onto the ring (Morfill & Goertz 1983). Color images of the spokes show that the particles are narrowly confined in size (radius $s = 0.6 \pm 0.2$ µm (Doyle & Grün 1990)). In most of their dynamical properties and temporal variations, the spokes of Saturn's B ring are unlike any other phenomenon observed in any planetary ring system.

Saturn's innermost (or D) ring was detected by the cameras of Voyagers 1 and 2. The ring is too faint to be detected by any occultation experiments or Earth-based imaging. It is composed of two major narrow ringlets and a set of fainter wave-like structures surrounding them (Showalter 1996). Dust particles are generally larger than ~ 10 µm. An excess of backscattered light indicates that as well as dust macroscopic bodies must also be present.

The narrow F ring, which is located just outside the A ring, has a dust optical depth of ~ 0.1, and is one of the dustiest of Saturn's rings. The F ring is composed primarily of sub-centimeter-sized particles, possibly down to 0.01 µm in size (Esposito et al. 1984). The ring was discovered with the imaging photopolarimeter on board Pioneer 10 (Gehrels et al. 1980). It was, at the time at least, the archetypical example of a "shepherded" ring, as it appears to be confined by the two nearby moons Pandora and Prometheus. The F ring shows a lot of internal structure including clumps, strands and kinks and the clumps vary on timescales of days to months. Unseen moonlets embedded in the rings may be responsible for these features. The clumps are perhaps produced by impacts of interplanetary meteoroids onto unseen macroscopic bodies. For this ring, the sweeping up of dust by the parent bodies and gravitational perturbations of the dust by these bodies are the dominant processes acting on the grains.

The G ring is a 7000 km wide ring located between the F and the E ring further away from Saturn. The earliest evidence for this ring came from Pioneer 11 in 1979, which detected a high-energy charged particle absorption signature in that region. The G ring was also detected with Voyager's unintended "dust detectors", the PWS and PRA instruments (Gurnett *et al.* 1983) and on Voyager images. The optical thickness of the ring is $\tau \sim 10^{-6}$ and the particle sizes are a few micrometers with a probable extension down to 0.03 μm (Showalter & Cuzzi 1993). Plasma drag is the dominant force for the evolution of G ring particles of these sizes.

The E ring is the outermost of Saturn's rings, spanning the orbits of Mimas, Enceladus, Tethys and Dione. This ring was discovered from Earth during the crossing of Saturn's ring plane in 1966 and confirmed during later ring plane crossings. Two dust impacts were recorded in this region by the Pioneer 11 dust detector (Humes 1980). Voyager imaging showed a distinct peak in brightness near the orbit of Enceladus. This moon probably serves as a source for the ring dust (Hamilton & Burns 1994). The ring has a significant vertical extension compared with most other rings, ranging from about 6000 km near its inner limit to about 20000 km near its outer edge. The ring is extremely faint with a peak optical thickness of $\tau \sim 10^{-5}$ (Showalter *et al.* 1991). Measurements with the Voyager 1 plasma instruments confirmed the existence of E ring dust. Contrary to the much smaller G ring particles, the dynamics of grains in the E ring is dominated by solar radiation pressure and electromagnetic interaction, which leads to large orbital eccentricities and inclinations.

Most of our present knowledge of Saturn's dust comes from the fly-bys of Pioneer 10/11 and Voyager 1/2 at the planet. Beginning in 2004, the Cassini spacecraft will study the Saturnian system and a wealth of new information about spatial distribution, dynamics and chemical composition of Saturnian dust grains can be expected from the Cosmic Dust Analyser carried on board (Section 3.2).

3.3.3. *Uranus*

The nine main Uranian rings – discovered in 1977 – are all very narrow and devoid of dust. One additional narrow ring, λ, was discovered in images obtained with Voyager 2 (Smith *et al.* 1986). Unlike the other narrow rings, this one brightened substantially in forward-scatter, indicating the presence of dust. The λ ring shows periodic longitudinal variations in brightness (Ockert *et al.* 1987; Showalter 1995). Optical depths are of the order of 10^{-5}. The grain size distribution is dominated by particles 1 μm in radius, and lifetimes of grains of this size are of the order of 100 years to reach the atmosphere of Uranus by exospheric drag. The dust ring is most likely to be maintained by meteoroid impacts onto larger bodies (Esposito & Colwell 1989*a,b*).

As well as the λ ring the Uranian system contains an extraordinary family of dust belts and gaps around and among all the better known narrow rings (Fig. 3.6). Radial structure is visible at a variety of scales from 50 km to more than 1000 km, with the lower limit set by the Voyager image resolution. The structure in this material may be connected with unseen shepherd satellites. An additional broad ring lies interior to all the other nine Uranian rings, has a radial width of about 5000 km and is most probably dominated by dust. Dust impacts were also reported by the plasma instruments on board Voyager during ring plane crossing outside the region of the main narrow rings and

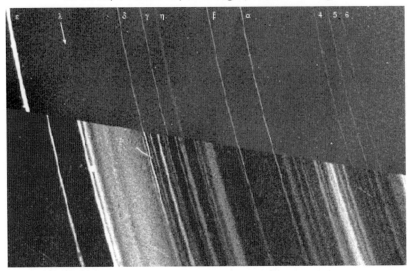

FIGURE 3.6. Two Voyager images of the Uranian ring system. The upper image was taken in backscattered light, the lower one in forwardscattered light. In forwardscattered light micrometer-sized dust particles are clearly visible, whereas in backscattered light only the narrow rings that contain macroscopic bodies can be seen. Courtesy of J. Kelly Beatty.

moons (Meyer-Vernet *et al.* 1986; Gurnett *et al.* 1987). Particles there are inferred to be micrometer-sized.

3.3.4. *Neptune*

After the discovery of the Uranian rings in 1977, Neptune's environs have been searched for analogous narrow rings, but with no initial success. In 1984, an occultation event was detected on one side of the planet that was not repeated on the other, indicating that incomplete arcs might be orbiting the planet (Hubbard *et al.* 1986). In 1989, Voyager 2 detected three to five slender and discontinuous ring arcs embedded within a narrow fainter Neptunian ring (Smith *et al.* 1989). Interspersed among these rings are a number of small moons.

Voyager images indicated that the Neptunian rings are dusty, being significantly dustier than the main rings of Saturn and Uranus. The grain size distribution probably extends from below 0.1 μm to at least several tens of microns and perhaps continuously to centimeter-sized and larger source bodies. The dust is broadly distributed and has a low optical depth (Smith *et al.* 1989). The large dust fractions, up to 50%, require vigorous collisions between source bodies to maintain the dust optical depth against sweep up by the parent bodies (Colwell & Esposito 1990), unlike the Uranian rings, which are maintained by meteoroid excavation.

As at Uranus, the Voyager plasma instruments detected dust impacts at ring plane crossing of Neptune, illustrating the presence of widely distributed dust beyond the prominent ring region (Gurnett *et al.* 1991). Particle sizes are probably in the range of micrometers and the grains may be ejecta from the moon Proteus (Colwell & Esposito 1990). Interestingly, the dust is not centred on the equatorial plane and is much more vertically extended than the main rings of Neptune. Considerable dust that has also been reported by the Voyager plasma instruments to exist at high latitudes must be transported there by electromagnetic processes or radiation pressure.

3.3.5. *Mars*

Thirty years ago it was suggested that a faint dust ring should exist around Mars near the orbits of Phobos and Deimos (Soter 1971). Despite several attempts, however, the putative ring could not yet be unambiguously detected. Imaging with the cameras of the Viking spacecraft placed an upper limit on the ring's normal optical depth of $\tau < 3 \times 10^{-5}$. When the Phobos-2 spacecraft crossed the orbit of Phobos the plasma and magnetic field instruments registered fluctuations of the magnetic field and plasma parameters (Dubinin *et al.* 1991). This so-called "Phobos event" remained inconclusive, however, the data were also compatible with a gaseous rather than a dusty torus (Dubinin 1993). A similar "Deimos event" during crossing of the Sun–Deimos line could not ultimately confirm the dust ring either (Sauer *et al.* 1995).

Despite the failure to detect the presumed Martian dust tori, they have been substantially studied theoretically. As a result, a rather detailed picture of the geometry, spatial distribution of dust, grain sizes and many other parameters has been developed. The dynamics of grains released from Phobos and Deimos can be broadly categorized according to particle size. The largest, millimeter-sized grains follow nearly Keplerian orbits close to their parent moons. They are concentrated in narrow tori around the orbits of Phobos and Deimos, and their main loss mechanism is re-accretion by their parent moons on timescales of several years (Kholshevnikov *et al.* 1993). Grains smaller than about 1 mm down to several tens of micrometers are expected to strongly dominate the dust belts. Their dynamics are governed by the combined action of the oblateness of Mars and solar radiation pressure (Hamilton 1996; Krivov & Hamilton 1997; Krivov & Jurewicz 1999). These grains form dust complexes that are vertically and azimuthally asymmetric as well as time-variable: Deimos ejecta form a ring displaced away from the Sun, whereas the ring of Phobos is displaced toward the Sun. Grain lifetimes are tens of years to thousands of years, which explains the dominance of these grains in the tori. Grains even smaller in the size range between several tens of micrometers and 1 μm radii are present in low number densities: they hit the Martian surface within less than a year after release from their parent body (Krivov & Hamilton 1997). The smallest grains, below 1 μm in radius, are strongly affected by electromagnetic forces and they are rapidly (within tens of days) swept into interplanetary space (Horányi & Burns 1991).

A systematic search for dust in the Martian environment will begin in 2003 when the Nozomi (Planet-B) spacecraft will arrive at Mars, carrying a dedicated in situ dust detector on board (Igenbergs *et al.* 1998) in order to test the hypothesis above. Recent remote observations with the Hubble Space Telescope (D. P. Hamilton, personal communication) failed to detect the Martian dust ring.

3.3.6. *Pluto*

It has been suggested that a dust ring exists around Pluto and Charon, at the outskirts of our solar system (Thiessenhusen *et al.* 2002). The ring is maintained mostly by Charon ejecta released by meteoroid erosion due to Edgeworth–Kuiper dust grains. The motion of the grains is dominated by the gravity of Pluto and Charon and the dust density is sufficient to be detectable with an in situ dust detector on a future space mission but too tenuous for remote sensing observations.

3.3.7. *Mercury, Venus and Earth*

Dust around the inner terrestrial planets Mercury and Venus is virtually negligible because of the absence of moons which could act as dust sources. Only at Mercury might a faint dust cloud be created by interplanetary impactors (Al Jackson, personal communication) because Mercury has no significant atmosphere.

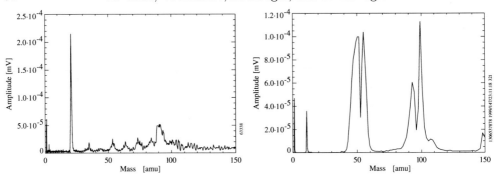

FIGURE 3.7. Left. Laboratory mass spectrum of an organic 0.75 micron dust particle (latex) impacting the target of Cassini's CDA with 17 km s^{-1} and a mass of 2.4×10^{-13} g. Right. Mass spectrum of an interplanetary dust particle with an impact speed of 24 km s^{-1} and a mass of 10^{-13} g. The peak at mass 103 reveals Rhodium in the target material of the sensor.

Dust around the Earth is mostly anthropogenic (Section 3.2) and is becoming an increasing hazard for space missions. Although the Earth moon is a potential source of dust via impact meteoroid erosion, direct in situ measurements close to the moon did not give definite results (Iglseder *et al.* 1996). Lunar meteorites ejected by other meteorite impacts, however, have been found on Earth.

3.4. Interplanetary Dust

3.4.1. *The Smooth Interplanetary Dust Cloud*

Zodiacal light was the first indication of the existence of the interplanetary dust cloud. It can be seen with the naked eye as a faint cone of light above the western horizon after sunset or above the eastern horizon before sunrise. The much better sensitivity of telescopic observations reveals that zodiacal light comes from every point on the celestial sphere, and that the spectrum of this emission matches that of the Sun, so that the orbit of the Earth is immersed in a swarm of particles scattering sunlight, an extension of the sporadic meteoroid background to small grain sizes.

Under certain assumptions on the scattering properties of the interplanetary particles, the spatial number densities of the dust may be inferred based on the observations of zodiacal light from different viewpoints and along different viewpaths though the dust cloud. Maps of zodiacal light as seen from the Earth were created based on the observations from the ground in Tenerife observatory (Levasseur-Regourd & Dumont 1980). The Helios satellite surveyed zodiacal light from its eccentric orbit with a perihelion at 0.3 AU (Leinert *et al.* 1981).

Infrared observations further facilitate the inversion of brightness measurements to number densities since unlike scattering of visible light the thermal emission of dust is isotropic, and the temperatures of dust particles have, at least, a well-predictable radial dependence (Röser & Staude 1978; Temi *et al.* 1989). IRAS and COBE satellites surveyed the sky in the infrared in a limited range of solar elongations from 60° to 135° due to instrument design constraints. These observations are unable, therefore, to probe zodiacal dust inside 0.86 AU. Midcourse Space Experiment (MSX) by the US Ballistic Missile Defence Organization includes an inspection of infrared background emission at lower and higher solar elongations (S.D. Price, personal communication), which extends the coverage of the infrared sky and replaces the old observations on board high-altitude rockets (Murdock & Price 1985).

Analysis of the data that have been obtained at visual and infrared wavelengths unveils a very smooth cloud of the normal optical depth of $\sim 5 \times 10^{-8}$ at 1 AU (Dermott *et al.* 2001). An inversion of IRAS observations (Clark *et al.* 1993) yielded the vertical distribution of the dust in the form

$$n(z) \propto \frac{w}{z^2 + w^2}, \tag{3.4.1}$$

with $w = 0.26 \pm 0.03$ AU at the Earth's distance from the Sun w is close to the half-width at half-maximum of the dust cloud. The width appears to be proportional to the heliocentric distance. The cloud has a symmetry plane which deviates from the ecliptic plane with the inclination $2°$ and the longitude of ascending node $77°$ (Kelsall *et al.* 1998).

The radial profile of the volume number density $n(r)$ of the zodiacal cloud is usually represented with

$$n(r) = \frac{1}{r^\alpha} \tag{3.4.2}$$

where the slope α has a weak dependence on the distance from the Sun r.

Inversion of zodiacal light observations yields $\alpha = 1.3$ at and less than the Earth's distance from the Sun. The same slope was obtained in Kelsall *et al.* (1998) based on COBE DIRBE observations, while Reach (1991) comes up with $\alpha = 1.1 \pm 0.3$ using IRAS observations and theoretical calculations of the infrared emission from silicate and graphite grains heated by the Sun. Zodiacal light observations at greater than 1 AU performed from the Pioneer 10 and 11 probes suggest a steeper slope up to the outskirts of the asteroid belt, marking a decrease of the number of sources (asteroids) with increasing distance from the Sun.

In situ flux measurements with the dust detectors provided different information on the interplanetary dust cloud, supplementary to that obtained with the remote observations. Early dust experiments in the interplanetary space on Pioneers 8 and 9 (see Table 3.1) yielded a discovery of β-meteoroids – the particles flying away from the Sun along hyperbolic paths (Zook & Berg 1975) and too small to be observed with remote optical systems.

Cosmic velocities are sufficient to vaporize and ionize the dust particles impacting on dust detector target, so that chemical analysis of the cosmic matter can be performed in situ. Dust detectors with built-in chemical analyzers were flown on board Helios spacecraft, and are operating now on Stardust and Cassini spacecraft. Two mass spectra obtained with Cassini's CDA are presented in Fig. 3.7.

In the outer solar system, outside the orbit of Jupiter, the only observations so far of dust have been performed in situ by the dust experiments on board Pioneers 10 and 11 (Humes 1980) and, indirectly, by the Voyager 1 and 2 plasma wave experiments (Gurnett *et al.* 1997). Results of the Pioneer dust experiments were surprising, suggesting a nearly constant number density between 5 AU and 18 AU. One has to note, however, that the two independent sensors of the Pioneer 11 dust detector gave very inconsistent results, and one of the sensors on Pioneer 10 did not work at all.

The interplanetary meteoroid flux (IMF) model by Grün *et al.* (1985) used results from several kinds of observations and resulted in a wide range of mass distribution of the meteoroid flux at 1 AU. It is shown in Fig. 3.8. The model defines the number of meteoroids hitting a plane spinning about the axis perpendicular to the ecliptic plane.

The key data set used in Grün *et al.* (1985) – lunar microcrater counts – does not give an unambiguous answer to the question of the velocities of the meteoroids that produced the craters. In the IMF model, the authors assumed an effective impact speed of 20 km s^{-1}

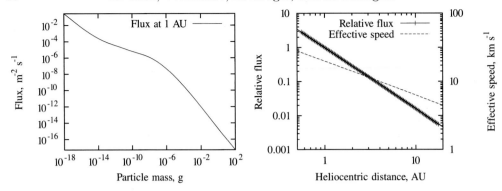

FIGURE 3.8. Interplanetary meteoroid flux model (Grün *et al.* 1985). Left: cumulative particle flux on a spinning flat plate at 1 AU distance from the Sun. The plate normal vector lies in and the spin axis is perpendicular to the ecliptic plane. Right: flux at other heliocentric distances and effective speed. The effective speed is the speed of meteoroids relative to each other or the speed of meteoroids relative to the target on a circular orbit.

at the orbit of the Earth, and in order to apply the model at other heliocentric distances, they introduced a flux modification factor $r^{-1.8}$ corresponding to the radial number density slope $r^{-1.3}$ and a velocity dispersion modification factor $r^{-0.5}$ which stems from the laws of Keplerian motion.

A sophisticated numerical model of the interplanetary meteoroid flux (Divine 1993) attempts to describe its directional properties anywhere from 0.05 to 18 AU based on the distributions of dust particles in mass and orbital elements, semimajor axis a, eccentricity e and inclination i, fitted to a variety of measurements. Zodiacal light observations, in situ flux measurements on board Helios, Galileo and Ulysses, radar meteors and the IMF model were used to adjust the parameters of the Divine model. Provision for calculation of various properties of the zodiacal cloud has been made. Fig. 3.9 shows the relative meteoroid flux distribution over the celestial sphere as measured by a flat plate. Directional structure of the fluxes is clearly seen.

An upgraded version of the meteoroid model (which includes software available to the public) was recently released and described in Garrett *et al.* (1999). Staubach *et al.* (1997) modified parameters of small particle distributions in order to account for more recent data from Galileo and Ulysses, and introduced a new population composed of interstellar particles (see also Grün *et al.* 1997).

3.4.2. *Minor Bodies and Sources of Dust*

The interplanetary dust we observe today is not a relict of the protoplanetary nebula. Estimates show that the lifetimes of interplanetary grains are, although strongly size-dependent, well below the age of the solar system. During formation of the planets, dust particles and gas stuck and formed major planets and minor bodies like comets and asteroids. The dust should therefore be continuously dispatched from these reservoirs to replenish the population of interplanetary grains.

Comets are an obvious source of dust in the present-day solar system. They move about the Sun on eccentric orbits and emit large amounts of gas and dust when they near the Sun. Tails of particularly bright comets can sometimes be observed with the naked eye (e.g. comet Hale–Bopp). The tails mostly consist of dust grains which scatter sunlight. Observations by ISO have shown that millimeter-sized and larger cometary dust gets distributed over large spatial regions along the comet's orbit, forming cometary trails observable in the infrared (Reach *et al.* 2000). If the Earth crosses such a cometary orbit,

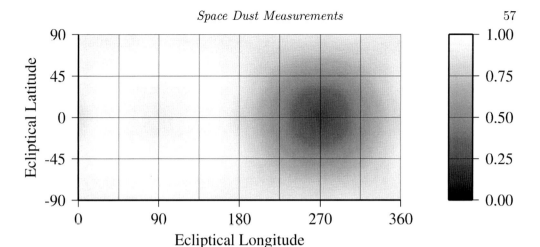

FIGURE 3.9. The relative distribution of the meteoroid flux as measured by a flat plate installed at 1 AU, moving about the Sun at the circular velocity of 30 km s^{-1} and turned in different directions specified by the ecliptic longitude and latitude. The prediction of the meteoroid model Divine (1993). Particles with the masses above 10^{-8} g are taken into account.

large dust grains with sizes up to one centimeter produce the well-known meteor streams, like the Leonid meteor shower caused by comet Tempel–Tuttle. The dust densities in the Leonid stream near its maximum are sufficient to be detectable with modern optical instruments (Nakamura *et al.* 2000).

Two classes of comets are designated based on their orbital elements and estimated lifetimes since their insertion in the low perihelion distance orbits. Comets on highly eccentric and randomly inclined orbits have also large semimajor axes and orbital periods, so that they are called long-period comets. They are usually brand new comets most recently arrived in the inner solar system from the Oort cloud – a hypothetical cometary reservoir tens of thousands astronomical units away from the Sun. For example, comet P/Halley is a long-period comet, its orbit has eccentricity 0.967 and inclination 162°; the orbital period is 76 years.

The second class is formed by the comets which have flown by Jupiter and got gravitationally pulled into a relatively flat swarm near the ecliptic plane. This class is also called Jupiter family. Certainly, there are Saturn family comets as well as comets trapped by other giant planets. Comet P/Schwassmann–Wachmann 1 is a short-period comet.

In addition to cometary dust, a significant fraction of meteors and dust grains in the zodiacal cloud has its origin in the asteroid belt between Mars and Jupiter. Although no single asteroid's emission of dust has been observed, these minor bodies are ubiquitous. Surveys report there are ~ 50000 asteroids with magnitude brighter than 15^m corresponding to diameters above ~ 200 m with the semimajor axes between 2 and 3.5 AU (taken from Durda & Dermott 1997). The abundance of asteroids suggests that dust may be produced, for example, through their mutual collisions, and indeed, infrared observations with the IRAS satellite confirmed that the dust is emitted by asteroids of several prominent families (Sykes 1990). The dust forms bands extending from the region occupied by the family asteroids towards the Sun.

Generation of dust by Edgeworth–Kuiper belt objects (EKBOs) orbiting the Sun at about 30 to 100 AU and discovered in the 1990s (Jewitt & Luu 1993) has not been confirmed by a direct observation; however, mutual collisions between EKBOs (Liou *et al.* 1996) and bombardment of their surfaces by interstellar meteoroids (Yamamoto & Mukai 1998) should produce ejecta forming bands analogous to the asteroidal dust bands. The presence of dust in the outer solar system is supported by the Pioneers 10 and 11 dust experiments (Humes 1980) as well as the measurements with Voyager 1 and 2 plasma wave instruments (Gurnett *et al.* 1997). Genetic relation of the observed dust to EKBOs (and not to the long-period comets, for example) remains an open question. EKBOs dwell in generally low-inclined, low-eccentric orbits.

In spite of a remarkably long list of visible objects capable of supplying dust in interplanetary space, the interplanetary dust could also have been created by a single event in the past, such as intrusion of an exceptionally active comet in the solar system or a collision of two large asteroids. The interplanetary dust complex may not be in a steady state now.

3.4.3. *Physical Modeling of Interplanetary Dust*

Although the dust sources listed in the previous section are already spread over large spatial extents in the solar system, dust itself gets further distributed by forces specific to small particles. Interplay between forces and other effects gives rise to interesting theoretical problems not all of which have been solved up to now.

In addition to the obvious gravity of the Sun, dust particles are affected by the gravity of major planets, which leads, depending on conditions of the interaction, to three distinct types of orbital behaviour. First, when a particle doesn't approach close to the planet and the period of the particle orbit is not a rational number multiplied by the orbital period of the planet, the orbit of the particle precesses under the perturbation from the planet, i.e., the argument of perihelion and the longitude of nodes of the dust grain orbits increase at a constant rate. This is called non-resonant perturbation of the particle orbit by the planet. The result of such interaction can readily be seen in the distribution of the arguments and nodes of the orbits of most asteroids.

Second, when the particle approaches the planet, the gravity of the latter may shortly exceed that of the Sun. Consequently, heliocentric orbit of the particle is modified on a very short time scale. This type of interaction is called gravitational scattering by planet, or distant encounter with planet. As a result, grain orbits are modified so that only the Tisserand criterion is kept constant:

$$T = \frac{a_\mathrm{p}}{a} + 2\sqrt{\frac{a}{a_\mathrm{p}}(1 - e^2)} \cos i \qquad (3.4.3)$$

where a, e and i are the semimajor axis, eccentricity and inclination of the particle orbit, a_p stands for the planetary orbit's semimajor axis. Most of the short-period comets flew by Jupiter and their orbital distributions have evolved into a flat, radially extended cloud.

Third, if the ratio of the periods of the grain and planet orbits is equal or close to a ratio of two integer numbers, the particle falls in a resonant interaction with the planet. Even very far from the planet, that is without any encounter, grain orbits are modified in a non-trivial way that cannot be predicted based on the averaged equations of the motion applicable in the first case. The order of the resonance, i.e. the numerator and denominator of the ratio of the orbital periods, is important for the outcome of the interaction. The particle may be locked in one resonance, and it may be expelled from another resonance. Different resonances with Jupiter form humps and gaps in the radial profile of the number density of asteroids. In resonance, a linear combination of orbital angles (longitude of node, argument of pericenter, true anomaly) of both particle and planet tends to stay nearly constant, coefficients of the combination are integer.

The dusty interplanetary space turns out to be "non-transparent" for particles of some sizes. Collisions with other particles erode or break up the grains. Collisional destruction probability depends on grain size distribution and velocity dispersion in the cloud. Fatal collision rate can be approximated by the integral (e.g. Grün *et al.* 1985; Ishimoto 2000)

$$\lambda(s_0) \sim \int_{s_{\min}(s_0)}^{\infty} \pi(s + s_0)^2 \langle v_{\mathrm{imp}} \rangle\, n(s) \mathrm{d}s. \tag{3.4.4}$$

The minimum impactor mass $s_{\min}$ sufficient to break up the grain of size s_0 provides the lower limit for the integral. Differential size distribution of the interplanetary dust $n(s)$ multiplied by the collision cross-section area $\pi(s + s_0)^2$ gives the inverse mean free path of the grain (between collisions) and the average impact velocity $\langle v_{\mathrm{imp}} \rangle$ converts the former into collision frequency. The integral (3.4.4) can be used to estimate grain lifetimes against collisions $\tau_{\mathrm{c}}(s) = 1/\lambda(s)$.

Large particles are not destroyed by collisions because there are few impactors in the interval $(s_{\min}, \infty)$ capable to break them up. All asteroids larger than ~ 10 km in size observed now survived from the time of formation of the solar system. The smaller is the particle, the higher is the fatal collision rate. Interplanetary dust having masses greater than $\sim 10^{-6}$ g is destroyed by collisions. The fragments feed in the population of smaller grains. Unless the fragments are very small, their heliocentric orbits are similar to the orbit of larger collided particles. This follows from the preservation of momentum direction in collision (the magnitude of momentum itself is changed when collision is inelastic).

However, solar radiation imposes pressure on the illuminated sides of dust particles, which is proportional to the area-to-mass ratio of the grains and therefore becomes important when small particles are considered. The strongest component of the radiation pressure is directed outward from the Sun and counteracts the solar gravity. Since both gravity and radiation pressure are proportional to the inverse square of the distance from the Sun, direct radiation pressure may be described as a reduction of the solar mass in the equations of the Keplerian motion about the Sun.

As a result of direct radiation pressure, small particles move more slowly than large particles even if their orbital paths are identical. This leads to radial displacement of the peaks and gaps driven by the resonances. The solar ring is an example. Fine dust particles trapped in 1:1 resonance with the Earth have semimajor axes slightly larger than the radius of the terrestrial orbit. As long as in situ flux measurements are concerned, incidence angles of small impactors will differ from incidence angles of large impactors when they all move along the same orbit. Finally, fine fragments of large particle collision may acquire orbits with higher eccentricities and larger semimajor axes compared with the elements of the parent particles.

The Poynting–Robertson effect is the other component of the radiation pressure. It is perpendicular to the outward component and is caused by the aberration of light due to the orbital velocity of the grain. Due to the Poynting–Robertson effect, grains spiral towards the Sun. An integral of the orbit-averaged equations of the motion was taken by Wyatt & Whipple (1950)

$$a\frac{1 - e^2}{e^{4/5}} = \mathrm{const} \tag{3.4.5}$$

The plane of the orbit is constant. Recently, Breiter & Jackson (1998) succeeded in solving the problem in the osculating orbital elements, i.e., exactly.

Solar radiation is responsible for a number of other long-term perturbations of the particle orbits, e.g. the windmill effects. The Lorentz force is a strong perturbation of submicrometer-sized particle orbits, especially close to the Sun. An overview (Burns *et al.* 1979) contains a description of these and the above-listed forces and provides plenty of useful constants.

Now let us see how the effects introduced above can be synthesized in comprehensive models of the zodiacal cloud capable of reproducing the most important trends revealed by the observations. In what follows, we put emphasis on analytical and semianalytical approaches to the problem.

Evolution of a single particle can be followed from its generation at a source until destruction or encounter with the Sun or a planet using the description of the motion given by the ordinary differential equations (ODEs) in appropriate variables. Numerical simulations usually choose Cartesian coordinates and velocity which make the right-hand sides of the ODEs easy to define for most perturbations. On the contrary, the integral of motion (3.4.5) was obtained from the ODEs of the motion perturbed by the Poynting–Robertson drag written in Keplerian orbital elements a and e and averaged over the revolution about the Sun.

However, particles in the interplanetary dust cloud start from an ensemble of sources, and their fates often depend on the other particles, as in the case of dust grains destroyed by collisions. Evolution of co-existing clouds of grains ejected at different times from different sources can be described in terms of the partial differential equations (PDEs), or Boltzman-type equations with integral terms. Indeed, evolution of the distribution of grains in size, semimajor axis, inclination and eccentricity $f(t; s, a, e, i)$ under the perturbations listed above can be written in the form (cf. Gor'kavyi et al. 1997)

$$\frac{\partial f}{\partial a}\frac{da}{dt} + \frac{\partial f}{\partial e}\frac{de}{dt} + \frac{\partial f}{\partial i}\frac{di}{dt} + f \cdot \left(\frac{\partial}{\partial a}\frac{da}{dt} + \frac{\partial}{\partial e}\frac{de}{dt} + \frac{\partial}{\partial i}\frac{di}{dt}\right) =$$

$$= \frac{\partial f}{\partial t} + N^+(s, a, e, i) - N^-(s, a, e, i) \qquad (3.4.6)$$

where da/dt, de/dt, di/dt come from the equations of the motion of a single particle, N^+ are sources of dust and N^- are sinks. Sources and sinks may be two aspects of the same process, like gravitational scattering which removes particles from (a, e, i) before the distant encounter and injects them in new (a', e', i') according to transition probability matrix. In a steady state $\partial f/\partial t = 0$.

The Boltzman equation for the full set of forces turns out to be rather complex. In the literature, several solutions are available for restricted problems. Leinert et al. (1983) derive the distribution of dust particles $f(a, e)$ spiraling towards the Sun under the Poynting–Robertson effect, and launched from the sources characterized by arbitrary distribution $\tilde{f}(a, e)$:

$$f(a, e) = \frac{2}{5C\sqrt{1 - e^2}e^{1/5}} \int_e^1 \tilde{f}(\tilde{a}, \tilde{e})\tilde{a}^2 \tilde{e}^{-4/5}(1 - \tilde{e}^2)d\tilde{e}, \qquad (3.4.7)$$

where C is a size-dependent factor proportional to the strength of the radiation pressure, $\tilde{a}$ and $\tilde{e}$ are the orbital elements of the sources, and $\tilde{a}$ is calculated using the integral (3.4.5) so that (a, e) and $(\tilde{a}, \tilde{e})$ must belong to the same trajectory. Collisional removal of the dust can be accounted for by multiplication of the integrand in (3.4.7) by the probability of safe transition of the dust particle from $(\tilde{a}, \tilde{e})$ to (a, e):

$$P(\tilde{a}, \tilde{e} \rightarrow a, e) = \exp\left[-\frac{8}{5}\frac{\tau_{PR}(s)}{\tau_c(s)}\ln\frac{\tilde{e}}{e}\right] \qquad (3.4.8)$$

with τ_{PR} being the Poynting–Robertson lifetime.

When converted into the number densities and represented with the power law (3.4.2), the distribution function (3.4.7) gives the slope $\alpha \approx 1.1$ for the asteroidal dust on circular orbits and $\alpha \approx 1.3$ for the cometary dust on eccentric orbits. This falls in a very good agreement with the results of inversion of the zodiacal irradiation in the visible and infrared wavelengths. The dust observed remotely from the Earth is composed of particles whose dynamics are mainly governed by the Poynting–Robertson effect.

The statement of the problem in Leinert *et al.* (1983) does not treat breakups of particles as an additional dust source. In fact, however, collisions destroy large particles and generate small grains. Ishimoto (2000) considered the radial drift of dust grains due to the Poynting–Robertson effect coupled with the collisional grinding of dust. He wrote the Boltzmann equation in the form

$$\frac{\partial n(m,r)}{\partial r} = -\frac{n(m,r)}{r}\left(2 + \frac{r}{|\bar{v}_r|}\frac{\partial|\bar{v}_r|}{\partial r}\right) + \frac{dt}{dr}\frac{dn'(m,r)}{dt} \qquad (3.4.9)$$

where m is the mass of the particle, $|\bar{v}_r|$ is the average speed of the radial drift of the grains due to the Poynting–Robertson effect, dn'/dt describes sources and sinks, including collisional breakups as well as dust production by asteroids and comets.

Numerical solution of the equation (3.4.9) showed that the relative mass distribution is not uniform in space. Although it can closely match the distribution in mass of the interplanetary meteoroid flux at 1 AU (Grün *et al.*, 1985), at other heliocentric distances the mass distributions are different.

Following Ishimoto (2000), we introduce the critical mass m_0 that separates the two regimes of particle evolution, the Poynting–Robertson drag and the collisional destruction. Small particles ($m \ll m_0$) evolving under the Poynting–Robertson effect and having negligible cross-section areas have a small probability of collision with other particles and their lifetimes are determined solely by the rate of the drag which is inversely proportional to the solar radiation pressure-to-gravity ratio $\beta \propto m^{-1/3}$, which in combination with the production rate ($\propto m^{-\gamma}$) results in

$$n(m) \propto m^{-\gamma+1/3}. \qquad (3.4.10)$$

Large particles ($m \gg m_0$) are destroyed by collisions before their semimajor axes can change noticeably due to the Poynting–Robertson effect and their lifetimes are constrained by collisions which have a frequency proportional to the inverse cross-sectional area ($\propto m^{-2/3}$). This yields, for the regions of continuous replenishment of the dust,

$$n(m) \propto m^{-\gamma-2/3} \qquad (3.4.11)$$

and where there is no source the number of large particles drops down abruptly (they cannot be supplied by other regions through the Poynting–Robertson drag).

At 1 AU the critical mass $m_0 \approx 10^{-6}$ g (see Fig. 3.8). Substitution of $\gamma = 5/3$ reproduces the cumulative mass distribution slopes of the interplanetary flux model between 10^{-14} g and 10^2 g. Closer to the Sun a shift of the critical mass towards smaller masses can be expected because the increase of the number density increases the loss of smaller particles due to collisions. Vice versa, farther from the Sun the critical mass becomes larger.

The formalism by Ishimoto (2000), however, doesn't take into account distant encounters and resonances with planets, so that it is good in the inner solar system only where the effects are not as strong as near the giant planets. Incorporation of the gravitational interaction with planets, direct solar radiation pressure and the Poynting–Robertson effect in numerical or semianalytical models were done by Liou *et al.* (1995), Liou & Zook (1997), Durda & Dermott (1997), Kortenkamp & Dermott (1998), Dermott *et al.* (2001). These works disregard collisions between grains.

In general, importance of gravitational interactions with planets for the orbital evolution has been shown. Distant encounters with planets shape the distributions of both large and small particles if their orbits allow for close approaches to the planets. Small particles spiraling towards the Sun from the outer solar system go through cascades of encounters and resonances. Resonances form azimuthally asymmetric structures in the zodiacal cloud like the leading and trailing dust blobs near the Earth's orbit.

3.5. Interstellar Dust

Cosmic dust phenomena are not restricted to the solar system. As early as in the 1920s it was recognized that the reddening of starlight is caused by galactic interstellar dust (Schalén 1929; Trümpler 1930; Öhman 1930). It was argued that the small interstellar grains scatter predominantly blue light, thereby removing it from the line of sight between the observer and the star. The red portion of the light is less affected, and consequently the stars appear reddened if a dust cloud is in front of them. The first model of the composition of interstellar grains was created in the 1940s (Oort & Van de Hulst 1946; Van de Hulst 1949). It described an icy dust population that consisted of hydrogenated compounds of the most abundant heavy elements (heavier than hydrogen and helium) O, C, and N. Mainly H_2O, CH_4, and NH_3 ice was believed to condense in interstellar clouds and to form conglomerates with typical sizes of about 0.3 μm. The model grain size distribution that was achieved in the equilibrium of grain growth and destruction was able to explain the observed extinction at that time (Struve 1937; Henyey & Greenstein 1941; Stebbins & Whitford 1943).

Just like interplanetary dust grains, interstellar dust emits most energy in the thermal spectral bands. The detection of interstellar dust at lower column densities, i.e. closer to our Sun, became possible after infrared astronomy was used more frequently to analyze interstellar phenomena. The spectroscopic analysis of the infrared light also allowed the identification of molecular bands that provided information on the chemical composition of interstellar dust. Prominent silicate (SiO) features near 10 and 20 μm, and the polycyclic aromatic hydrocarbon (PAH) features at 3.3, 6.2, 7.7, and 11.3 μm were detected. It was the European Infrared Space Observatory (ISO) that provided for the first time high resolution spectra in the wavelength range from 2 to 200 μm. With the ISO data it was possible to obtain an inventory of molecular ice features of H_2O, CO, CO_2, and CH_3OH. Early infrared observations had already shown, however, that interstellar absorption spectra of lines-of-sight through the diffuse medium lacked the O–H stretching feature of H_2O ice at 3.08 μm that was predicted by the icy dust model (Danielson et al. 1965; Knacke et al. 1969). Consequently, new models of the interstellar dust composition called for refractory grains, mainly consisting of graphite (Hoyle & Wickramasinghe 1962), SiO_2 (Kamijo 1963), and SiC (Friedemann 1969). Recent models that also take into account the cosmic abundance of chemical elements as well as polarization measurements suggest grains of various sizes and compositions: small graphite grains, silicate grains, and composite grains containing carbon (amorphous, hydrogenated, or graphitic), silicates, and oxides (Mathis 1996). Also, grains consisting of a silicate core with an organic refractory mantle as well as nanometer-sized grains, or large molecules, of polycyclic aromatic hydrocarbons are discussed (Li & Greenberg 1997).

For *interplanetary* cosmic dust grains the sources are well known: comets and asteroids continuously release dust grains as they disintegrate due to solar heating or as they collide with each other. But what is the origin of interstellar grains? It is known that evolved stars continuously lose mass. About 90% of the stellar mass loss is provided by cool high-luminosity stars, in particular by asymptotic giant branch (AGB) and post-AGB stars. As the ejected material cools in the expanding stellar wind, solid particles condense out of the gas phase (for reviews see Whittet (1989); Whittet et al. (1992); Sedlmayr (1994)). This so-called "stardust" provides the seeds for grains that grow in cool interstellar clouds by accretion of atoms and molecules and by agglomeration (Ossenkopf & Henning 1994). From observations of the interstellar medium it is known that dust is not only created there, but also destroyed. Shocks in the interstellar medium are mainly caused by supernova explosions and supersonic stellar winds. It is in these shocks that dust grains are ground down by the plasma and mutual collisions (Jones *et al.* 1994). Additionally, dust grains lose their volatile constituents when they are exposed to the interstellar UV radiation (Greenberg *et al.* 1995). Ultimately, an interstellar dust grain can be incorporated and destroyed in a newly forming star, or it becomes part of a planetary system. This way cosmic dust is repeatedly recycled through the galactic evolution process (Dorschner & Henning 1995). More locally, dust is contained in the interstellar clouds that surround the solar system. Our Sun is located in one of these clouds that was found to be a small-scale (≈ 1 pc) part of the warm medium. The local interstellar medium is believed to be part of a fragmented super-bubble shell created by repeated supernova explosions in the nearby Scorpius–Centaurus star-forming region (Frisch 1995, 1998).

From the measurement of the Doppler-shift of the absorption lines found in lines of sight towards nearby stars it is known that our Sun moves relative with respect to the local cloud at 26 km s^{-1}. Assuming that there is a dust component embedded in the local cloud, we can expect that interstellar cosmic dust traverses the solar system. Early attempts to detect interstellar dust in the solar system with in situ instruments on board the Pioneers 8 and 9 spacecraft failed or were inconclusive (McDonnell & Berg 1975). The non-detection of interstellar dust by Pioneers 8 and 9 was interpreted to be caused by the electrostatic charge that grains acquire in the interplanetary environment. This charge causes the grains to interact with the solar wind magnetic field that diverts them out of the heliosphere. A more detailed analysis of the proposed diversion effect (Gustafson & Misconi 1979; Morfill & Grün 1979) showed that the solar wind magnetic field is not only capable of diverting small interstellar grains but also of concentrating them towards the solar equator plane, depending on the phase of the solar cycle. More than a decade after these discussions, interstellar dust grains were unambiguously detected in the solar system with in situ instruments (Grün *et al.* 1993). After its fly-by of Jupiter, the highly sensitive dust detector on board the joint ESA/NASA spacecraft Ulysses detected impacts predominantly from a direction that was opposite from the expected impact direction of interplanetary dust grains on low-inclination and low-eccentricity orbits. It was found that on average the impact velocities exceeded the local solar system escape velocity, even if radiation pressure effects were neglected (Grün *et al.* 1994a). Once it became evident that galactic interstellar dust is accessible to in situ detection and even sample return to Earth, NASA selected the Stardust mission to analyze and return to Earth samples of interstellar dust at asteroid belt distances (Brownlee *et al.* 1997). Stardust was launched on 7 February 1999. While the primary mission of Stardust is the encounter with comet Wild-2 in January 2004, its secondary goal is to collect and detect the interstellar grains that have been discovered by Ulysses. With its time-of-flight mass spectrometer Stardust is capable of analyzing the chemical composition of local interstellar dust. Stardust returns to Earth in January 2006, where it sends a capsule into the Earth's atmosphere that contains the collected dust samples.

From the analysis of the Ulysses and Galileo data it was found that the motion of interstellar dust through the solar system is parallel to the stream of neutral interstellar gas that was also detected by Ulysses, which means that local interstellar dust and gas are nearly in rest with respect to each other (Witte *et al.* 1993; Frisch *et al.* 1999). Figure 3.10 shows the determination of the ecliptic upstream direction of interstellar dust by fitting a mono-directional flow model to the Ulysses data. The flux of interstellar dust grains,

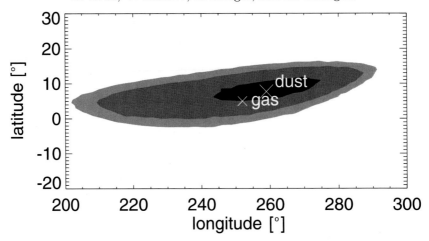

FIGURE 3.10. Best-fit analysis of the dust direction using the Ulysses in situ data. The black, dark, and light gray areas show the 1σ, 2σ, and 3σ confidence levels of the interstellar dust upstream direction. The best-fit direction (indicated by the large cross) is at $8°$ ecliptic latitude and $259°$ ecliptic latitude. The upstream direction of interstellar neutral Helium is shown by the small cross. Both directions are indistinguishable within the 1σ confidence level.

which is continuously measured by Ulysses, began to decrease in mid-1996. The observed decrease by a factor of three was interpreted to be the result of the electro-magnetic interaction of the grains with the solar wind magnetic field (Landgraf 2000). The polarity of the solar wind magnetic field changes with the 22-year solar cycle. During the solar maximum in 1991 the field polarity became north-pointing. The azimuthal component of the corresponding radially expanding Parker (1958) spiral field deflects interstellar grains in the northern hemisphere to the north and grains in the southern hemisphere to the south. Thus, the net effect of the interplanetary magnetic field during the 1991 polarity cycle was to divert interstellar dust out of the solar system. During the 2000/2001 solar maximum the field polarity reversed again. Then, interstellar grains in the northern hemisphere were deflected to the south and grains in the southern hemisphere to the north. This focusing effect will increase the flux of interstellar grains in the inner Solar System significantly, starting in 2005.

The in situ measurements as well as the detection of interstellar meteors (Taylor *et al.* 1996) show that interstellar dust can be measured at 1 AU. The analysis of data obtained by the dust detector on board the Hiten satellite, which is on a highly eccentric orbit about the Earth, indicates that micron- and sub-micron-sized interstellar grains indeed reach 1 AU. Measurements by the Cosmic Dust Analyzer (CDA) instrument on board the Cassini spacecraft also suggest the presence of interstellar dust in the inner solar system. Modeling of the Ulysses data furthermore implies that up to 30% of the flux of cosmic dust grains larger than 10^{-13} g at 1 AU is of interstellar origin (Grün *et al.* 1997). The presence of interstellar dust in Earth's vicinity gives us the opportunity to obtain information about stellar evolution processes in our galaxy that is otherwise inaccessible to in situ measurements with today's technology.

The analysis of the interstellar dust stream by means of a dust telescope in Earth orbit is desirable, because a dust telescope allows the precise determination of the orbit for grains larger than 10^{-11} g. From the Ulysses results, one large grain is expected to hit a target area of 1 m^2 every 12 days. Thus, a 1 m^2 dust telescope can already detect more than 50 large interstellar grains in a 2-year mission. These relatively large interstellar grains are particularly interesting, because their trajectory can be traced back to their origin, be it a nearby dusty interstellar cloud, or another planetary system. When the dust telescope is pointed into the interstellar stream direction, more than 10 small ($m < 10^{-11}$ g) interstellar grains are expected to hit the target per square meter per day. The high-resolution mass spectroscopy capability of the dust telescope allows the determination not only of the chemical but also the isotopic composition of the impacting grain.

A comparison of the isotopic properties of interstellar dust measured by a dust telescope to analyses of sub-micron grains embedded in cosmic interplanetary dust particles (IDPs) collected in the Earth's atmosphere gives important information about the connection between the ancient solar nebula cosmic dust and contemporary interstellar dust. An in situ measurement also provides "ground" truth for the laboratory analysis of IDPs. The most direct evidence for the interstellar origin of constituents of IDPs comes from their infrared spectrum. The spectra of silicate phases in IDPs match spectra of circumstellar and interstellar material measured by astronomical observations (Bradley *et al.* 1999). The physical properties of these silicate phases that are dubbed GEMS (Glass with Embedded Metal and Sulfides) are exotic (for example, they contain super-paramagnetic metal inclusions) but similar to those found throughout interstellar and circumstellar space (Bradley 1994). Isotopic anomalies of H, N, O in IDPs indicate a presolar or interstellar origin of other constituents of IDPs (Tielens 1997; Zinner 1998). In some IDPs hotspots of extreme isotopic anomalies up to a factor of 50 have been found (Messenger & Walker 1997). Due to the small size of these spots it is difficult, however, to identify the carriers of the anomaly.

Cosmic dust has also been observed in other galaxies. Its most prominent effect is the emission of electromagnetic radiation in the 100 to 1000 μm wavelength region (far infrared to sub-millimeter). At least the lower end of this spectral range is inaccessible for ground-based observations, because the Earth's atmosphere is optically thick. The use of space-based infrared telescopes, especially the Infrared Space Observatory (ISO), increased our capability for mapping and analyzing the dust components of other galaxies. From the infrared observations the fraction of the mass of the observed galaxy that is in the form of gas and the fraction that is in the form of dust can be derived. The average value for the gas-to-dust ratio of our Milky Way is approximately 100. The value derived for other galaxies is pretty much uncertain, especially due to the uncertainty in the amount of molecular gas and the emission efficiency of the dust (Papadopoulos & Allen 2000; Alton *et al.* 2000). For example, the gas-to-dust ratio found for the distant galaxy NGC 891 is in the order of 260. The question arises as to whether deviation arises from uncertainties in the measurements or from significant variation in the dust content from galaxy to galaxy. Another dust phenomenon that has been observed in distant, as well as nearby galaxies is the appearance of two separate components of dust (Haas *et al.* 1998). One of these components is very cold (10 to 20 K) and the other is "warm" (≈ 45 K). The interpretation of this observation is that the cold dust is embedded in optically thick molecular clouds, while the warm dust component is exposed to the interstellar radiation field. It was found that the warm component becomes more important (in terms of the mass fraction) closer to the galactic center (Trewhella *et al.* 2000).

3.6. Summary

Space dust (size less than 0.1 mm) is the small particle extension of meteor particles. Because of their abundance and the large cross-section-to-mass ratio clouds of dust particles can be observed by their scattered sunlight. Comets, planetary rings, and the faint zodiacal cloud are examples of clouds of space dust. Dust particles in space have different dynamics than their larger parent bodies. While large objects in interplanetary space are affected only by gravitation, small particles feel solar radiation pressure and even the electromagnetic interactions with the ambient magnetic field. These forces rapidly disperse dust in space so that they can be found at large distances from their sources within short time intervals.

Micron-sized dust grains are generally too small to observe the meteor phenomenon upon entry into planetary atmospheres but the interaction with solid surfaces like those of space dust detectors can be recorded. In situ dust detectors of less than 1 m^2 sensitive area that were mounted on interplanetary space probes have found dust everywhere in the planetary system. These instruments do not only detect space dust, but they can provide a wealth of information on individual particles. There are modern dust analyzers that determine the mass, velocity, chemical composition and electrical charge of micron-sized dust grains in space. Because of their versatility in situ detectors are carried on most space missions to unknown territory in the planetary system.

Satellite and ring systems of the outer planets are models for the (early) solar system with satellites and ring particles in intimate interaction. These systems were studied by optical observations from spacecraft but additional information was gained from in situ studies of dust in the jovian system. Clouds of ejecta particles in the vicinity of satellites have been observed that were emitted upon the constant hail of interplanetary meteoroids impacting the satellites. These ejecta particles are the material from which dust rings around planets form. Even exotic dust sources like the volcanos on Jupiter's moon Io have been identified.

Comets are sources of interplanetary dust particles. This has been suggested by optical zodiacal light, comet, and meteor observations. Space missions to comet Halley found dust emitted over a wide size range. Recent observations of the thermal emission from the zodiacal cloud confirmed that asteroids are a significant source of interplanetary dust as well. The relative contributions from these different sources are still under debate. The size distribution of interplanetary dust particles was derived from analyses of microcraters on lunar rocks and on satellite surfaces that were exposed for some time to the interplanetary dust flux. Synthesis of different observations of interplanetary dust is provided by dust models like that of Divine (1993). This and similar models provide an empirical description of the size and orbital element distribution of interplanetary dust particles from which other properties of interplanetary dust like spatial densities, dust fluxes, line-of-sight brightnesses, etc. can be derived. New attempts are under way which describe the interplanetary dust complex by a physical model that starts from the sources and follows the evolution of dust in space and time. The key processes are planetary scattering of meteor and larger particles and the Poynting–Robertson effect of small dust grains. The larger particles evolve in space while mutual collisions grind them down until they reach the size of dust. These dust grains are then removed from the solar system by the dynamic effects that act on small particles.

The study of interstellar dust has long been the domain of astronomers. They observed the obscuration of starlight by foreground dust and derived some compositional information from solid state features through spectroscopic observations. For about ten years interstellar dust has been found accessible by in situ measurements within the planetary system. Interstellar grains sweep through the solar system in a collimated stream that is modulated by combined effects of solar gravity, radiation pressure and electromagnetic interaction with the solar wind magnetic field. Space missions are on their way to better characterize this basic material from which stars and planet form.

The field of space dust research is far from being complete. Despite the great advances made in the last years of the understanding of the interplanetary and planetary dust environments, there remain many important questions to be answered. Starting close to the Sun, nature separates meteoroid material according to its volatility. Analysis of the spatial distribution of matter close to the Sun will immediately give us information on the volatility of its constituents.

Closer to home, the dust environment of the Earth is of interest, because humans are affecting this environment due to their space activities. However, the natural dust environment is also of technological and scientific interest. Hazards from meteor streams (like the Leonids) will require continuous attention. Meteoroids that pass the Earth are of scientific interest because they are messengers from distant worlds, like asteroids, comets and even interstellar space. Once we know where dust grains originate from, compositional analysis of grains can tell us many things about those worlds. Therefore, the goal of such dust studies is to identify the sources of dust particles together with their in-depth analysis. The future of dust measurements in Earth orbit lies in three areas: (1) environmental monitoring, (2) use of dust telescopes to separate and analyze dust populations of different origin, and (3) collection and sample return of dust from various sources for in-depth analysis in laboratories.

Analysis of particulates from comet Halley brought us new and important information that has relevance to the understanding of the formation of our planetary system. Currently the Stardust mission is on its way to analyze, collect and return dust from comet Wild-2. This will give us a second example from the large variety of comets. It is the analysis of this variety that will tell us about the spatial and compositional variations in the protoplanetary nebula through the comets that may have sampled different regions of this nebula.

Enhanced dust densities in the Martian environment have long been suspected. The comparison of future dust observations at Mars with those in the dusty rings of Saturn and the other giant planets will tell us what effects solar radiation pressure (strongest at Mars) and planetary magnetospheres (negligible at Mars) have.

All giant planets have ring systems which display a variety of different "dusty" phenomena. While Saturn's ring system is of high complexity, the rings of Jupiter, Uranus and Neptune show other features that have not yet been found elsewhere. The Pluto–Charon system may have rings with yet other features. To understand the common characteristics of all these rings and the reasons why they are so different requires detailed measurements of the rings and their environments.

Detection of Edgeworth–Kuiper belt objects (EKBOs) of diameters of up to a few hundred kilometers confirmed the existence of objects beyond the orbits of the planets. Mutual collisions among EKBOs as well as impacts of interstellar grains generate dust locally. The action of the Poynting–Robertson effect together with resonances with the outer giant planets, interaction with the solar wind and neutral interstellar gas, may have lead to radial and azimuthal structure of the distribution of dust at the edge of the planetary system. The detection of infrared excess at main sequence stars started renewed interest in the outer extensions of our own solar system dust cloud. In particular, the observation of dust disks around β Pictoris stimulated this interest. Observations of the Edgeworth–Kuiper dust belt in our solar system can, therefore, be used as a model for extra-solar dust clouds and can help to reveal information about other planetary systems.

Effects of the solar-cycle-dependent heliosphere reach into interstellar space out to about 300 AU from the Sun where it interacts with the small ($\leq 0.1\,\mu$m) particles entering the heliosphere and modulates their flow. Their origin, however, may be different from that of larger grains that are accessible in Earth orbit. We know that evolved stars continuously lose mass. This "stardust" provides the seeds for ISD grains that grow in cool interstellar clouds by accretion of atoms and molecules and by agglomeration. An unbiased look into this interstellar dust factory will provide us with information on processes that are difficult to quantify by astronomical observations alone. Thereby, in situ dust analysis will be an important method when automated probes will leave our solar system.

REFERENCES

ALTON, P. B., XILOURIS, E. M., BIANCHI, S., DAVIES, J. & KYLAFIS, N. 2000 Dust properties of external galaxies; NGC 891 revisited. *Astronomy and Astrophysics* **356**, 795–807.

AUER, S. & SITTE, K. 1968 Detection technique for micrometeoroids, using impact ionization. *Earth and Plan. Sci. Let.* **4**, 178–183.

BERG, O. E. & RICHARDSON, F. F. 1969 The Pioneer 8 cosmic dust experiment. *Rev. Sci. Instrum.* **40**, 1333–1337.

BRADLEY, J. P. 1994 Chemically anomalous preaccretionally irradiated grains in interplanetary dust from comets. *Science* **265**, 925–929.

BRADLEY, J. P., KELLER, L. P., SNOW, T. P., HANNER, M. S., FLYNN, G. J., GEZO, J. C., CLEMETT, S. J., BROWNLEE, D. E. & BOWEY, J. E. 1999 An infrared spectral match between GEMS and interstellar grains. *Science* **285**, 1716–1718.

BREITER, S. & JACKSON, A. A. 1998 Unified analytical solutions to two-body problems with drag. *Monthly Notice of the Royal Astron. Soc.* **299**, 237–243.

BROWNLEE, D. E., TSOU, P., BURNETT, D., CLARK, B., HANNER, M. S., HORZ, F., KISSEL, J., McDONNELL, J. A. M., NEWBURN, R. L., SANDFORD, S., SEKANINA, Z., TUZZOLINO, A. J. & ZOLENSKY, M. 1997 The STARDUST mission: Returning comet samples to Earth. *Meteoritics & Planetary Science* **32**, A22.

BURNS, J. A., HAMILTON, D. P. & SHOWALTER, M. R. 2001 Dusty rings and circumplanetary dust: observations and simple physics. In *Interplanetary Dust* (ed. E. Grün, B. A. S. Gustafson, S. F. Dermott & H. Fechtig), pp. 641–725. Springer Verlag, Berlin.

BURNS, J. A., LAMY, P. L. & SOTER, S. 1979 Radiation forces on small particles in the solar system. *Icarus* **40**, 1–48.

BURNS, J. A., SCHAFFER, L. E., GREENBERG, R. J. & SHOWALTER, M. R. 1985 Lorentz resonances and the structure of the Jovian ring. *Nature* **316**, 115–119.

BURNS, J. A., SHOWALTER, M. R., HAMILTON, D. P., NICHOLSON, P. D., DE PATER, I., OCKERT-BELL, M. E. & THOMAS, P. C. 1999 The formation of Jupiter's faint rings. *Science* **284**, 1146–1150.

CLARK, F. O., TORBETT, M. V., JACKSON, A. A., PRICE, S. D., KENNEALY, J. P., NOAH, P. V., GLAUDELL, G. A. & COBB, M. 1993 The out-of-plane distribution of zodiacal dust near the Earth. *Astronomical Journal* **105**, 976–979.

COLWELL, J. E. 1996 Size distributions of circumplanetary dust. *Advances in Space Research* **17**, 161–170.

COLWELL, J. E. & ESPOSITO, L. W. 1990 A model of dust production in the Neptune ring system. *Geophysical Research Letters* **17**, 1741–1744.

COLWELL, J. E., HORÁNYI, M. & GRÜN, E. 1998 Capture of interplanetary and interstellar dust by the Jovian magnetosphere. *Science* **280**, 88–91.

DANIELSON, R. E., WOOLF, N. J. & GAUSTAD, J. E. 1965 A search for interstellar ice absorption in the infrared spectrum of μ Cephei. *Astrophysical Journal* **141**, 116–125.

DE PATER, I., SHOWALTER, M. R., BURNS, J. A., NICHOLSON, P. D., LIU, M. C., HAMILTON, D. P. & GRAHAM, J. R. 1999 Keck infrared observations of Jupiter's ring system near Earth's 1997 ring plane crossing. *Icarus* **138**, 214–223.

DERMOTT, S. F., GROGAN, K., DURDA, D. D., JAYARAMAN, S., KEHOE, T. J. J., KORTENKAMP, S. J. & WYATT, M. C. 2001 Orbital evolution of interplanetary dust. In *Interplanetary Dust*, (ed. E. Grün, B. A. S. Gustafson, S. F. Dermott & H. Fechtig), pp. 569–639. Springer Verlag, Berlin.

DIETZEL, H., EICHHORN, G., FECHTIG, H., GRÜN, E., HOFFMANN, H. J. & KISSEL, J. 1973

The IEOS 2 and HELIOS micrometeoroid experiments. *Journal of Physics and Scientific Instruments* **6**, 209–217.

DIVINE, N. 1993 Five populations of interplanetary meteoroids. *Journal of Geophysical Research* **98**, 17029–17048.

DORSCHNER, J. & HENNING, T. 1995 Dust metamorphis in the galaxy. *Astron. Astrophs. Rev.* **6**, 271–333.

DOYLE, L. R. & GRÜN, E. 1990 Radiative transfer modeling constraints on the size of the spoke particles in Saturn's rings. *Icarus* **85**, 168–190.

DUBININ, E. M., PISSARENKO, N. F., BARABASH, S. V., ZAKHAROV, A. V. & LUNDIN, R. 1991 Plasma and magnetic field effects associated with Phobos and Deimos tori. *Planetary and Space Science* **39**, 113–121.

DUBININ, E. M. 1993 The Phobos and Deimos effects. *Advances in Space Research* **13(10)**, 271–290.

DURDA, D. D. & DERMOTT, S. F. 1997 The collisional evolution of the asteroid belt and its contribution to the zodiacal cloud. *Icarus* **130**, 140–164.

ESPOSITO, L. W. & COLWELL, J. E. 1989*a* Creation of the Uranus rings and dust bands. *Nature* **339**, 605–607.

ESPOSITO, L. W. & COLWELL, J. E. 1989*b* Errata: Creation of the Uranus rings and dust bands. *Nature* **340**, 322.

ESPOSITO, L. W., CUZZI, J. N., HOLBERG, J. B., MAROUF, E. A., TYLER, G. L. & PORCO, C. C. 1984 Saturn's rings: structure, dynamics and particle properties. In *Saturn* (ed. T. Gehrels and M. S. Matthews), pp. 463–545. University of Arizona Press.

FECHTIG, H., GRÜN, E. & KISSEL, J. 1978 Laboratory simulation. In *Cosmic Dust* (ed. J. A. M. McDonnel), pp. 607–669. Wiley-Interscience, Chichester, UK.

FECHTIG, H., LEINERT, C. & BERG, O. E. 2001 Historical perspectives. In *Interplanetary Dust* (ed. E. Grün, B. A. S. Gustafson, S. F. Dermott & H. Fechtig), pp. 1–56. Springer Verlag, Berlin.

FRIEDEMANN, C. 1969 Evolution of silicon carbide particles in the atmospheres of carbon stars. *Physica* **41**, 139–143.

FRIICHTENICHT, J. F. 1962 Two-million-volt electrostatic accelerator for hypervelocity research. *Rev. Sci. Instr.* **33**, 209–212.

FRISCH, P. C. 1995 Characteristics of nearby interstellar matter. *Space Science Reviews* **72**, 499–592.

FRISCH, P. C. 1998 The local bubble, local fluff, and heliosphere. In *Lecture Notes in Physics* (ed. D. Breitschwerdt, M. J. Freyberg & J. Trumper), vol. 506, pp. 269–278. Springer Verlag, Berlin.

FRISCH, P. C., DORSCHNER, J., GEISS, J., GREENBERG, J. M., GRÜN, E., LANDGRAF, M., HOPPE, P., JONES, A. P., KRÄTSCHMER, W., LINDE, T. J., MORFILL, G. E., REACH, W. T., SLAVIN, J., SVESTKA, J., WITT, A. & ZANK, G. P. 1999 Dust in the local interstellar wind. *Astrophysical Journal* **525**, 492–516.

GARRETT, H. B., DROUILHET, S. J., OLIVER, J. P. & EVANS, R. W. 1999 Interplanetary meteoroid environment model update. *Journal of Spacecraft and Rockets* **36** (1), 124–132.

GEHRELS, T., BAKER, L. R., BESHORE, E., BLENMAN, C., BURKE, J. J., CASTILLO, N. D., DACOSTA, B., DEGEWIJ, J., DOOSE, L. R., FOUNTAIN, J. W., GOTOBED, J., KENKNIGHT, C. E., KINGSTON, R., MCLAUGHLIN, G., MCMILLAN, R., MURPHY, R., SMITH, P. H., STOLL, C. P., STRICKLAND, R. N., TOMASKO, M. G., WIJESINGHE, M. P., COFFEEN, D. L. & ESPOSITO, L. W. 1980 Imaging photopolarimeter on Pioneer Saturn. *Science* **207**, 434–439.

GOR'KAVYI, N. N., OZERNOY, L. M. & MATHER, J. C. 1997 A new approach to dynamical evolution of interplanetary dust. *Astrophysical Journal* **474**, 496–502.

GRAPS, A. L., GRÜN, E., SVEDHEM, H., KRÜGER, H., HORÁNYI, M., HECK, A. & LAMMERS, S. 2000 Io as a source of the Jovian dust streams. *Nature* **405**, 48–50.

GREENBERG, J. M., LI, A., MENDOZA-GOMEZ, C. X., SCHUTTE, W. A., GERAKINES, P. A. & GROOT, M. D. 1995 Approaching the interstellar grain organic refractory component. *Astrophysical Journal Letters* **455**, L177–L180.

GRÜN, E., BAGUHL, M., SVEDHEM, H. & ZOOK, H. A. 2001 In situ measurements of cosmic dust. In *Interplanetary Dust* (ed. E. Grün, B. A. S. Gustafson, S. F. Dermott & H. Fechtig), pp. 295–346. Springer Verlag, Berlin.

GRÜN, E., FECHTIG, H., HANNER, M. S., KISSEL, J., LINDBLAD, B. A., LINKERT, D., MAAS, D., MORFILL, G. E. & ZOOK, H. A. 1992a The Galileo dust detector. *Space Science Reviews* **60**, 317–340.

GRÜN, E., FECHTIG, H., KISSEL, J., LINKERT, D., MAAS, D., McDONNELL, J. A. M., MOR-FILL, G. E., SCHWEHM, G., ZOOK, H. A. & GIESE, R. H. 1992b The Ulysses dust experiment. *Astronomy and Astrophysics Supplement Series* **92**, 411–423.

GRÜN, E., GUSTAFSON, B. Å. S., MANN, I., BAGUHL, M., MORFILL, G. E., STAUBACH, P., TAYLOR, A. & ZOOK, H. A. 1994a Interstellar dust in the heliosphere. *Astronomy and Astrophysics* **286**, 915–924.

GRÜN, E., HAMILTON, D., BAGUHL, M., RIEMANN, R., HORÁNYI, M. & POLANSKEY, C. 1994b Dust streams from comet Shoemaker–Levy 9? *Geophysical Research Letters* **21**, 1035–1038.

GRÜN, E., HAMILTON, D. P., RIEMANN, R., DERMOTT, S., FECHTIG, H., GUSTAFSON, B. A., HANNER, M. S., HECK, A., HORÁNYI, M., KISSEL, J., KIVELSON, M., KRÜGER, H., LIND-BLAD, B. A., LINKERT, D., LINKERT, G., MANN, I., McDONNELL, J. A. M., MORFILL, G. E., POLANSKEY, C., SCHWEHM, G., SRAMA, R. & ZOOK, H. A. 1996 Dust measurements during Galileo's approach to Jupiter and Io encounter. *Science* **274**, 399–401.

GRÜN, E., KRÜGER, H., GRAPS, A., HAMILTON, D. P., HECK, A., LINKERT, G., ZOOK, H., DERMOTT, S., FECHTIG, H., GUSTAFSON, B., HANNER, M., HORÁNYI, M., KISSEL, J., LINDBLAD, B., LINKERT, G., MANN, I., McDONNELL, J. A. M., MORFILL, G. E., POLANSKEY, C., SCHWEHM, G. & SRAMA, R. 1998 Galileo observes electromagnetically coupled dust in the jovian magnetosphere. *Journal of Geophysical Research* **103**, 20011–20022.

GRÜN, E., STAUBACH, P., BAGUHL, M., HAMILTON, D. P., ZOOK, H. A., DERMOTT, S., GUSTAFSON, B. A., FECHTIG, H., KISSEL, J., LINKERT, D., LINKERT, G., SRAMA, R., HANNER, M. S., POLANSKEY, C., HORÁNYI, M., LINDBLAD, B. A., MANN, I., McDON-NELL, J. A. M., MORFILL, G. E. & SCHWEHM, G. 1997 South–north and radial traverses through the interplanetary dust cloud. *Icarus* **129**, 270–288.

GRÜN, E., ZOOK, H. A., BAGUHL, M., BALOGH, A., BAME, S. J., FECHTIG, H., FORSYTH, R., HANNER, M. S., HORÁNYI, M., KISSEL, J., LINDBLAD, B. A., LINKERT, D., LINKERT, G., MANN, I., McDONNELL, J. A. M., MORFILL, G. E., PHILLIPS, J. L., POLANSKEY, C., SCHWEHM, G., SIDDIQUE, N., STAUBACH, P., SVESTKA, J. & TAYLOR, A. 1993 Discovery of Jovian dust streams and interstellar grains by the Ulysses spacecraft. *Nature* **362**, 428–430.

GRÜN, E., ZOOK, H. A., BAGUHL, M., BALOGH, A., BAME, S. J., FECHTIG, H., FORSYTH, R., HANNER, M. S., HORÁNYI, M., KISSEL, J., LINDBLAD, B.-A., LINKERT, D., LINKERT, G., MANN, I., McDONNELL, J. A. M., MORFILL, G. E., PHILLIPS, J. L., POLANSKEY, C., SCHWEHM, G., SIDDIQUE, N., STAUBACH, P., SVESTKA, J. & TAYLOR, A. 1993 Discovery of jovian dust streams and interstellar grains by the Ulysses spacecraft. *Nature* **362**, 428–430.

GRÜN, E., ZOOK, H. A., FECHTIG, H. & GIESE, R. 1985 Collisional balance of the meteoritic complex. *Icarus* **62**, 244–272.

GURNETT, D. A., ANSHER, J. A., KURTH, W. S. & GRANROTH, L. J. 1997 Micron-sized dust particles detected in the outer solar system by Voyager 1 and 2 plasma wave instruments. *Geophysical Research Letters* **24**, 3125–3128.

GURNETT, D. A., GRÜN, E., GALLAGHER, D., KURTH, W. S. & SCARF, F. L. 1983 Micron-sized particles detected near Saturn by the Voyager plasma wave instrument. *Icarus* **53**, 236–254.

GURNETT, D. A., KURTH, W. S., GRANROTH, L. J., ALLENDORF, S. C. & POYNTER, R. L. 1991 Micron-sized particles detected near Neptune by the Voyager 2 plasma wave instrument. *Journal Geophysical Research Supplement* **96**, 19177–19186.

GURNETT, D. A., KURTH, W. S., SCARF, K. L., BURNS, J. A. & CUZZI, J. N. 1987 Micron-sized particle impacts detected near Uranus by the Voyager 2 plasma wave instrument. *Journal of Geophysical Research* **92**, 14959–14968.

GUSTAFSON, B. S. & MISCONI, N. Y. 1979 Streaming of interstellar grains in the solar system. *Nature* **282**, 276–278.

HAAS, M., LEMKE, D., STICKEL, M., HIPPELEIN, H., KUNKEL, M., HERBSTMEIER, U. &

MATTILA, K. 1998 Cold dust in the Andromeda Galaxy mapped by ISO. *Astronomy and Astrophysics* **338**, L33–L36.

HAMILTON, D. & BURNS, J. 1994 Origin of Saturn's E ring: Selfsustained – naturally. *Science* **264**, 550–553.

HAMILTON, D. P. 1996 The asymmetric time-variable rings of Mars. *Icarus* **119**, 153–172.

HAMILTON, D. P. & BURNS, J. A. 1993 Ejection of dust from Jupiter's gossamer ring. *Nature* **364**, 695–699.

HANNER, M. S. & WEINBERG, J. L. 1973 Gegenschein observations from Pioneer 10. *Sky and Telescope* **45**, 217–218.

HANSEN, D. O. 1968 Mass analysis of ions produced by hypervelocity impact. *Appl. Phys. Letters* **13**, 89.

HENYEY, L. G. & GREENSTEIN, J. L. 1941 Diffuse radiation in the galaxy. *Astrophysical Journal* **93**, 70–83.

HORÁNYI, M. 1996 Charged dust dynamics in the solar system. *Annual Review of Astronomy and Astrophysics* **34**, 383–418.

HORÁNYI, M. & BURNS, J. A. 1991 Charged dust dynamics – Orbital resonance due to planetary shadows. *Journal of Geophysical Research* **96**, 19283.

HORÁNYI, M., GRÜN, E. & HECK, A. 1997 Modeling the Galileo dust measurements at Jupiter. *Geophysical Research Letters* **24**, 2175–2178.

HORÁNYI, M., MORFILL, G. & GRÜN, E. 1993a Mechanism for the acceleration and ejection of dust grains from Jupiter's magnetosphere. *Nature* **363**, 144–146.

HORÁNYI, M., MORFILL, G. & GRÜN, E. 1993b The dusty ballerina skirt of Jupiter. *Journal of Geophysical Research* **98**, 221–245.

HOYLE, F. & WICKRAMASINGHE, N. C. 1962 On graphite particles as interstellar grains. *Monthly Notices of the Royal Astronomical Society* **124**, 417–433.

HUBBARD, W. B., BRAHIC, A., SICARDY, B., ELICER, L., ROQUES, F. & VILAS, F. 1986 Occultation detection of a Neptunian ring-like arc. *Nature* **319**, 636–640.

HUMES, D. H. 1976 The Jovian meteoroid environment. In *Jupiter* (ed. T. Gehrels), pp. 1052–1067. Univ. of Arizona Press, Tucson.

HUMES, D. H. 1980 Results of Pioneer 10 and 11 meteoroid experiments: interplanetary and near-Saturn. *Journal of Geophysical Research* **85**, 5841–5852.

HUMES, D. H., ALVAREZ, J. M., O'NEAL, R. L. & KINARD, W. H. 1974 The interplanetary and near-Jupiter meteoroid environment. *Journal of Geophysical Research* **79**, 3677–3684.

IGENBERGS, E. & KUCZERA, H. 1979 Micrometeoroid and dust simulation. In *Comet Halley Micrometeoroid Hazard Workshop*, pp. 109–114.

IGENBERGS, E., SASAKI, S., MÜNZENMAYER, R., OHASHI, H., FÄRBER, G., FISCHER, F., FUJIWARA, A., GLASMACHERS, A., GRÜN, E., HAMABE, Y., IGLSEDER, H., KLINGE, D., MAIYAMOTO, H., MUKAI, T., NAUMANN, W., NOGAMI, K.-I., SCHWEHM, G., SVEDHEM, H. & YAMAKOSHI, K. 1998 Mars dust counter. *Earth Planets Space* **50**, 241–245.

IGLSEDER, H., UESUGI, K. & SVEDHEM, H. 1996 Cosmic dust measurements in lunar orbit. *Advances in Space Research* **17(12)**, 177–182.

ISHIMOTO, H. 2000 Modeling the number density distribution of interplanetary dust on the ecliptic plane within 5 AU of the Sun. *Astronomy and Astrophysics* **362**, 1158–1173.

JESSBERGER, E. K., STEFAN, T., ROST, D., ARNDT, P., MAETZ, M., STADERMANN, F. J., BROWNLEE, J. P., BRADLEY, J. P. & KURAT, G. 2001 Properties of interplanetary dust: Information from collected samples. In *Interplanetary Dust* (ed. E. Grün, B. A. S. Gustafson, S. F. Dermott & H. Fechtig), pp. 253–294. Springer Verlag, Berlin.

JEWITT, D. & LUU, J. 1993 Discovery of the candidate Kuiper belt object 1992 QB1. *Nature* **362**, 730–732.

JONES, A. P., TIELENS, A. G. G. M., HOLLENBACH, D. J. & McKEE, C. F. 1994 Grain destruction in shocks in the interstellar medium. *Astrophysical Journal* **433**, 797–810.

KAMIJO, F. 1963 A theoretical study of the long period variable stars III. Formation of solid or liquid particles in the circumstellar envelope. *Publ. Astron. Soc. Japan* **15**, 440–448.

KELSALL, T., WEILAND, J. L., FRANZ, B. A., REACH, W. T., ARENDT, R. G., DWEK, E., FREUDENREICH, H. T., HAUSER, M. G., MOSELEY, S. H., ODEGARD, N. P., SILVERBERG, R. F. & WRIGHT, E. L. 1998 The COBE diffuse infrared background experiment search for the cosmic infrared background. II. model of the interplanetary dust cloud. *Astrophysical Journal* **508**, 44–73.

KHOLSHEVNIKOV, K. V., KRIVOV, A. V., SOKOLOV, L. L. & TITOV, V. B. 1993 The dust torus around Phobos orbit. *Icarus* **105**, 351–362.

KISSEL, J. 1986 The Giotto particulate impact analyzer. In *European Space Agency Spec. Publ. ESA SP-1077*, pp. 67–83.

KNACKE, R. F., CUDABACK, D. D. & GAUSTAD, J. E. 1969 Infrared spectra of highly reddened stars: A search for interstellar ice grains. *Astrophysical Journal* **158**, 151.

KORTENKAMP, S. J. & DERMOTT, S. F. 1998 Accretion of interplanetary dust particles by the earth. *Icarus* **135**, 469–495.

KRIVOV, A. V. 1994 On the dust belts of Mars. *Astronomy & Astrophysics* **291**, 657–663.

KRIVOV, A. V. & HAMILTON, D. P. 1997 Martian dust belts: Waiting for discovery. *Icarus* **128**, 335–353.

KRIVOV, A. V. & JUREWICZ, A. 1999 The ethereal dust envelopes of the Martian moons. *Planetary and Space Science* **47**, 45–56.

KRIVOV, A. V., KRÜGER, H., GRÜN, E., THIESSENHUSEN, K.-U. & HAMILTON, D. P. 2002 A tenuous dust ring of Jupiter formed by escaping ejecta from the Galilean satellites. *Journal of Geophysical Research*. In Press.

KRIVOV, A. V., WARDINSKI, I., SPAHN, F., KRÜGER, H. & GRÜN, E. 2002 Dust on the outskirts of the Jovian system. *Icarus*, In Press.

KRÜGER, H. & GRÜN, E. 2002 Dust en-route to Jupiter and the Galilean satellites. In *Proceedings of the IAU Colloquium 181 and COSPAR Colloquium 11 Dust in the Solar System and in Other Planetary Systems held at University of Kent at Canterbury in April 2000* (ed. S. Green). In Press.

KRÜGER, H., GRÜN, E., GRAPS, A. & LAMMERS, S. 1999a Observations of electromagnetically coupled dust in the Jovian magnetosphere. In *Proceedings of the VII. International Conference on Plasma Astrophysics and Space Physics, held in Lindau in May 1998* (ed. E. M. J. Büchner, I. Axford & V. Vasyliunas), vol. 264, pp. 247–256. Kluwer Academic Publishers, Dordrecht, The Netherlands.

KRÜGER, H., KRIVOV, A. V. & GRÜN, E. 2000 A dust cloud of Ganymede maintained by hypervelocity impacts of interplanetary micrometeoroids. *Planetary and Space Science* **48**, 1457–1471.

KRÜGER, H., KRIVOV, A. V., HAMILTON, D. P. & GRÜN, E. 1999b Detection of an impact-generated dust cloud around Ganymede. *Nature* **399**, 558–560.

LANDGRAF, M. 2000 Modeling the motion and distribution of interstellar dust inside the heliosphere. *Journal of Geophysical Research* **105** (A5), 10,303–10,316.

LEINERT, C. & GRÜN, E. 1990 Interplanetary dust. In *Physics of the Inner Heliosphere I* (ed. R. Schwenn & E. Marsch), pp. 207–275. Springer Verlag, Berlin.

LEINERT, C., RICHTER, I., PITZ, E. & PLANCK, B. 1981 The zodiacal light from 1.0 to 0.3 A.U. as observed by the HELIOS space probes. *Astronomy and Astrophysics* **103**, 177–188.

LEINERT, C., ROSER, S. & BUITRAGO, J. 1983 How to maintain the spatial distribution of interplanetary dust. *Astronomy and Astrophysics* **118**, 345–357.

LEVASSEUR-REGOURD, A. C. & DUMONT, R. 1980 Absolute photometry of zodiacal light. *Astronomy and Astrophysics* **84**, 277–279.

LI, A. & GREENBERG, J. M. 1997 A unified model of interstellar dust. *Astronomy and Astrophysics* **323**, 566–584.

LIOU, J. & ZOOK, H. A. 1997 Evolution of interplanetary dust particles in mean motion resonances with planets. *Icarus* **128**, 354–367.

LIOU, J., ZOOK, H. A. & DERMOTT, S. F. 1996 Kuiper belt dust grains as a source of interplanetary dust particles. *Icarus* **124**, 429–440.

LIOU, J.-C., ZOOK, H. A. & JACKSON, A. A. 1995 Radiation pressure, Poynting–Robertson drag, and solar wind drag in the restricted three-body problem. *Icarus* **116**, 186–201.

MATHIS, J. 1996 Dust models with tight abundance constraints. *Astrophysical Journal* **472**, 643–655.

MCDONNELL, J. A. M. & BERG, O. E. 1975 Bounds for the interstellar to solar system microparticle flux ratio over the mass range 10^{11}–10^{13} g. In *Space Research XV* (ed. M. J. Rycroft), pp. 555–563. Akademie-Verlag, Berlin.

MESSENGER, S. & WALKER, R. M. 1997 Evidence for molecular cloud material in meteorites and interplanetary dust. In *Astrophysical Implications of the Laboratory Study of Presolar Materials* (ed. T. J. Bernatowicz & E. Zinner), vol. CP402, pp. 523–544. New York: American Institute of Physics.

MEYER-VERNET, N., AUBIER, M. G. & PEDERSEN, B. M. 1986 Voyager 2 at Uranus – Grain impacts in the ring plane. *Geophysical Research Letters* **13**, 617–620.

MORFILL, G. E. & GOERTZ, C. K. 1983 Plasma clouds in Saturn's rings. *Icarus* **55**, 111–123.

MORFILL, G. E. & GRÜN, E. 1979 The motion of charged dust particles in interplanetary space ii. interstellar grains. *Planetary and Space Science* **27**, 1283–1292.

MORFILL, G. E., GRÜN, E. & LEINERT, C. 1986 The interaction of solid particles with the interplanetary medium. In *The Sun and the Heliosphere in Three Dimensions*, pp. 455–474.

MURDOCK, T. L. & PRICE, S. D. 1985 Infrared measurements of zodiacal light. *Astronomical Journal* **90**, 375–386.

NAKAMURA, R., FUJII, Y., ISHIGURO, M., MORISHIGE, K., YOKOGAWA, S., JENNISKENS, P. & MUKAI, T. 2000 The discovery of a faint glow of scattered sunlight from the dust trail of the Leonid parent comet 55P/Tempel–Tuttle. *Astrophysical Journal* **540**, 1172–1176.

OCKERT, M. E., CUZZI, J. N., PORCO, C. C. & JOHNSON, T. V. 1987 Uranian ring photometry – Results from Voyager 2. *Journal of Geophysical Research* **92**, 14,969–14,978.

OCKERT-BELL, M. E., BURNS, J. A., DAUBAR, I. J., THOMAS, P. C., VEVERKA, J., BELTON, M. J. S. & KLAASEN, K. P. 1999 The structure of Jupiter's ring system as revealed by the Galileo imaging experiment. *Icarus* **138**, 188–213.

ÖHMAN, Y. 1930 *Spectrophotometric Studies of B, A and F-type Stars*. Acta Regiae Societatis Scientiarum Upsaliensis, Uppsala: AlmQvist & Wiksell.

OORT, J. H. & VAN DE HULST, H. C. 1946 Gas and smoke in interstellar space. *Bull. Astron. Inst. Netherlands* **10**, 187.

OSSENKOPF, V. & HENNING, T. 1994 Dust opacities for protostellar cores. *Astronomy and Astrophysics* **291**, 943–959.

PAPADOPOULOS, P. P. & ALLEN, M. L. 2000 Gas and Dust in NGC 7469: Submillimeter imaging and CO J=3-2. *Astrophysical Journal* **537**, 631–637.

PARKER, E. N. 1958 Dynamics of interplanetary gas and magnetic fields. *Astrophysical Journal* **128**, 664–676.

RATCLIFF, P. R., McDONNELL, J. A. M., FIRTH, J. G. & GRÜN, E. 1992 The cosmic dust analyzer. *J. Br. Interplanet. Soc.* **45**, 375–380.

REACH, W. T. 1991 Zodiacal emission. ii Dust near ecliptic. *Astrophysical Journal* **369**, 529–543.

REACH, W. T., SYKES, M. V., LIEN, D. & DAVIES, J. K. 2000 The formation of Encke meteoroids and dust trail. *Icarus* **148**, 80–94.

RÖSER, S. & STAUDE, H. J. 1978 The zodiacal light from 1500 Å to 60 micron – Mie scattering and thermal emission. *Astronomy and Astrophysics* **67**, 381–394.

SAUER, K., DUBININ, E., BAUMGÄRTEL, K. & BOGDANOV, A. 1995 Deimos: an obstacle to the solar wind. *Science* **269**, 1075–1078.

SCHALÉN, C. 1929 Untersuchungen über Dunkelnebel. *Medd. Astron. Obs.* **58**.

SEDLMAYR, E. 1994 From molecules to grains. In *Molecules in the Stellar Environment VIII* (ed. U. G. Jorgensen), *Proceedings of the IAU Colloquium* 146, p. 163. Springer Verlag, Berlin.

SHOWALTER, M., CUZZI, J. & LARSON, S. 1991 Structure and particle properties of Saturn's E ring. *Icarus* **94**, 451–473.

SHOWALTER, M. R. 1995 Arcs and clumps in the Uranian λ ring. *Science* **267**, 490–493.

SHOWALTER, M. R. 1996 Saturn's D ring in the Voyager images. *Icarus* **124**, 677–689.

SHOWALTER, M. R., BURNS, J. A., CUZZI, J. N. & POLLACK, J. B. 1985 Discovery of Jupiter's 'gossamer' ring. *Nature* **316**, 526–528.

SHOWALTER, M. R., BURNS, J. A., CUZZI, J. N. & POLLACK, J. B. 1987 Jupiter's ring system – New results on structure and particle properties. *Icarus* **69**, 458–498.

SHOWALTER, M. R. & CUZZI, J. N. 1993 Seeing ghosts – Photometry of Saturn's G ring. *Icarus* **103**, 124–143.

SIMPSON, J. A. & TUZZOLINO, A. J. 1985 Polarized polymer films as electronic pulse detectors of cosmic dust particles. *Nucl. Instrum. and Methods* **A 236**, 187–202.

SMITH, B. A., SODERBLOM, L., BEEBE, R. F., BOYCE, J. M., BRIGGS, G., BUNKER, A., COLLINS, S. A., HANSEN, C., JOHNSON, T. V., MITCHELL, J. L., TERRILE, R. J., CARR, M. H., COOK, A. F., CUZZI, J. N., POLLACK, J. B., DANIELSON, G. E., INGERSOLL, A. P., DAVIES, M. E., HUNT, G. E., MASURSKY, H., SHOEMAKER, E. M., MORRISON, D., OWEN, T., SAGAN, C., VEVERKA, J., STROM, R. & SUOMI, V. E. 1981 Encounter with Saturn – Voyager 1 imaging science results. *Science* **212**, 163–191.

SMITH, B. A., SODERBLOM, L. A., BANFIELD, D., BARNET, C., BEEBE, R. F., BAZILEVSKII, A. T., BOLLINGER, K., BOYCE, J. M., BRIGGS, G. A., BRAHIC, A., BRIGGS, G. A., BROWN, R. H., CHYBA, C., COLLINS, S. A., COLVIN, T., COOK II, A. F., CRISP, D., CROFT, S. K., CRUIKSHANK, D., CUZZI, J. N., DANIELSON, G. E., DAVIES, M. E., E., D., DONES, L., GODFREY, D., GOGUEN, J., GRENIER, I., HAEMMERLE, V. R., HAMMEL, H., HANSEN, C. J., HELFENSTEIN, C. P., HOWELL, C., HUNT, G. E., INGERSOLL, A. P., V., J. T., KARGEL, J., KIRK, R., KUEHN, D. I., LIMAYE, S., MASURSKY, H., McEWEN, A., D., M., OWEN, T., OWEN, W., B., P. J., PORCO, C. C., RAGES, K., ROGERS, P., RUDY, D., SAGAN, C., SCHWARZ, J., A., S. L., STOKER, C., STROM, R. G., SUOMI, V. E., SYNOTT, S. P., TERRILE, R. J., THOMAS, P., THOMPSON, W. R., VERBISCER, A. & VEVERKA, J. 1989 Voyager 2 at Neptune – imaging science results. *Science* **246**, 1422–1449.

SMITH, B. A., SODERBLOM, L. A., BEEBE, R., BLISS, D., BROWN, R. H., COLLINS, S. A., BOYCE, J. M., BRIGGS, G. A., BRAHIC, A., CUZZI, J. N., DANIELSON, G. E., DAVIES, M. E., DOWLING, T. E., GODFREY, D., HANSEN, C. J., HARRIS, C., HUNT, G. E., INGERSOLL, A. P., V., J. T., KRAUS, R. J., MASURSKY, H., D., M., OWEN, T., OWEN, W., PLESCIA, J. B., B., P. J., PORCO, C. C., RAGES, K., SAGAN, C., SHOEMAKER, E. M., A., S. L., STOKER, C., STROM, R. G., SUOMI, V. E., SYNNOTT, S. P., TERRILE, R. J., THOMAS, P., THOMPSON, W. R. & VEVERKA 1986 Voyager 2 in the Uranian system – imaging science results. *Science* **233**, 43–64.

SOTER, S. 1971 The dust belts of Mars. *Tech. Rep.*. Center for Radiophysics and Space Research Report No. 462.

SRAMA, R. & GRÜN, E. 1996 The cosmic dust analyzer for the cassini mission to saturn. In *Physics, Chemistry and Dynamics of Interplanetary Dust, ASP Conference Series* (ed. B. A. S. Gustafson & M. S. Hanner), vol. 104, pp. 227–232. Astronomical Society of the Pacific, San Francisco, CA.

STAUBACH, P., GRÜN, E. & JEHN, R. 1997 The meteoroid environment near earth. *Advances in Space Research* **19**, 301–308.

STEBBINS, J. & WHITFORD, A. E. 1943 Six-color photometry of stars i. The low of space reddening from the colors of O and B stars. *Astrophysical Journal* **98**, 20–32.

STRUVE, O. 1937 On the interpretation of the surface brightness of diffuse galactic nebulae. *Astrophysical Journal* **85**, 194–212.

SYKES, M. V. 1990 Zodiacal dust bands – their relation to asteroid families. *Icarus* **85**, 267–289.

TAYLOR, A. D., BAGGALEY, W. J. & STEEL, D. I. 1996 Discovery of interstellar dust entering the earth's atmosphere. *Nature* **380**, 323–325.

TEMI, P., DE BERNARDIS, P., MASI, S., MORENO, G. & SALAMA, A. 1989 Infrared emission from interplanetary dust. *Astrophysical Journal* **337**, 528–535.

THIESSENHUSEN, K.-U., KRIVOV, A. V., KRÜGER, H. & GRÜN, E. 2002 A dust cloud around Pluto and Charon. *Planetary and Space Science*, **50**, 79–87.

THIESSENHUSEN, K.-U., KRÜGER, H., SPAHN, F. & GRÜN, E. 2000 Dust grains around Jupiter – The observations of the Galileo Dust Detector. *Icarus* **144**, 89–98.

TIELENS, A. G. G. M. 1997 Deuterium and interstellar chemical processes. In *Astrophysical Implications of the Laboratory Study of Presolar Materials* (ed. T. J. Bernatowicz & E. Zinner), vol. CP402, pp. 523–544. New York: American Institute of Physics.

TREWHELLA, M., DAVIES, J. I., ALTON, P. B., BIANCHI, S. & MADORE, B. F. 2000 ISO Long wavelength spectrograph observations of cold dust in galaxies. *Astrophysical Journal* **543**, 153–160.

TRÜMPLER, R. 1930 Preliminary results on the distances, dimensions, and space distribution of open star clusters. *Lick Obs. Bull.* **420**, 214.

TUZZOLINO, A. J. 1992 PVDF co-polymer dust detectors: particle response and penetration characteristics. *Nucl. Instrum. and Methods* **A 316**, 223.

VAN DE HULST, H. C. 1949 The solid particles in interstellar space. *Rech. Astron. Obs. Utrecht* **11** (2).

WEINBERG, J. L., HANNER, M. S., BEESON, D. E., DESHIELDS II, L. M. & A., G. B. 1974 Background starlight observed from Pioneer 10. *Geophys. Res.* **121**, 750–770.

WHIPPLE, F. L. 1967 On maintaining the meteoritic complex. In *NASA SP-150: The Zodiacal Light and the Interplanetary Medium* (ed. J. L. Weinberg), pp. 409–426. U. S. Govt. Printing Office, Washington, DC.

WHIPPLE, F. L. 1987 The cometary nucleus: current concepts. *Astronomy and Astrophysics* **187**, 852–858.

WHITTET, D. C. B. 1989 The composition of dust in stellar ejects. In *Interstellar Dust, Proceedings of the IAU Colloquium* 135 (ed. L. J. Allamandola & A. G. G. M. Tielen), p. 455. Kluwer Academic Publishers, Dordrecht

WHITTET, D. C. B., MARTIN, P. G., HOUGH, J. H., ROUSE, M. F., DAILEY, J. A. & AXON, D. J. 1992 Systematic variations in the wavelength dependence of interstellar linear polarization. *Astrophysical Journal* **386**, 562–577.

WITTE, M., ROSENBAUER, H., BANASZKIEWICZ, H. & FAHR, H. 1993 The Ulysses neutral gas experiment – determination of the velocity and temperature of the interstellar neutral helium. *Advances in Space Research* **13**, (6)121–(6)130.

WYATT, S. P. & WHIPPLE, F. L. 1950 The Poynting–Robertson effect on meteor orbits. *Astrophysical Journal* **111**, 134–141.

YAMAMOTO, S. & MUKAI, T. 1998 Dust production by impacts of interstellar dust on Edgeworth–Kuiper Belt objects. *Astronomy and Astrophysics* **329**, 785–791.

ZINNER, E. 1998 Stellar nucleosynthesis and the isotopic composition of presolar grains from primitive meteorites. *Annual Review of Earth and Planetary Sciences* **26**, 147–188.

ZOOK, H., GRÜN, E., BAGUHL, M., HAMILTON, D. P., LINKERT, G., LINKERT, D., LIOU, J.-C., FORSYTH, R. & PHILLIPS, J. L. 1996 Solar wind magnetic field bending of Jovian dust trajectories. *Science* **274**, 1501–1503.

ZOOK, H. A. & BERG, O. E. 1975 A source for hyperbolic cosmic dust particles. *Planetary and Space Science* **23**, 183–203.

4
EXTRATERRESTRIAL DUST IN THE NEAR-EARTH ENVIRONMENT

By GEORGE J. FLYNN[†]

Departments of Physics & Mathematics, State University of New York–Plattsburgh,
Plattsburgh, 12901 NY, USA

Meteors are extraterrestrial particles that vaporize during atmospheric entry because of their size and velocity. Interplanetary dust particles (IDPs), micrometeorites, and meteorites are extraterrestrial objects that survive atmospheric entry. These survivors bracket the size-range of the meteors, and some of these survivors are likely to sample the same parent bodies as the meteors. Thus, the chemical, physical, and mineralogical properties of the IDPs, micrometeorites, and meteorites, all of which have been subjected to extensive laboratory analysis, provide indications of the properties of the meteors. The stream meteors, which are generally associated with comets, are likely to have a chemical composition similar to that of the subset of IDPs that are believed to be from comets. Those porous, anhydrous IDPs are enriched in the volatile elements (Na, K, P, Zn, Cu, Ga, Ge, and Se) and carbon compared with even the most volatile-rich meteorites. Both the IDPs and unweathered stone meteorites exhibit significant porosity, frequently in the 10% to 25% range, suggesting meteors are also porous, and may fragment either during ejection from their parent bodies or during atmospheric entry. Fragmentation can result in chemically and mineralogically distinct meteors from the same source. Pulse heating of extraterrestrial materials, and the comparison of heated with unheated IDPs, indicates that volatile elements, such as S, Zn and Ge, are lost prior to the onset of silicate melting. In addition, both unmelted and melted micrometeorites collected from the polar ices are depleted in moderately volatile elements compared to their presumed starting composition, providing further evidence for volatile loss during atmospheric entry. Thus, meteor trails may begin with the release of the most volatile elements or compounds (Na, K, and, possibly, water and some organics) followed by the moderately volatile elements (S, Zn and Ge), prior to the onset of silicate vaporization.

4.1. Introduction

Spacecraft impact measurements indicate that, in the current era, about 30,000 tons of extraterrestrial dust in the size range of 20 to 400 microns is incident on the Earth annually [Love & Brownlee (1993); Taylor et al. (1998)]. This value is in general agreement with the long-term average accretion rate of extraterrestrial material, 37,000 tons/year, derived from the Os content of marine sediment deposited over the past 80 million years [Peucker-Ehrenbrink (1996)]. Much of the extraterrestrial dust ablates or vaporizes during atmospheric entry. Taylor et al. (1998) estimate that only $\sim 4\%$ of the extraterrestrial particles in the 50 to 700 μm size range that are incident on the Earth's atmosphere settle to the Earth's surface as identifiable particles. The vaporized material is likely to recondense and settle to the Earth's surface [Hunten et al. (1980)].

The size-frequency distribution of extraterrestrial material incident on the Earth's atmosphere is shown in Figure 4.1. The dust particles accrete onto the Earth in a continuous, planet-wide "rain" of extraterrestrial material, while the huge impactors ($\sim 10^{15}$ grams) strike the Earth at time intervals greater than a million years. The meteors are at

[†] Email: george.flynn@plattsburgh.edu

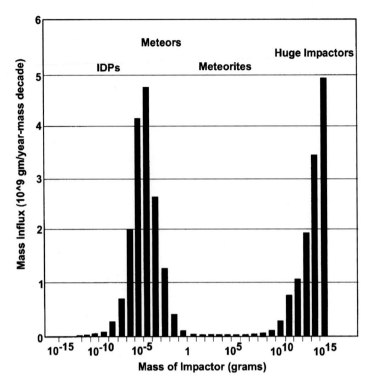

FIGURE 4.1. Mass influx per decade of mass vs. mass of the impacting object for dust accreting onto the Earth. Figure based on data in Kyte & Wasson (1986) and Love & Brownlee (1993).

the peak of the mass influx, while the meteorites (∼1 gram to thousands of kilograms), which have provided most of the information on the chemistry, mineralogy, and physical properties of extraterrestrial material, constitute a relatively small fraction of the extraterrestrial mass incident on the Earth (see Figure 4.1).

Extraterrestrial particles accelerate, due to the Earth's gravity, as they approach the Earth, resulting in a minimum atmospheric entry velocity of about 11 km/s. A hypervelocity particle entering the atmosphere of the Earth is heated during its deceleration. The degree of heating and the ultimate fate of each particle depend strongly on the velocity, size, density, composition and entry angle of that particle. Large particles, generally those greater than a few centimeters in size, slow to their terminal velocity before all of their mass ablates away, and the surviving mass strikes the Earth as a meteorite. Meteors are extraterrestrial particles that vaporize, forming a trail of excited ions during atmospheric entry. Meteors generally range from about 100 microns up to centimeters in size. Smaller particles, some up to several hundred microns in diameter, because of their larger ratio of surface area to volume (or mass), are able to radiate heat away more efficiently than the meteors [Whipple (1950)]. These small particles, called interplanetary dust particles (IDPs) and micrometeorites, can survive atmospheric entry. They are decelerated to their settling velocities, and gently settle to the Earth's surface.

The meteors span only a small range of the total mass distribution of extraterrestrial material incident on the Earth. The division of particles into survivors (the IDPs, micrometeorites, and meteorites) or meteors, which are vaporized on atmospheric entry, is both a size selection effect and a velocity selection effect. This division into survivors or meteors is Earth-specific, since the heating of a particle depends on the scale height of the atmosphere and the acceleration the particle receives due to the planet's gravity. Flynn & McKay (1990a) have shown that Mars is a more favorable site for the gentle deceleration of extraterrestrial dust because of its significantly lower gravity and slightly larger atmospheric scale height. Their modeling indicates that, on average, particles ~100 μm in size experience similar heating entering the Martian atmosphere as particles ~10 μm in size experience entering the Earth's atmosphere. Thus, unless the Earth's gravity and atmosphere are remarkably well-matched to the properties of extraterrestrial dust, it is unlikely that the separation into survivors and meteors proceeds in a manner that efficiently segregates all particles of one type into the meteor group and particles of other types into the survivors.

A significant fraction of the extraterrestrial dust is believed to be produced by the collisional erosion or collisional destruction of parent bodies, mechanisms that give rise to power-law size-frequency distributions. Thus, at least with respect to the size selection effect, some IDPs and micrometeorites are likely to be smaller samples of the material that in larger sizes produces meteors, and some of the smallest meteorites are likely to be samples of the material that in smaller sizes produces meteors. Laboratory analyses of the physical, chemical, and mineralogical properties of the IDPs, micrometeorites, and meteorites, which bracket the size of the meteors, can be used to understand and constrain the properties of the meteors and their interactions with the Earth's atmosphere.

4.2. Collection of Extraterrestrial Particles

One of the earliest efforts to collect extraterrestrial dust was undertaken on the H. M. S. Challenger expedition in 1873, by sorting extraterrestrial from terrestrial particles based on their magnetic properties. A few magnetic spherules were collected on that expedition [Murray & Renard (1891)]. Such separations, however, rely on the production of magnetite due to heating during atmospheric entry, and thus identify only severely heated, and thermally altered, extraterrestrial dust. The unheated extraterrestrial particles cannot be sorted from terrestrial particles based on magnetic properties. Such separations can only be accomplished by examining the chemical, mineralogical, or other properties of each particle in a collection. To maximize the efficiency of identification of extraterrestrial particles it is necessary to collect samples in environments where the abundance of terrestrial particles is low. Two such terrestrial environments have been identified: the polar ices, and the Earth's stratosphere.

Micrometeorites have been recovered from the sea floor by magnetic separation [Brownlee (1978)], and from the polar ices [Maurette *et al.* (1991)] including the South Pole water well [Taylor *et al.* (2000)] by hand separation based on color and morphology followed by chemical characterization. The recovered micrometeorites, many of which partially or completely melt on entry (see Figure 4.2), typically range from ~25 μm (the minimum size trapped by the filters that have been used to separate these particles from the melted ice samples) up to a few hundred micrometers in diameter. Even smaller particles, generally ranging from ~5 to ~50 μm in size, settle through the atmosphere slowly enough that they are present in high enough concentrations to permit collection from the Earth's stratosphere by NASA stratospheric sampling aircraft [Brownlee (1978)]. The extraterrestrial particles recovered from the stratosphere are called interplanetary dust particles (IDPs). Many of these IDPs do not melt on atmospheric entry, thus they provide relatively pristine samples of the extraterrestrial dust striking the Earth.

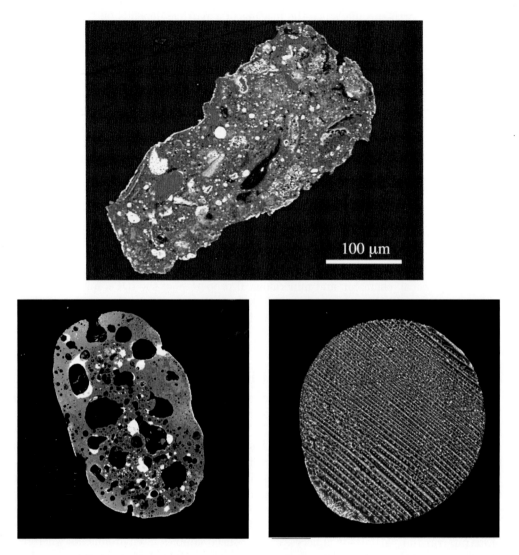

FIGURE 4.2. Micrometeorites recovered from the South Pole Water Well exhibit a variety of degrees of heating ranging from irregular, unmelted particles (upper), through scoriaceous particles exhibiting bubbles, presumably resulting from the vaporization of volatile material during atmospheric entry heating (lower left), and finally melted spherules (lower right). (Photos provided by S. Taylor and R. Harvey.)

4.3. Identification of Extraterrestrial Particles

The first criterion to separate terrestrial from extraterrestrial particles is the chemical composition. Since the Earth is a differentiated planet, its indigenous surface materials are, on average, depleted in elements such as Ni, Pt, and Ir, which are concentrated in the Earth's core. The most common meteorites, the chondrites, are from primitive (i.e. undifferentiated) parent bodies, and they have higher concentrations of Ni, and the Pt–Ir group metals. Thus, chemical composition provides an efficient way to sort primitive chondritic dust particles from terrestrial dust.

Chemical similarity, for the major rock-forming elements, to the chondritic composition does not, in itself, provide an absolute identification of a particle as extraterrestrial. But, since extraterrestrial dust particles generally spend long times in space, observing characteristics that result from long space exposure, particularly effects of exposure to solar radiation, provides compelling evidence that a particle is extraterrestrial. Heavy solar flare ions, such as Fe, produce radiation damage trails, called "tracks," when they pass through many types of insulating material. These tracks can be imaged directly in the Transmission Electron Microscope (TEM), or they can be chemically etched until they are wide enough to be viewed by optical microscopy. In the IDPs the observation of tracks in the silicate minerals, in abundances much larger than can be accounted for by fission fragments from decaying radioactive material in a particle, demonstrates that the particle was outside of the Earth's magnetic field [Bradley *et al.* (1984)]. Lower energy solar wind ions, which penetrate only tens of nanometers, produce radiation-damaged rims that can also be imaged by TEM (see Figure 4.3). Solar wind gases are implanted into interplanetary particles, and the detection of noble gases in the isotopic and elemental ratios characteristic of the solar wind provides identification that a particle has been exposed to solar radiation [Hudson *et al.* (1981)]. Evidence of preserved interstellar material of exotic isotopic compositions can also identify a particular particle as extraterrestrial [McKeegan (1987)].

4.4. Sources of Interplanetary Dust

Every solar system object is a potential source of small particles, emitted by cratering, collisional disruption, volcanism, levitation by gas emission, or other mechanisms. However, it requires significantly higher velocities to eject material into space from the deep gravity wells of the planets and their moons than from the weak gravity wells of the asteroids and comets. Infrared emission is a particularly efficient way to detect small particles and dust in space. The InfraRed Astronomical Satellite (IRAS) detected dust bands associated with three families of main belt asteroids [Low *et al.* (1984)], which are in nearly circular heliocentric orbits between Mars and Jupiter, and dust trails associated with many comets [Sykes *et al.* (1986)]. The IRAS results demonstrate that both asteroids and comets are significant sources of small particles in the Solar System.

Four physical mechanisms affect the delivery of interplanetary particles to the Earth: Poynting–Robertson radiation drag, gravitational perturbations, gravitational focusing, and collisional disruption. Particles that are large enough to escape ejection from the Solar System due to solar radiation pressure [Burns *et al.* (1979)] initially enter orbits that are generally similar to the orbit of their parent asteroid or comet. However, the orbits of small particles evolve relatively rapidly because of Poynting–Robertson radiation drag, which causes the particles to spiral towards the Sun [Wyatt & Whipple (1950)]. Particles from main belt asteroids are in nearly circular orbits when they reach 1 AU, while cometary particles are generally in orbits having considerable eccentricity when an Earth encounter opportunity occurs [Flynn (1989a)].

Modeling by Flynn (1989a) and Jackson & Zook (1992) indicates that dust particles from the main belt asteroids generally have geocentric velocities ~ 1 to 5 km/s, while dust particles from typical comets have significantly higher geocentric velocities. For example, comet Encke provides dust with a geocentric velocity of $\sim$20 km/s and Comet Temple–Tuttle provides dust with a geocentric velocity of $\sim$70 km/s. The higher geocentric velocity for the cometary particles results in a lower Earth collection probability

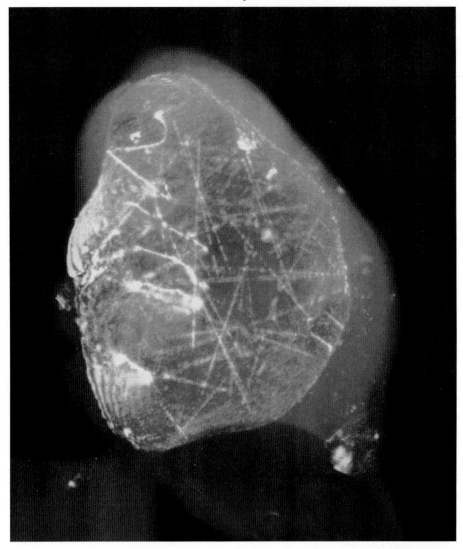

FIGURE 4.3. Transmission electron microscope image of an ultramicrotome section of an IDP, showing the amorphous, radiation-damaged thin white rim on the exterior of silicate crystals in the IDP. The straight, narrow lines are solar flare ion tracks. (Photo provided by J. P. Bradley.)

[Flynn (1990)] and a higher degree of heating on atmospheric entry [Flynn (1989a)], both of which bias of all Earth-based collection efforts in favor of asteroidal particles.

Öpik (1951) showed that the effective cross-section (C_{eff}) for capture of a particle by a planet having a physical cross section C_{act} ($= \pi R^2$) depends on the relative velocity of that particle (v_{g}) with respect to that planet:

$$C_{\text{eff}} = C_{\text{act}}(1 + v_{\text{e}}^2/v_{\text{g}}^2) \qquad (4.4.1)$$

where v_{e} is the escape velocity from the surface of the planet (11.2 km/s in the case of the Earth). Thus, the Earth preferentially accretes particles that have low geocentric velocities.

Equation (4.4.1) indicates that dust particles from main belt asteroids are accreted onto the Earth with an efficiency ∼10 to 100 times greater than the dust particles from typical comets [Flynn (1990)]. The result is that the IDPs and micrometeorites entering the Earth's atmosphere are expected to be dominated by main belt asteroidal particles [Flynn (1989a); Flynn (1990)]. Micrometeorites, which are generally larger than IDPs, experience more severe heating during atmospheric deceleration. As a result, Love & Brownlee (1991) and Flynn (1992) have concluded that only those extraterrestrial particles that have the lowest atmospheric entry velocities (generally asteroidal particles) can produce micrometeorites. Measurements by Nishiizumi *et al.* (1991) of the abundances of cosmic ray induced nuclides in micrometeorites several hundred microns in size, give space exposure histories consistent with a main belt asteroidal origin for these particles.

The time required for a particle from the main belt to reach 1 AU under the influence of Poynting–Robertson radiation drag increases in proportion to the particle diameter. As the cross-sectional area of the particle increases so does the probability that the particle will experience a catastrophic collision with another dust particle. Thus, at some size, the catastrophic collision lifetime becomes shorter than the time it takes for the particle's orbit to evolve to 1 AU, and most particles larger than that crossover size are destroyed before they can reach Earth from the main belt under the influence of Poynting–Robertson radiation drag. Modeling, under various different assumptions, gives different crossover sizes, but most modeling suggests the crossover size is near 50 μm [Dohnanyi (1978); Grün *et al.* (1985)]. However, a significant change in the slope of the size-frequency distribution of particles incident on the Earth is seen at $\sim 10^{-5}$ grams [Love & Brownlee (1993)]. This may suggest a change in the delivery mechanism at $\sim 10^{-5}$ grams (about 200 μm in diameter for an assumed density of 2.5 g/cm^3). Both the modeling results and the change in slope of the size-frequency distribution are at sizes small enough to indicate that few visible meteors are likely to be delivered to the Earth by Poynting–Robertson radiation drag.

The delivery of main belt asteroidal objects larger than this cutoff size to the Earth is believed to result from gravitation perturbations, by which objects in orbits near the gravitational resonances with Jupiter are ejected from the asteroid belt [Wetherill (1974)]. Meteorites are believed to be delivered from the main belt to the Earth by these gravitational resonances [Wetherill (1974)]. The sporadic meteors, although their specific parent bodies are unknown, should include some objects ejected from the main belt by the same gravitational resonances that eject the meteorites. Thus, asteroidal meteors should have properties similar to the common meteorites, as well as the micrometeorites, and the asteroidal IDPs.

Since comets are observed to contribute dust to the interplanetary medium, there must be some cometary dust among the IDPs. Because cometary dust enters the Earth's atmosphere with a significantly higher mean geocentric velocity than main belt asteroidal dust, cometary dust particles are, on average, heated more severely on atmospheric entry than asteroidal particles of the same size and density [Flynn (1989a)]. Thus, Flynn (1994) and Love & Brownlee (1994) have shown that the degree of heating can be used to identify a cometary subset of the IDPs collected from the Earth's stratosphere. Using this heating criterion, Brownlee *et al.* (1993) have identified a cometary subset of the IDPs, using the release temperature of implanted solar He as a monitor of the peak temperature experienced by an individual particle on atmospheric entry. They indicate that the cometary subgroup is dominated by chondritic IDPs that are porous and anhydrous [Brownlee *et al.* (1993); Joswiak *et al.* (2000)].

The stream meteors, whose orbits are associated with comets, are likely to be similar to the subgroup of IDPs (porous, chondritic, anhydrous IDPs) which have been associated with comets. No identified meteorites have yet been convincingly associated with cometary parent bodies; however, some researchers have suggested that the porous, volatile-rich CI carbonaceous chondrite meteorites, usually assigned an asteroidal origin, might possibly be from comets [Campins (1996); Campins & Swindle (1998)].

Interstellar grains also contribute to the flux of extraterrestrial material at Earth, particularly when the Solar System passes through a dense region of the interstellar medium [Frisch *et al.* (1999)]. Interstellar grains sweep through the solar system due to the motion of the Sun through the local interstellar medium. The velocity of the interstellar grains at Earth is modulated by the Earth's motion around the Sun, reaching a maximum when the Earth's orbital motion is directed along the Sun's motion [Flynn (1997)]. The result is that in January and February the velocity of interstellar grains is a maximum, resulting in the formation of meteors by these very high velocity particles as small as 15 to 20 μm [Taylor *et al.* (1996)]. Six months later, when the atmospheric entry velocity of these interstellar grains is at a minimum, the velocity is low enough to allow their survival and collection [Flynn (1997)]. This phenomenon provides the unique opportunity to observe the characteristics of the interstellar particles as meteors in one season and to recover and analyze surviving particles of similar size from the same source in another season.

4.5. Chemical Compositions

The CI carbonaceous chondrite meteorites have an average chemical composition which is quite similar to that measured for the photosphere of the Sun, except for H, He and the noble gases, which are depleted in the CI meteorites compared with the Solar photosphere [Anders & Grevesse (1989)]. These authors suggest that, except for the elements that are gases at 300 K, the CI meteorites represent the average Solar System composition. However, many of the moderately volatile trace elements are poorly determined or not measured in the solar photosphere [Anders & Grevesse (1989)] and the CI chondritic meteorites provide the only inference of the abundances of these elements in the Solar Nebula. The chemical compositions of other meteorites are generally referenced to this standard CI composition.

Meteorites are classified as "falls," objects that were observed to fall from the sky, and "finds," objects that are discovered on the ground. Selection effects complicate statistical studies of the finds. For example, meteorites that deteriorate rapidly are less likely to be found than meteorites (such as irons) that are very durable. In addition, meteorites that are quite distinct from the local rocks are more likely to be identified as unusual and "found" than are meteorites which are similar to local rocks. Because of these biases in the survival and identification of the finds, statistical studies of meteorites are frequently confined to the falls.

Among the falls, the most common type of meteorite is the ordinary chondrite. In the ordinary chondrites, the refractory elements (including the major elements Al, Mg, Si, O, Ti, Cr, Fe, and Ni) are generally within a factor of 2 of the CI composition, but the volatile elements (those elements with nebular condensation temperatures lower than about 1000 K) and carbon are significantly depleted compared to CI (see Figure 4.4).

Those sporadic meteors, which are ejected from the main belt by gravitational effects, might be expected to have a chemical composition similar to the most common meteorites, the ordinary chondrites. Thus, these sporadic meteors would be expected to have a refractory content similar to the CI meteorites, but should be depleted in the volatile elements and carbon compared with the CI meteorites. It is important to note, however, that the carbonaceous chondrite meteorites are significantly more friable than the ordinary chondrites, and the Earth's atmosphere may act as a filter, causing breakup of the

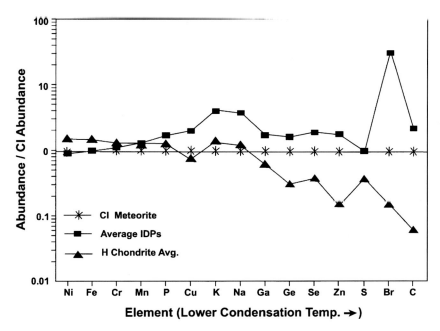

FIGURE 4.4. Average CI normalized element abundances in a group of IDPs [data from Flynn *et al.* (1996)], the CI carbonaceous chondrite meteorites, and the H-type ordinary chondrite meteorites. Elements are plotted with increasing volatility to the right.

most friable objects. It may be more difficult for large pieces of carbonaceous chondrite material to pass through the Earth's atmosphere without fragmentation and disruption than it is for ordinary chondrite material. Thus, the dominance of ordinary chondrites among the falls may be determined by their survival during passage through the atmosphere rather than their abundance at the top of the atmosphere. If the carbonaceous chondrite meteorites dominate at the top of the atmosphere, then the sporadic meteors from asteroids should have a more CI-like chemical composition.

Detailed chemical analyses of the IDPs and micrometeorites give a somewhat different composition for these smaller objects than for the meteorites. The refractory elements are again within a factor of two of the CI composition [Schramm *et al.* (1989)], but, on average, the ∼10 µm IDPs are enriched in the volatile elements (Na, K, P, Zn, Cu, Ga, Ge, and Se) by factors of 2 to 4 [Flynn *et al.* (1996)] and in carbon by a factor of 2 or 3 [Schramm *et al.* (1989); Thomas *et al.* (1994)] compared with the CI meteorite composition (see Figure 4.4). This volatile enrichment is particularly pronounced among the anhydrous, porous subgroup of the IDPs [Flynn *et al.* (1993)], a group which includes IDPs of the type that Joswiak *et al.* (2000) associate with comets. Although some of the carbon in IDPs is elemental (graphitic or amorphous) [Keller *et al.* (1994)], carbon X-ray Absorption Near Edge Structure spectroscopy [Flynn *et al.* (2000)] and Fourier Transform Infrared Spectroscopy [Flynn *et al.* (2000)] demonstrate the presence of organic compounds, which are likely to be relatively volatile compared with the silicate minerals.

The direct analysis of dust particles in the coma of Comet Halley, by dust analyzers on the Giotto and Vega spacecraft, indicated a very high carbon content (C/Mg = 11.6 times CI) as well as enrichments in some other volatiles (N/Mg = 7.5 times CI and Na/Mg = 1.9 times CI) over CI meteorite concentrations [Grün & Jessberger (1990)]. Based on the high volatile contents observed in astronomical measurements of Comet Hale–Bopp and Comet Hyakutake compared with the assumed composition of meteors, Steel (1998) has suggested that Leonid stream meteors, particularly those recently released from the parent comet, may significantly enriched in volatile compounds compared with an assumed CI meteorite composition. Rietmeijer (1999) has suggested, based on the high Na in IDPs, that Na might be enriched in comets compared with the CI meteorite composition. The laboratory analyses of porous chondritic IDPs [Flynn *et al.* (1993)] as well as the in situ dust analyses of comet Halley dust particles [Grün & Jessberger (1990)] suggest the stream meteors whose orbits are associated with comets should have a refractory element content similar to the chondritic meteorites but should be enriched, possibly by a factor of 2 to 4, in volatile elements and carbon over the CI carbonaceous chondrite composition.

4.6. Physical Properties, Heterogeneity, and Fragmentation

Most 5 to 25 μm IDPs are aggregates, usually of sub-micron mineral grains and glass in a non-crystalline matrix [see reviews by Mackinnon & Rietmeijer (1987); Rietmeijer (1998)]. In many IDPs the mineral grains and glass are coated with sub-micron-thick layers of carbonaceous material [Keller *et al.* (1994); Flynn *et al.* (1998b)] that appears to be the "glue" that holds the subunits together. Meteors composed of this material would be well-mixed aggregates on the millimeter scale. However, most stone meteorites are heterogeneous objects at the millimeter to centimeter scale. The chondritic stone meteorites typically consist of crystalline material, including mineral fragments and "chondrules" (typically millimeter-sized spherical units of olivine and/or pyroxene), in a porous, volatile-rich matrix. These chondritic meteorites contain chemically and mineralogically distinct subunits close to the size of the meteors. The ordinary chondrites consist of about 70% chondrules, 10% mineral fragments, 10% matrix, and the remainder Fe–Ni metal and sulfides [Scott & Newsom(1989)]. The rarer carbonaceous chondrites also include refractory (Ca–Al-rich) inclusions, and have a higher content of matrix and lower contents of chondrules and metal than the ordinary chondrites [Scott & Newsom(1989)].

The IDPs have unusually low densities compared with the densities of the minerals that dominate their structure. Flynn & Sutton (1991) reported a mean density of 1.1 g/cm^3 on a set of 25 IDPs, while Love & Brownlee (1993) found that roughly one-half of a group of ∼100 IDPs examined had densities below 2.0 g/cm^3. Since the typical silicates observed in IDPs all have densities >2.2 g/cm^3, these low densities suggest that IDPs exhibit considerable porosity. The porous chondritic IDPs, which Brownlee *et al.* (1993) suggest include the cometary subgroup, have unusually high porosities, visible as reentrant structures in Scanning Electron Microscope images of these particles. Significant porosity has also been observed in TEM examination of ultramicrotome sections of IDPs [Rietmeijer (1993)].

Unweathered ordinary chondrite meteorites have porosities ranging from 5 to 25%, while carbonaceous chondrites have porosities ranging from 10 to 35% [Britt & Consolmagno (1997) ; Flynn & Klöck (1998a); Flynn *et al.* (1999)]. Thus, meteors derived from either comets or stone asteroids are likely to exhibit considerable porosity. This is consistent with inferences of relatively low densities for some meteors based on atmospheric deceleration [Verniani (1969)].

Porosity may affect the interaction of meteors with the atmosphere, possibly causing breakup during atmospheric entry. Fragmentation may also occur due to vaporization of volatile phases, some of which may form the "glue" which holds the crystalline phases

TABLE 4.1. Compositions of some subunits of chondritic meteorites

	SiO_2	Al_2O_3	TiO_2	FeO	MgO	CaO	Na_2O	K_2O	Cr_2O_3	MnO	NiO
Matrix[a]	30.0	2.4	0.06	33.6	17.8	2.4	0.30	0.04	0.37	0.20	1.46
Chondrule[b]	33.9	1.9	0.09	29.6	24.7	0.8	1.2	0.13	0.60	0.31	<0.1
Refractory[c]	29.1	29.6	1.3	6.6	10.2	28.8	0.18	0.01	0.04	n.d.	0.06

[a] Allende CV3 matrix data from Scott *et al.* (1988).
[b] Chondrule bulk analysis from Scott *et al.* (1984).
[c] Refractory inclusion data from MacPherson *et al.* (1988).

together. This atmospheric breakup is likely to separate a particle into its subunits. The different types of materials (matrix, chondrules, and refractory inclusions) in chondritic meteorites show distinctly different compositions. Representative compositions of each type of material are given in Table 4.1. If a chondritic object fragments into its chemically distinct subunits, then some meteors from a single parent might be dominated by volatile-rich matrix, others by olivine or pyroxene, and others by Ca–Al-rich minerals.

Fragmentation may also occur during ejection from the parent body. To simulate the disruption of the asteroidal parent of a chondritic meteorite, Durda & Flynn (1999) impacted three ~300 grams pieces of Hawaiian porphyritic olivine basalt with $\frac{1}{4}$ inch aluminum projectiles shot at ~5 km/s by the NASA Ames Vertical Gun. The porphyritic olivine basalt contained millimeter-size olivine phenocrysts in a fine-grained vesicular matrix. In the primary debris, which was captured in aerogel, all nine fragments less than 100 μm in size were matrix while the majority of the largest fragments (>200 μm in size) were olivine [Durda & Flynn (1999)]. Larger debris was collected from the floor of the gun chamber. In the millimeter size range a large number of isolated olivine crystals were found, indicating the target experienced preferential failure along the phenocryst–matrix boundaries. All three shots showed distinct changes in the slopes of the mass–frequency distribution near 0.4 grams (about 5 mm in diameter), the size of typical olivine phenocrysts. This suggests that the mechanical failure of the material was affected by the presence of the phenocrysts. If the fragmentation of chondritic meteorites is similar to that of this Hawaiian basalt, then millimeter-size debris from parent bodies like the stone meteorites might experience chemical and mineralogical segregation during the fragmentation process. Meteors from these parent bodies might demonstrate compositional variations reflecting millimeter-scale heterogeneity in the parent body, with some fragments consisting of olivine or pyroxene chondrules or chondrule pieces, others consisting of Ca–Al-minerals, others sampling metal or sulfides, and still others sampling the more volatile-rich matrix. The refractory Ca–Al inclusions are enriched by more than an order of magnitude in Ca and Al compared with the matrix and the chondrules, while Ni is enriched by more than an order of magnitude in the matrix compared to the chondrules and refractory inclusions (see Table 4.1). Thus, the fragmentation of material from a single source could produce meteors with a diverse range of chemical compositions.

Although little is known about the heterogeneity of comets on the millimeter size scale, the dust analyzers on the Giotto and Vega spacecraft detected several distinct particle types (silicates, carbon–hydrogen–nitrogen–oxygen-rich (CHON) particles, and mixtures of the two) in the ~ 10^{-12} to 10^{-14} gram dust particles in the comet Halley coma [Grün & Jessberger (1990)]. Thus, particles from comets are heterogeneous, at least at the nanometer scale.

4.7. Atmospheric Heating and Element Loss

Long duration (hours to days) heating experiments demonstrate that volatile elements are lost from stone meteorites at temperatures well below the melting point of the silicates, the dominant mineral in those meteorites [Lipschutz & Wollum (1988)]. Experiments on the Allende carbonaceous chondrite establish loss temperatures for a variety of elements including carbon (700 to 1000 °C), Ga (900 to 1200 °C), In (600 to 1200 °C), Se (600 to 1300 °C), and Zn (700 to 1400 °C) [Lipschutz & Wollum (1988)].

The duration of the heating pulse experienced by particles entering the Earth's atmosphere is typically only a few seconds [Flynn (1989b); Love & Brownlee (1991)]. Thus, the long duration heating experiments on meteorites provide no information on the element loss from particles entering the Earth's atmosphere. Until recently, only a few short-duration, pulse-heating experiments had been performed to monitor the chemical and mineralogical alteration of meteoritic materials. Frauendorf et al. (1982) reported that IDPs lose the element S during a 20-second pulse heating to 1000 °C, and Sandford (1986) presented evidence for water loss from hydrated IDPs at temperatures near 600 °C.

More recently, Klöck et al. (1994) heated ∼100 μm size fragments of the CI carbonaceous chondrites Orgueil and Alais, and Greshake et al. (1996, 1998) heated 50 to 100 μm size fragments of Orgueil and Alais as well as 30 to 60 μm size Fe-sulfide (pyrrhotite) grains hand-selected from Orgueil to temperatures ranging from 600° to 1200 °C for times from 10 to 40 seconds. They studied the loss of the volatile minor elements S, Zn, Ga, Ge and Se by measuring the element/Fe ratios in the individual grains both before and after the heating pulse was applied. The Ni concentration in each sample was also measured to monitor loss of refractory elements. The Ni/Fe ratio remained constant, while the S/Fe, Se/Fe, Ga/Fe, Ge/Fe and Zn/Fe ratios dropped with increasing temperature. In Alais, the 50% loss temperatures of S (∼900 °C), Se (∼900 °C), Ga (∼1200 °C), and Ge (∼1200 °C) were determined, and only an ∼30% decrease in the Zn/Fe ratio occurred at 1200 °C [Greshake et al. (1998)]. The fragments from Orgueil showed similar loss temperatures [Greshake et al. (1996)]. The study of pyrrhotite from Orgueil showed lower 50% loss temperatures, ∼700 °C for both Se and S [Greshake et al. (1998)]. These results demonstrate that small meteoritic particles can lose significant fractions of their volatile elements at temperatures well below the melting temperatures of most silicate minerals. However, these results are likely to depend on both the grain size and the mineral host of the volatile elements.

The recovery of micrometeorites and IDPs which have experienced heating, sometimes to the melting temperature of the major silicates, but which have not completely vaporized provides a method of monitoring element loss prior to complete vaporization. Significant mass-dependent isotopic fractionation, the pattern expected for mass-loss by vaporization, has been measured for Ni and Fe (up to 32 per mil per atomic mass unit) in some melted spheres recovered from the sea floor [Davis & Brownlee (1993)]. This demonstrates that these spherules lost a significant fraction of their pre-atmospheric mass during atmospheric deceleration. However, the recovery of surviving refractory material demonstrates that some particles deposit volatiles into the upper atmosphere while retaining a significant fraction of their refractory material.

The majority of the micrometeorites are spherical, suggesting they melted during atmospheric deceleration. However, some are irregularly shaped and contain fine-grained mineral phases, indicating they did not melt on entry. Even these unmelted micrometeorites have chemical abundance patterns indicating volatile loss during atmospheric entry. Greshake et al. (1995) analyzed 20 fine-grained micrometeorites ranging from 200 to 250 μm in diameter from the Greenland collections and 14 fine-grained micrometeorites from 100 to 150 μm in diameter from the Antarctic collections. These particles are depleted in Ca, Ni, Cu, Ge, Se, Zn and S. While the depletions in Ni, S, and Se are attributed to the leaching of sulfides or sulfates in the ice water, the depletions in Cu, Ge, and Zn are consistent with loss during atmospheric deceleration [Greshake *et al.* (1995)]. Thus, particles a few hundred microns in diameter are believed to deposit volatile elements in the Earth's upper atmosphere without reaching the melting temperature of the silicate minerals in the particles.

Magnetite is produced on the exposed surfaces of some IDPs by the high-temperature interaction of atmospheric oxygen with silicates such as olivine and Fe-sulfides. Flynn *et al.* (1993) compared the chemical compositions (for the elements from Cr to Br) of two sets of chondritic, anhydrous IDPs: one group of eight particles identified as having experienced severe atmospheric entry heating based on the observations of magnetite rims, and a second group of five particles that were unequilibrated and had no magnetite rims, indicating much less severe heating. The heated group was depleted in Zn, Br, and Ge compared with the less severely heated group, indicating that, even among the smallest particles which are not melted, some moderately volatile elements are lost during the atmospheric entry heating pulse.

The observation that elements are lost from IDPs and micrometeorites, even at temperatures well below the silicate melting temperature, suggests that meteors may also experience the loss of volatile elements during the initial phase of their atmospheric deceleration, before they reach the silicate vaporization temperature. McNeil *et al.* (1998) have proposed "differential ablation" to explain the higher concentration of the more volatile species in the mesosphere. The degree to which volatile elements can be lost from a meteor may be size dependent, governed by both the rate of heat flow into the core of the meteoroid and the rate of diffusion of the volatile element out of the host mineral or the bulk meteoroid. Thus, meteor trails may begin with the loss of the most volatile material, including Na, K, water and possibly organic matter, followed by the moderately volatile material, including S, Zn, Se, and Ge, prior to complete silicate vaporization.

4.8. Collection of Particles from Specific Meteor Streams

The compositional diversity among the meteorites as well as the differences in composition between IDPs and meteorites suggest there are a wide variety of parent bodies that contribute to small objects in the Solar System. Thus, individual meteors may sample compositionally distinct parent bodies. Although the sporadic meteors have a wide variety of orbital elements, suggesting they sample a variety of sources, most meteor streams are associated with a particular parent comet. Thus, it would be desirable to associate some individual IDPs collected from the stratosphere with specific meteor showers, providing information on the properties of the meteors in that shower as well as these properties of the parent comet.

The IDPs recovered from the Earth's stratosphere at any given time are a mixture of material from several different sources, including both comets and asteroids. A typical meteor shower produces 10 to 20 visually observable meteors per hour, a significant increase over the sporadic flux. However, the duration of each meteor shower is relatively short, at most one to three days with the highest flux lasting only a few hours. If we assume that smaller particles accompany these stream meteors, then the flux of ~10 μm particles would increase during a meteor shower, and some of these might survive atmospheric entry as IDPs. However, the time it takes for an ~10 μm IDP to settle from the deceleration altitude (~90 to 110 kilometers) to the collection altitude (~20 km) is long compared with the duration of a meteor shower. The contribution of a meteor shower to the IDPs at the 20 km collection altitude is significantly diluted by the sporadic (non-shower) IDPs accumulating over an interval comparable to the settling time. Thus a single meteor shower is not likely to dominate the IDPs from any stratospheric collection. As a consequence, unless the IDPs from a particular meteor shower have an unusual composition, mineralogy, or morphology, it may not be possible to associate specific particles recovered from the stratosphere with their parent bodies based simply on the time of year of their collection.

The collection of IDPs from a specific meteor stream is further complicated by the fact that very small dust particles are quickly separated from a meteor stream by solar radiation pressure. Modeling by Burns et al. (1979) indicates that smallest particles emitted near perihelion by many comets are rapidly ejected from the Solar System by solar radiation pressure. The maximum size of particles emitted near perihelion that are ejected from the Solar System varies with the orbital parameters of the comet. Typical values for this maximum diameter are 5 μm from Comet Kopff, 15 μm from Comet Encke, and 70 μm from Comet Halley [Flynn (1989a); Burns et al. (1979)]. Even particles that are large enough to enter an orbit are subject to Poynting–Robertson drag, which quickly alters the orbits of the smallest particles in a stream [Wyatt & Whipple (1950)]. Thus, small dust particles from a specific comet do not generally arrive at Earth at the same time as the larger particles in the meteor stream from that comet, and the smallest of the particles from any comet are ejected from the Solar System.

However, solar radiation pressure does not preclude the collection of the smallest dust from a comet immediately after the emission of these particles from the comet, when they are still in close proximity to the comet nucleus and the larger particles in the stream. The best opportunity to recover IDPs that can be identified with a specific parent comet is during meteor storms. These meteor storms occur just before or just after the parent comet passes close to the Earth. Meteor storms offer two advantages for particle collection: the smallest particles have had little time to separate from the stream, and the meteor flux in the storm is very high.

Meteor storms produce much higher meteor fluxes than the normal meteor streams. In 1966, when Comet Temple–Tuttle passed close to the Earth, observers reported up to 150,000 Leonid meteors per hour. Even though the Leonid storm lasts only a few hours, when this contribution is integrated over a period of a few weeks these Leonid storms significantly perturb the total particle accretion onto the Earth. Thus, meteor storms of this magnitude offer the opportunity to collect IDPs that can be associated with a specific parent body.

Because of the 33-year orbital period of Comet Temple–Tuttle, the parent of the Leonid meteor stream, Leonid storms occur every 33 years. However, the extremely high atmospheric entry velocity of Leonid particles (more than 70 km/s) makes these Leonid storms unfavorable for the recovery of thermally unaltered IDPs, since most 10 μm particles that enter the Earth's atmosphere at 70 km/s are heated above the silicate melting temperature [Flynn (1989a)].

Jenniskens (1995) tabulates a number of observed meteor storms, associated with the return of the parent comet to perihelion. Many of these storms, including the 1933 and 1946 Draconid outbursts, have significantly lower geocentric velocities (some near 20 km/s) than the Leonid storm but have comparable rates of visual meteors. These lower velocity meteor storms are likely to contribute a significant number of unmelted IDPs to the Earth's stratosphere.

Because several days to weeks elapse between the observation of a meteor storm and the settling of IDPs to the collection altitude, advance prediction of the meteor storm is not necessary. The collection of IDPs from a meteor storm requires only a planned effort to schedule IDP collections after the observation of major meteor storm.

4.9. Conclusions

Shower meteors, which are generally associated with comets, are likely to have chemical compositions that are very similar to the CI meteorite composition for the refractory elements, but carbon and the volatile elements are likely to be enriched relative to the CI meteorite composition, possibly by factors of 2 to 4.

Those sporadic meteors that are derived from the asteroid belt, should, on average, have a composition similar to the most common meteorites, the ordinary chondrites, which are depleted in the volatile elements compared to the CI meteorites. In the millimeter size range, fragmentation can separate the chondritic meteoritic material into its mineralogical sub-units (matrix, olivine or pyroxene chondrules, and Ca–Al-rich inclusions), resulting in particles with distinctly different compositions coming from a single parent body.

Meteors from both asteroidal and cometary parent bodies may exhibit significant porosity, which could aid in breakup during atmospheric deceleration.

Volatile-depleted chondritic spheres are recovered from the polar ices, and volatile depleted but unmelted IDPs are recovered from the stratosphere. This demonstrate that atmospheric entry heating results in the release of some of the most volatile components (including Na, K, S, Zn, Ge, water and organics) prior to the onset of silicate vaporization, particularly in the smallest samples where diffusion effects are minimized. Thus, we might expect chemical differences along the trail of a meteor, with the most volatile elements or compounds being released at the highest altitudes and the less volatile material vaporizing at lower altitudes.

By timing stratospheric collection missions to follow the arrival of dust from significant meteor storms, it may be possible to associate specific IDPs with their cometary parent bodies, and, thus, to determine the physical, chemical, and mineralogical properties of particular stream meteors and their cometary parents.

Acknowledgement: This work was supported by a NASA Cosmochemistry Grant NAG5-4843.

REFERENCES

ANDERS E. & GREVESSE N. 1989. Abundance of the elements: meteoritic and solar, *Geochim. Cosmochim. Acta*, **53**, 194–214.

BRADLEY J. P., BROWNLEE D. E. & FRAUNDORF P. 1984. Discovery of nuclear tracks in interplanetary dust particles, *Science*, **226**, 1432–1434.

BRITT D. T. & CONSOLMAGNO G. J. 1997. The porosity of meteorites and asteroids: results from the Vatican collection, *Lunar Planet. Sci.*, **XXVIII**, 159–160.

BROWNLEE D. E. 1978. Microparticle studies by sampling techniques, in *Cosmic Dust*, edited by J. A. M. McDonnell, Wiley, New York, pp. 295–336.

BROWNLEE D. E., JOSWIAK D. J., LOVE S. G., NIER A. O., SCHLUTTER D. J. & BRADLEY J. P. 1993. Identification of cometary and asteroidal particles in stratospheric IDP collections *Lunar Planet. Sci.*, **XXIV**, 205–206.

BURNS J. A., LAMY P. L. & SOTER S. 1979. Radiation forces on small particles in the Solar System, *Icarus*, **40**, 1–48.

CAMPINS H. 1996. Meteorites from comets? Recent observational developments, in *Physics, Chemistry, and Dynamics of Interplanetary Dust*, edited by B. Å. S. Gustafson and M. S. Hanner, ASP Conference Series, Vol. 104, Astronomical Society of the Pacific, pp. 395–398.

CAMPINS H. & SWINDLE T. D. 1998. Expected characteristics of cometary meteorites, *Meteoritics Planet. Sci.*, **33**, 1201–1211.

DAVIS A. M. & BROWNLEE D. E. 1993. Iron and nickel isotopic mass fractionation in deep-sea spherules, *Lunar Planet. Sci.*, **XXIV**, 373–374.

DOHNANYI J. S. 1978. Particle dynamics, in *Cosmic Dust*, edited by J. A. M. McDonnell, Wiley, New York, pp. 527–605.

DURDA D. D. & FLYNN G. J. 1999. Experimental study of the impact disruption of a porous, inhomogeneous target, *Icarus*, **142**, 46–55.

FLYNN G. J. 1989. Atmospheric entry heating: a criterion to distinguish between asteroidal and cometary sources of interplanetary dust, *Icarus*, **77**, 287–310.

FLYNN G. J. 1989. Atmospheric entry heating of micrometeorites, *Lunar Planet. Sci.*, **XIX**, 673–682.

FLYNN G. J. 1990. The near-earth enhancement of asteroidal over cometary dust, *Lunar Planet. Sci.*, **XX**, 363–371.

FLYNN G. J. & McKAY D. S. 1990. An assessment of the meteoritic contribution to the Martian soil, *J. Geophys. Res.*, **95**, 14,497–14,509.

FLYNN G. J. & SUTTON S. R. 1991. Cosmic dust particle densities: evidence for two populations of stony meteorites, *Lunar Planet. Sci.*, **XXII**, 171–184.

FLYNN G. J. 1992. Atmospheric entry survival of large micrometeorites: implications for their sources and for the cometary contribution to the Zodiacal Cloud, in *Asteroids, Comets, and Meteors 1991*, edited by A. W. Harris and E. Bowell, Lunar and Planetary Institute, Houston, TX, pp. 195–199.

FLYNN G. J., SUTTON S. R., BAJT S., KLÖCK W., THOMAS K. L. & KELLER L. P. 1993. The volatile content of anhydrous interplanetary dust, *Meteoritics*, **28**, 349–350.

FLYNN G. J. 1994. Cometary dust: a thermal crietrion to identify cometary samples among the interplanetary dust collected from the stratosphere, in *Analysis of Interplanetary Dust*, edited by M. E. Zolensky, T. L. Wilson, F. J. M. Rietmeijer, and G. J. Flynn, American Institute of Physics, New York, AIP Conf. Proc. 310, pp 223–230.

FLYNN G. J., BAJT S., SUTTON S. R., ZOLENSKY M., THOMAS K. L. & KELLER L. P. 1996. The abundance pattern of elements having low nebular condensation temperatures in interplanetary dust particles: Evidence for a new type of chondritic material, in *Physics, Chemistry, and Dynamics of Interplanetary Dust*, edited by B. Å. S. Gustafson and M. S. Hanner, ASP Conference Series, Vol. 104, Astronomical Society of the Pacific, pp. 291–294.

FLYNN G. J. 1997. Collecting interstellar dust grains, *Nature*, **387**, 248.

FLYNN G. J. & KLÖCK W. 1998. Densities and porosities of meteorites: implications for the porosities of asteroids, *Lunar Planet. Sci.*, **XXII**, CD-ROM, Abstract No. 1112.

FLYNN G. J., KELLER L. P., JACOBSEN C. & WIRICK S. 1998. Carbon and potassium mapping and carbon bonding state measurements on interplanetary dust, *Meteoritics*, **33** A50.

FLYNN G. J., MOORE L. B. & KLÖCK W. 1999. Density and porosity of stone meteorites: implication for the density, porosity, cratering, and collisional disruption of asteroids, *Icarus*, **142**, 97–105.

FLYNN G. J., KELLER L. P., JACOBSEN C., WIRICK S. & MILLER M. A. 2000. Organic carbon in interplanetary dust particles, in *A New Era in Bioastronomy*, edited by G. Lemarchand and K. Meech, ASP Conf. Series, Vol. 213, pp. 191–194.

FRAUNDORF P., BROWNLEE D. E. & WALKER R. M. 1982. Laboratory studies of interplanetary dust, in *Comets*, edited by L. Wilkening, University of Arizona Press, Tucson, pp. 383–409.

FRISCH P. C., DORSCHNER J. M., GEISS J., GREENBERG J. M., GRÜN E., LANDGRAF M., HOPPE P., JONES A. P., KRÄTSCHMER W., LINDE T. J., MORFILL G. E., REACH W., SLAVIN J. D., SVESTKA J., WITT A. N. & ZANK G. P. 1999. Dust in the local interstellar wind, *Ap. J.*, **525**, 492–516.

GRESHAKE A., KLÖCK W., FLYNN G. J., ARNDT P., MAETZ M. & BISCHOFF A. 1995. Volatile element abundances in micrometeorites: evidence for the loss of copper, germanium, and zinc during atmospheric entry, *Lunar Planet. Sci.*, **XXVI**, 509–510.

GRESHAKE A., KLÖCK W., ARNDT P., MAETZ M. & BISCHOFF A. 1996. Pulse-heating of fragments from Orgueil (CI): simulation of atmospheric entry heating of micrometeorites, in *The Interstellar Dust Connection*, edited by J. M. Greenberg, Kluwer Academic Publishers, Dordrecht, Netherlands, pp. 303–311.

GRESHAKE A., KLÖCK W., ARNDT P., MAETZ M., FLYNN G. J., BAJT S. & BISCHOFF A. 1998. Heating experiments simulating atmospheric entry heating of micrometeorites: clues to their parent body sources, *Meteoritics Planet. Sci.*, **33**, 267–290.

GRÜN E., ZOOK H. A., FECHTIG H. & GIESE R. H. 1985. Collisional balance of the meteoritic complex, *Icarus*, **62**, 244–272.

GRÜN E. & JESSBERGER E. K. 1990. Dust, in *Physics and Chemistry of Comets*, edited by W. F. Huebner, Springer-Verlag, 113–176.

HUDSON R. S., FLYNN G. J., FRAUNDORF P., HOHENBERG C. M. & SHIRCK J. 1981. Noble gases in stratospheric dust particles: confirmation of extraterrestrial origin, *Science*, **211**, 383–386.

HUNTEN D. M., TURCO R. P. & TOON O. B. 1980. Smoke and dust particles of meteoric origin in the mesosphere and stratosphere, *J. Atm. Sci.*, **32**, 1342–1357.

JACKSON A. & ZOOK H. A. 1992. Orbital evolution of dust particles from comets and asteroids, *Icarus*, **97**, 70–84.

JENNISKENS P. 1995. Meteor stream activity: II. Meteor outbursts, *Astron. Astrophys.*, **295**, 206–235.

JOSWIAK D. J., BROWNLEE D. E., PEPIN R. O. & SCHLUTTER D. J. 2000. Characteristics of asteroidal and cometary IDPs obtained from stratospheric collectors: summary of measured He release temperatures, velocities and descriptive mineralogy, *Lunar Planet. Sci.*, **XXXI**, CD-ROM, Abstract No.1500.

KELLER L. P., THOMAS K. L. & McKAY D. S. 1994. Carbon in primitive interplanetary dust particles, in *Analysis of Interplanetary Dust*, edited by M. E. Zolensky, T. L. Wilson, F. J. M. Rietmeijer, and G. J. Flynn, American Institute of Physics, New York AIP Conf. Proc. Vol. 310, pp. 159–164.

KLÖCK W., FLYNN G. J., SUTTON S. R., BAJT S. & NEUKING K. 1994. Heating experiments simulating atmospheric entry of micrometeorites, *Lunar Planet. Sci.*, **XXV** 713–714.

KYTE F. & WASSON J. T. 1986. Accretion rate of extraterrestrial matter: iridium deposited 33 to 67 million years ago, *Science*, **232**, 1225–1228.

LIPSCHUTZ M. E. & WOLLUM D. S. 1988. Highly labile elements, in *Meteorites and the Early Solar System*, edited by J. F. Kerridge and M. S. Matthews, University of Arizona Press, Tucson, Arizona, pp. 462–487.

LOVE S. G. & BROWNLEE D. E. 1991. Heating and thermal transformation of micrometeoroids entering the Earth's atmosphere, *Icarus*, **89**, 26–43.

LOVE S. G. & BROWNLEE D. E. 1993. A direct measurement of the terrestrial mass accretion rate of cosmic dust, *Science*, **262**, 550–553.

LOVE S. G., JOSWIAK D. J. & BROWNLEE D. E. 1993. Densities of 5 to 15 mm interplanetary dust particles, *Lunar Planet. Sci.*, **XXIV**, 901–902.

LOVE S. G. & BROWNLEE D. E. 1994. Peak atmospheric entry temperatures of micrometeorites, *Meteoritics*, **29**, 69–71.

LOW F. J., BEINTEMA D. A., GAUTIER T. N., GILLETT F. C., BEICHMAN C. A., NEUGEBAUER G., YOUNG E., AUMANN H. H., BOGGESS N., WALKER R. G. & WESSELIUS P. R. 1984. Infrared cirrus: new components of the extended infrared emission, *Ap. J.*, **278**, L19–L22.

MACKINNON I. D. R. & RIETMEIJER F. J. M. 1987. Mineralogy of chondritic interplanetary dust particles, *Rev. Geophys.*, **25**, 1527–1553.

MACPHERSON G. J., WARK D. A. & ARMSTRONG J. T. 1988. Primitive material surviving in chondrites: refractory inclusions, in *Meteorites and the Early Solar System*, edited by J. F. Kerridge and M. S. Matthews, University of Arizona Press, Tucson, Arizona, pp. 746–807.

MAURETTE M., OLINGER C., CHRISTOPHE MICHEL-LEVY M., VINCENT C. & KURAT G. 1994. A collection of diverse micrometeorites recovered from 100 tonnes of Antarctic blue ice, *Nature*, **351**, 44–47.

McKEEGAN, K. D. 1987. Oxygen isotopes in refractory stratospheric dust particles: proof of extraterrestrial origin, *Science*, **237**, 1468–1471.

McNEIL W. J., LAI S. T. & MURAD E. 1998. Differential ablation of cosmic dust and implications for the relative abundances of atmospheric metals, *J. Geophys. Res.*, **103**, 10,899–10,911.

MURRAY J. & RENARD A. F. 1891. in *Report on the Scientific Results of the H. M. S. Challenger During the Years 1873–76*, Vol. 4, Neill and Company, Edinburgh, Scotland, pp. 327–336.

NISHIIZUMI K., ARNOLD J. R., FINK D., KLEIN J., MIDDLETON R., BROWNLEE D. E. & MAURETTE M. 1991. Exposure histories of individual cosmic particles, *Earth Planet. Sci. Lett.*, **104**, 315–324.

ÖPIK E. J. 1951. Collisional probabilities with planets and the distribution of interplanetary matter, *Proc. R. Irish Acad.*, **54**, Sect A., 165–199.

PEUCKER-EHRENBRINK B. 1996. Accretion of extraterrestrial matter during the last 80 million years and its effect on the marine osmium isotope record, *Geochim. Cosmochim. Acta*, **60**, 3187–3196.

RIETMEIJER F. J. M. 1998b. Size distributions in two porous chondritic micrometeorites, in *Advanced Mineralogy*, edited by A. S. Marfunin, Vol. 3, Springer-Verlag, Berlin, pp. 22–28.

RIETMEIJER F. J. M. 1998. Interplanetary dust particles, *Earth Planet. Sci. Lett.*, **117**, 609–617.

RIETMEIJER F. J. M. 1999. Sodium tails of comets: Na/O and Na/Si abundances in interplanetary dust particles, *Ap. J.*, **514**, L125–L127.

SANDFORD S. A. 1986. Laboratory heating of an interplanetary dust particle: comparisons with an atmospheric entry model, *Lunar and Planet. Sci.*, **XVII**, 754–755.

SCHRAMM L. S., BROWNLEE D. E. & WHEELOCK M. M. 1989. Major element composition of stratospheric micrometeorites, *Meteoritics*, **24**, 99–112.

SCOTT E. R. D., RUBIN A. E., TAYLOR G. J. & KEIL K. 1984. Matrix material in type 3 chondrites: occurrence, heterogenity, and relationship with chondrules, *Geochim. Cosmochim. Acta*, **48**, 1741–1757.

SCOTT E. R. D., BARBER D. J., ALEXANDER C. M., HUTCHISON R. & PECK J. A. 1988. Primitive material surviving in chondrites: matrix, in *Meteorites and the Early Solar System*, edited by J. F. Kerridge and M. S. Matthews, University of Arizona Press, Tucson, pp. 718–745.

SCOTT E. R. D. & NEWSOM H. E. 1989. Planetary compositions: clues from meteorites and asteroids, *Z. Naturforsch.*, **44a**, 924–934.

STEEL D. 1998. The Leonid meteors: compositions and consequences, *Astron. Geophys.*, **39**, 24–26.

SYKES M. V., LOBOFSKY L. A., HUNTEN D. M. & LOW F. 1986. The discovery of dust trails in the orbits of periodic comets, *Science*, **232**, 1115–1117.

TAYLOR A. D., BAGGALEY W. J. & STEEL D. I. 1996. Discovery of interstellar dust entering the Earth's atmosphere, *Nature*, **380**, 323–325.

TAYLOR S., LEVER J. H. & HARVEY R. P. 1998. Accretion rate of cosmic spherules measured at the South Pole, *Nature*, **392**, 899–903.

TAYLOR S., LEVER J. H. & HARVEY R. P. 2000. Numbers, types, and composition of an unbiased collection of cosmic spherules, *Meteoritics Planet. Sci.*, **35**, 651–666.

THOMAS K. L., KELLER L. P., BLANFORD G. E. & McKAY D. S. 1994. Quantitative analysis of carbon in anhydrous and hydrated interplanetary dust particles, in *Analysis of Interplanetary Dust*, edited by M. E. Zolensky, T. L. Wilson, F. J. M. Rietmeijer, and G. J. Flynn, AIP Press, New York, AIP Conf. Proc. 310, pp. 165–172.

VERNIANI F. 1969. Structure and fragmentation of meteoroids, *Space Sci. Revs.*, **10**, 230–261.

WETHERILL, G. W. 1974. Solar system sources of meteorites and large meteoroids, *Ann. Rev. Earth Planet. Sci.*, **2**, 303–331.

WHIPPLE F. L. 1950. The theory of micro-meteorites: Part 1. In an isothermal atmosphere, *Proc. Natl. Acad. Sci. USA*, **36**, 687–695.

WYATT S. P. & WHIPPLE F. L. 1950. The Poynting–Robertson effect on meteor orbits, *Astrophys. J.*, **111**, 134–141.

Part 2
OBSERVATIONS IN THE EARTH'S ATMOSPHERE

5

DETECTION AND ANALYSIS PROCEDURES FOR VISUAL, PHOTOGRAPHIC AND IMAGE INTENSIFIED CCD METEOR OBSERVATIONS

By R O B E R T   L.   H A W K E S[†]

Department of Physics, Mount Allison University, Sackville, NB E4L 1E6, Canada

This chapter provides a description of visual, photographic and image intensified video techniques for the observation of meteors. Included are both descriptions of the detection techniques and an overview of routines for positional, photometric and triangulation analyses for optical data. Some of the main results obtained are briefly reviewed. Visual observations provide a wealth of valuable data for determining the rate profiles for major showers. Visual rates are converted to a ZHR (zenithal hourly rate), which is the number of meteors per hour which a single visual observer would see with a limiting astronomical magnitude of +6.5 if the radiant of the shower was at the observer's zenith. Visual observers have incomplete capture of meteors even for several magnitudes brighter than this, and this must be taken into account when computing fluxes from visual observations. Visual observations can be used to obtain the population index, r, which is the ratio of the number of meteors of magnitude $M + 1$ to the number of magnitude M. With intensified video techniques it is conventional to integrate the luminous trajectory in order to obtain the photometric mass of the meteor, and then use the number of meteors of different masses to determine the mass distribution index. Compared to image intensified CCD observations photographic methods offer much better spatial resolution, but lowered sensitivity. A typical small camera photographic system is limited to meteors of about +1 on the astronomical magnitude scale, whereas an image intensified CCD system can detect stars down to about +9 magnitude with the effective limit for meteors being several magnitudes brighter than this. Both photographic and optical techniques can be extended to multiple stations, from which precise atmospheric trajectories and orbits can be determined. An important advance in recent years has been the development of software which will detect and characterize meteors from image intensified CCD systems in real time. Optical techniques provide a representation of the entire ablation profile, and therefore are particularly valuable for providing evidence for the physical structure of meteoroids. A transmission diffraction grating can be added to photographic or image intensified television systems in order to obtain meteor spectra. At least in the visible, near infrared and near ultraviolet spectral regions the predominant meteor radiation comes from line emission from atomic and ionic species, with Ca, Na and Mg lines being the dominant in a majority of meteors. In the next few years we expect the spatial resolution of optical observing techniques to improve, both through employing systems with larger apertures and longer focal lengths, and by migrating to high definition video systems. A recent extension of optical meteor observing techniques has been the utilization of CCD cameras connected to telescopes to view meteoroid impacts on the Moon.

[†] Email: rhawkes@mta.ca

5.1. Introduction

Much of what we have learned about metoroids and meteor streams has come from observations of the optical emission produced when meteoroids ablate in the Earth's atmosphere. Most meteors ablate within the height range from 140 km to 70 km, with higher velocity meteors ablating at greater heights. In this chapter we consider the various techniques (visual observer, photographic, video and spectral) which can be used to study this optical emission, restricting our coverage to roughly the sensitivity region for the human eye (400 to 700 nm). One huge advantage of optical means to study meteors, when compared with radar, is that information on the entire ablation profile is provided for individual meteor events. Additional detail on many of the topics covered in this chapter is provided in the recent comprehensive review by Ceplecha *et al.* (1998).

When the meteoroid enters the Earth's atmosphere it rapidly heats to several thousand degrees kelvin, and meteoroid material begins to ablate. The meteoric atoms which have been released then undergo atomic collisions with atmospheric constituents at approximately the geocentric entry speed of the meteor (11 to 72 km/s). These collisions drive the meteoric atoms, and some atmospheric atoms, into excited states. In most cases these excited states have lifetimes shorter than 1 μs, so that the light is produced almost instantaneously with ablation from the meteoroid surface.

The physical theory of atmospheric ablation of meteors [McKinley (1961)] indicates that the luminous intensity I of meteor radiation in the visual region of the spectrum is given by the following relationship.

$$I = \tau_I \frac{1}{2} \frac{dm}{dt} v^2 \tag{5.1.1}$$

Since a typical small meteor decelerates by no more than a few percent over the visible part of the atmospheric trajectory, the luminous intensity is essentially proportional to the instantaneous ablation rate of the meteor. The luminous efficiency factor τ_I is the fraction of the released kinetic energy of motion which appears in the form of electromagnetic radiation in the visible region of the spectrum. Both the absolute value of this parameter, and its dependence on velocity, are poorly known (although it is widely assumed that a higher fraction of the energy goes into visible radiation for higher velocity meteors). Ceplecha *et al.* (1998) provide a review of the relevant literature on the luminous efficiency relationship. Essentially one can make estimates based on a theoretical computation of the excitation and decay processes, by studying the light produced by re-entry of artificial meteors projected from spacecraft or by comparison of the luminous profiles of meteorite producing fireballs for which dynamical masses can be estimated. Important contributions to this field include Öpik (1958), Verniani (1965), Revelle & Rajan (1979) and Ceplecha (1996). Probably the single greatest obstacle to definitive interpretation of optical meteor data is uncertainty regarding the luminous efficiency factor.

The astronomical magnitude scale is used for optical meteor work. The relationship between magnitudes (m) of two objects relative to the luminous flux (I) of optical radiation received from them is given by the following.

$$m_1 - m_2 = 2.5 \log \left(\frac{I_2}{I_1} \right) \tag{5.1.2}$$

This is based upon the proposal made by N. R. Pogson in 1856 that a 5 magnitude difference be defined to correspond precisely to a factor of 100 difference in luminous flux from the two sources (with the brighter object having the smaller magnitude value). These magnitudes are apparent magnitudes. An absolute meteor magnitude scale, which corrects for the distance of the meteor from the observer, has been defined for optical meteor work, with magnitudes being standardized to a distance of 100 km from the observer (capital M is used for absolute meteor magnitudes). A useful relationship between absolute meteor magnitude M and the intensity radiated in watts P by the meteor in the visible region of the spectrum is quoted by McKinley (1961).

$$M = 6.8 - 2.5 \log{(P)} \qquad (5.1.3)$$

For example, a 0 magnitude meteor radiates about 500 W in the visible region of the spectrum.

5.2. Visual Observing

A detailed treatment of visual observing techniques is beyond the scope of this article, and the reader is referred to the excellent books on the subject by Bone (1993) and Rendtel *et al.* (1995) as well as the brief introduction provided by Hawkes (2001). The correction procedures in these and other recent studies are derived from the original work done by Znojil (1966). During the past two decades a broadly based and highly skilled group of serious amateur meteor astronomers have provided a wealth of visual observations of the major showers. More than one and a half million visual meteor records (from the last 15 years) are in the Visual Meteor Database of the International Meteor Organization (IMO01).

The zenithal hourly rate (ZHR) is defined as the number of meteors per hour which would be viewed from a meteor shower which was located at the observer's zenith, assuming a limiting visual magnitude of +6.5. The formula for computation of ZHR from visual observations is given below.

$$\text{ZHR} = F \frac{N}{T_{\text{eff}}} \frac{1}{\sin{(e_{\text{R}})}} r^{(6.5 - m_{\text{l}})} \qquad (5.2.1)$$

In this relationship N is the number of meteors recorded in a time T_{eff} while F is a correction factor for any time when a portion of the observer's field of view was covered by cloud or other obstructions. The local elevation angle of the radiant for the meteor shower is represented by e_{R}. The limiting magnitude (on the astronomical apparent magnitude scale) of the observer is given by m_{l} while r is the population index for the meteor shower. This relationship is only valid for elevation angles of more than about 10 degrees.

The population index, r, is defined as the ratio of the number of meteors of one magnitude value to the number one magnitude brighter.

$$r = \frac{N_{(m+1)}}{N_{(m)}} \qquad (5.2.2)$$

The mass distribution index is defined by the following relationship, where $\mathrm{d}N$ is the number of meteors of mass between m and $m + \mathrm{d}m$.

$$\mathrm{d}N = Cm^{-s}\,\mathrm{d}m \qquad (5.2.3)$$

The population index r can be related to the mass distribution index s by the following relationship if the shape of the meteor light curve does not vary over the interval being studied. For typical visual observations it has been suggested by Koschak & Rendtel (1990a) that the 2.5 in the following relationship should be replaced with a 2.3. Brown *et al.* (2000) have found that in the range studied by image intensified video detectors

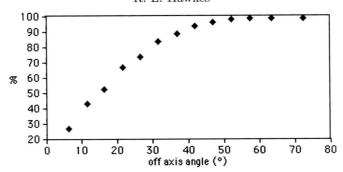

FIGURE 5.1. The percentage of meteors which are not observed by a visual observer as a function of the angle between the center of the field of view and the meteor. From Koschak & Rendtel (1990a).

there is not a strong dependence of meteor light curve shape with mass.

$$s = 1 + 2.5 \log (r) \tag{5.2.4}$$

Visual observations provide much of our information on the detailed activity profiles, the flux rates and the population index for showers. Indicative of the sort of analysis which can be done on large collections of visual data are the following articles on the Geminid by Rendtel & Arlt (1997), the Perseid shower by Arlt (1998), the June Bootid shower by Rendtel et al. (1998), the η-Aquarid shower by Rendtel (1997) and the Leonid shower by Arlt & Gyssens (2000a).

It is essential for visual observers to allow dark adaptation of the eye before commencement of the formally recorded observation period. While the pupil size adjusts quickly, full photochemical adaptation of the human eye takes at least 25 minutes to achieve. It is critical not to lose this adjustment and therefore to use only a dim red light during the observing period. Also, since the ZHR correction formula depends so critically on the limiting magnitude, this needs to be accurately estimated. The most accurate way to do this is to count carefully the number of visible stars in an extended region, and then to use a desktop planetarium program or another reference source to adjust the limiting magnitude until the same number of stars is recorded in the region. Generally speaking it is suggested that visual observers center their field of view from 20 to 40 degrees away from the radiant, and that the elevation angle for their field of view be at least 45 degrees.

An observer will see bright fireballs from nearly the entire visible hemisphere, but will observe fainter meteors only when they are near the center of the field of view. Results adapted from Koschak & Rendtel (1990a) and displayed in Figure 5.1 show the percentage of meteors missed as a function of the angle from the center of the observer's field of view. Clearly this percentage (which is of the total number of meteors of all magnitude classes for that observer) will vary somewhat with meteor shower, since the different meteor showers have different population indices. An effective visual observer field of view is usually assumed to be about 40 to 50 degrees, although it is clear that many faint meteors are lost before this is reached. The effect of this magnitude dependent field of view is particularly important when magnitude classes are assigned and population indices determined. The probability of perception, p, is defined as the ratio of the number of meteors observed N to the true number appearing ϕ.

$$p = \frac{N}{\phi} \tag{5.2.5}$$

TABLE 5.1. Perception probability (p) as a function of difference in magnitude Δm between limiting magnitude and the meteor apparent magnitude

Δm	0.5	1.0	1.5	2.0	2.5	3.0	3.5	4.0	5.0	5.5	6.0
p	0.01	0.03	0.10	0.16	0.26	0.36	0.54	0.63	0.74	0.81	0.89

The probability of perception is dependent on the difference between the observer's limiting magnitude and the magnitude of the meteors. We have created Table 5.1 based on the data presented by Koschak & Rendtel (1990a). It shows that the probability of one observer detecting a meteor only reaches 0.5 for a meteor which is between 3.5 and 4.0 astronomical magnitudes brighter than the limiting magnitude. Again the geocentric velocity and population index of the shower will affect these results. As well as counts and magnitude distributions, visual observers can plot meteor trails. From this the radiant can be approximately determined. McKinley (1961) quotes values on typical errors of trained teams of visual observers, with the angle of the trail typically being accurate within 3 to 14 degrees, the endpoints within 3 to 7 degrees, and the magnitude estimates being accurate to about one-half magnitude. It is probable that the highly experienced observers of the International Meteor Organization and other national meteor organizations have in recent years improved on these results.

Visual observers occasionally use binoculars or wide field telescopes in order to extend the observations to much fainter magnitudes. However, the very small field of view of these instruments result in very low hourly meteor rates, although the precision of plotting is much better. Kresakova (1978) discusses the relative efficiency of different types of telescopic optics for meteor observation.

5.3. Photographic Techniques

Despite the strides which have been made in making the observations of visual observers objective and consistent, nevertheless a more objective method to record meteor phenomena is desired for many types of research analysis. Photographic methods have been in use for many decades. Any camera which is capable of time exposures can be used. The camera is set on a tripod or similar mounting, usually a medium field lens (such as 50 mm focal length) is used, and time exposures from 5 to 20 minutes (depending on the quality of the observing location, the lens speed and the film speed). Most often films with ISO 400 are used. Typically a total of one to two hours of time exposures are needed per recorded meteor even during a moderate annual shower. The limiting sensitivity for meteor detection of these small cameras is usually about +1 astronomical apparent magnitude (the apparent stellar sensitivity goes to much fainter magnitudes). If a rotating shutter is used in front of the objective lens (typically 30 to 60 breaks per second) then angular velocity information can be obtained from the photographic record. The Dutch Meteor Society is currently one of the most active groups in meteor photography, and Betlem *et al.* (1999) provide an example of the precision which can be achieved in modern photographic meteor observations.

In the early 1950s Perkin–Elmer Corporation constructed six Super-Schmidt cameras specifically for meteor observation. These cameras had a focal length of just over 20 cm with a very fast effective f/ratio of 0.65. To achieve this remarkable performance, these cameras required huge optics (they weighed about 3 tons). Nevertheless, with a field of view of about 55 degrees and a limiting sensitivity of better than +3 these provided one of the most historically valuable optical meteor datasets [Jacchia *et al.* (1967)].

Camera patrol networks utilizing banks of small cameras have been set up in several countries in order to provide orbital characteristics for meteorite-dropping fireballs, as well as to provide influx statistics on large objects. Such networks have been reviewed by McCrosky & Ceplecha (1968) and Ceplecha (1996) with the Canadian MORP system described by Halliday *et al.* (1978). Halliday *et al.* (1983) has reviewed meteorite orbit information provided by photographic fireball networks. A recent example of very precise trajectory and orbit information from multi-station photographic observations is given by Betlem *et al.* (1999).

Before concluding this section, at least brief mention should be made of the method of instantaneous exposure [Babadzanov & Kramer (1968)] in which a long (by meteor standards) focal length lens (750 mm) is combined with a rapidly rotating shutter (0.6 ms effective exposure) in order to obtain excellent temporal and spatial resolution. With this equipment it was possible to provide effective spatial resolution of the order of 10 m. This has proven particularly valuable for estimating the sizes of the constituent grains from dustball meteors Simonenko (1968).

5.4. Video Observation of Meteors

The topic of video detection of meteors has been the source of several reviews, including Hawkes & Jones (1986), Hawkes *et al.* (1993) and Molau *et al.* (1997). While video observation of meteors was first reported in 1960, it has only been during the last two decades that this technique has played a major role in observational meteor astronomy. The CCD (charge coupled device) invented in 1970 is now the dominant detector for virtually all video-based meteor astronomy. The sensor is a MOS (metal oxide semiconductor) solid state device which can be regarded as a matrix of capacitive charge wells. The presence of light during the integration period generates charge in these wells through hole–electron pair creation and electronic diffusion. In order to reduce the number of connecting lines, which would for example be several hundred thousand for a typical video CCD, the charge is shifted from well to well and read out in a manner similar to an electronic shift register. A typical CCD used for video work employs about 500×800 active pixel elements. Most video-based CCDs use interline transfer, which means that there is a duplicate set of MOS capacitive cells adjacent to those which detect the light. At the end of the exposure the charges recorded in the primary cells are rapidly ($\ll 1$ μs) shifted to these interline cells. Then while the next image is being integrated the existing video image is shifted out from this electronic shift register. We will see below that the image intensifier required for video work introduces its own limitations on the resolution which can be obtained with moderate cost.

CCD chips offer distinctive advantages over photographic plates for several reasons. One is that the detection quantum efficiency, ϵ, defined as the ratio of the average number of detected photons per pixel per second to the average number of incident photons per pixel per second, is much higher than for photographic plates.

$$\varepsilon = \frac{n_{\mathrm{d}}}{n_{\mathrm{i}}} \tag{5.4.1}$$

A premium CCD chip can have a detection quantum efficiency of 0.80 or more in the middle of its sensitivity curve (compared with 0.04 for typical photographic films). The system quantum efficiency will be less than the detection quantum efficiency because of other losses in the system. The response of the CCD will cease for wavelengths longer than 1100 nm because these infrared photons do not have sufficient energy to create hole–electron pairs. At the blue end of the spectrum the photons increasingly do not penetrate sufficiently to the depletion region of the CCD where the charge pairs need to be created to be recorded. CCDs are constructed in two basic ways: frontside and backside. In frontside illuminated CCDs the light first crosses semi-transparent electrodes and silicon dioxide gate region before creating electron–hole pairs in the doped silicon substrate. Frontside CCDs are generally insensitive to violet and ultraviolet radiation. This lack of blue response can be partially overcome by coating the CCD with films which are sensitive to near ultraviolet light, and which will fluoresce in the visible region in response. In a backside CCD the incident light arrives directly in the substrate. The second reason that CCDs are preferred to photographic film for astronomical use is that they are linear over many orders of magnitude. The response of photographic film is highly nonlinear with saturation taking place.

The response of a CCD is often expressed in terms of amperes of current flow per watt of incident energy (A/W). A useful expression by Buil (1991) relates the response R (expressed in amperes per watt of incident radiation) to the quantum efficiency, where λ is the wavelength of the light, h is Planck's constant, c is the speed of light and e is the value of the fundamental electronic charge.

$$\varepsilon = R\frac{hc}{e\lambda} \tag{5.4.2}$$

To provide a measure of typical performance features for a scientific monochrome CCD used in video meteor work, we provide the characteristics for the Cohu 4910 scientific video cameras. The RS170 version of these cameras have $768(H) \times 492(V)$ active pixels, resulting in a total image area of 6.4×4.8 mm. The effective resolution is 580 horizontal television lines, and 350 vertical television lines. The sensitivity at the faceplate is 0.65 lux for full video and 0.02 lux for 80 applications.

There are several sources of noise within the CCD. Thermal noise is caused by the generation of charge pairs through thermal energy. Transfer noise is associated with lack of consistency in the transfer of charges from element to element in the CCD. Reset noise is associated with the process of reading the output capacitor of the shift register. If the signal is converted to digital format there is a quantification noise. The light arriving from the meteor are quantized into photons which appears as a noise for very low light levels. There are always additional noise contributions from other electronic circuitry and lines. In typical astronomical applications where exposure integrations from tens of seconds to several minutes or more are common, thermal noise predominates. However, for effective exposures of less than 0.1 s or so (i.e. for video recording) other sources of noise are most important, and little is achieved by cooling of the CCD through liquid nitrogen or Pelletier electronic coolers. The most serious of these in most video applications, is readout noise, which is defined as the noise measured from a darkened CCD when thermal signal is removed (or largely so). In essence readout noise is a combination of the several nonthermal noise sources noted earlier. Early CCDs had readout noise of typically 20 to 50 electronic charges, but that value has now been reduced to just a few electronic charges in premium CCDs.

When a commercial video rate CCD is pointed to the night sky the results are disappointing. Typical limiting stellar sensitivity is about +2 astronomical magnitude, and one would require hours of recording for a single meteor (except during a strong shower). Considering the improvements in readout noise which have been achieved during the past decade in premium CCDs, it is not absolutely clear that a CCD camera could not be

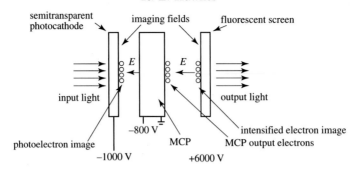

FIGURE 5.2. The principle of operation of a Gen II/III microchannel plate image intensifier. The energy of the incident photons is used to release electrons from the photocathode. They are accelerated in a strong electrical field in the microchannel plate, with electron multiplication taking place. The electrons strike a phosphor, recreating a light image which is many thousands of times brighter than the original image. Adapted from Murphy (1998).

optimized to become a useful meteor detector directly, but the approach which has been used in meteor research up to this point is to interface the CCD to an image intensifier.

Image intensifiers are essentially light multiplication imaging devices. They have an input window and an output phosphor, with the image on the output phosphor being much brighter than that which is incident on the input window. The early (Gen I) image intensifiers were essentially similar to electronic pinhole cameras. Photoelectrons were released when light fell on a photocathode, and these were accelerated in vacuum towards the point of a cone-shaped anode which was held at a high positive potential (typically 15,000 V). However, the rapidly moving electrons streamed through a small hole in the anode cone and struck the phosphor, creating a light pulse. Since the energy of the electrons had been increased many times over in the accelerating electric field, a much brighter output image is created. These Gen I image intensifiers could achieve a luminous gain of typically 75, and several (usually three) were used in sequence to produce a useful device with a gain of some tens of thousands. However, these Gen I devices were susceptible to severe blooming, which refers to the expansion of a bright image into neighboring pixels.

Gen II image intensifiers employ a microchannel plate (MCP). Again electrons are released at a photoelectrode and are accelerated towards a high positive potential. However now these optically released electrons are guided down the many tiny microchannel plates, and a cascade effect is created when a high energy electron strikes the MCP wall and generates several secondary electrons. The process is similar to the multi-stage photomultiplier tube, except in the case of a MCP image intensifier the image characteristics are retained. With Gen II image intensifiers gains of typically 35,000 can be achieved in a single stage. Although blooming is still present, it is not nearly as pronounced as in Gen I image intensifiers. Gen III intensifiers use a similar operating principle to Gen II, but have improved photocathode materials which extend both the lifetime of the device and also the spectral response. Figure 5.2 illustrates the operation of a microchannel plate image intensifier. Most image intensifers come in integrated packages which includes an oscillator, high voltage transformer and voltage multiplication chain so that only a low DC voltage (typically a few volts) need be provided by the user. The most common sizes of image intensifiers have diameters of 18 mm and 25 mm. Essentially what the image intensifier is doing for us in a meteor detection system is that it is increasing the light level incident on the CCD so that we can detect near the quantum limit, despite readout and other noise sources in the CCD detector itself. In order to make the image intensi-

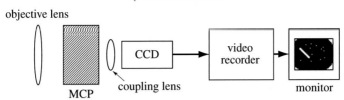

FIGURE 5.3. A block diagram of a lens–coupled microchannel plate image intensified CCD meteor observation system.

fier useful we need to optically couple it to the CCD. There are two main approaches possible. One option is to use a fiber-optic connector and optical cement to physically couple the two directly (usually a reducing taper is needed on the fiber-optic element, to map the 18 mm active surface of the image intensifier down to the 6 mm or so active surface of the CCD). The other approach is to use a coupling lens to image optically the output of the image intensifer on the input of the CCD. Clearly the fiber-optic approach is more light efficient, but the interface may actually degrade resolution. Furthermore, the optical gain of the image intensifier is such that we don't need nearly the entire gain possible in order to make a bright enough image to overcome the readout noise in the CCD, and therefore the light loss may not be significant. For these reasons, in recent years most image intensified CCD systems, and certainly almost all of those used for meteor detection, have adopted lens coupling.

A quality image intensifier has a resolution of about 28 line pairs per mm, or 504 line pairs across 18 mm. However, at most about 400 line pairs can be supported in a video format with a 4:3 dimension ratio, or 800 pixels of resolution. This would be further degraded by the CCD and other components of the overall system. Hence it is clear that one must go to higher resolution, or larger format, image intensifiers to support the higher resolution CCDs which will soon be applied to meteor research.

A complete video-based MCP image intensified CCD meteor observation system is pictured in Figure 5.3. Figure 5.4 shows a typical stellar and meteor video frame observed with such a system, which is sensitive to stars of nearly +9 magnitude. In North America and Japan NTSC video standards are used and each 1/30 s video frame is made up of two interlaced video fields, while the PAL video standard is used in most other areas, with a 1/50 s per video field. In Figure 5.4 one can clearly distinguish between the even and odd video field recording of the meteor. Direct digital recording of data in real time has only become feasible in the last few years with the commercial development of digital video tape recorders (a typical monochrome video system with a resolution of 640 × 480 pixels, using NTSC video with 30 frames per second, and with one byte per pixel photometric resolution will generate about 33 GB of data per hour of recording if no compression is used). These digital recording systems offer many advantages including very stable time base and the ability to be used to make multiple lossless copies. Furthermore, they are less susceptible to noise and signal degradation. Now many desktop computers have digital video (IEEE 1394) interfaces so that the digital data can be transferred to the computer directly without the need for additional frame grabbing circuitry. We show in Figure 5.4 a typical night sky image with one frame of a meteor trail recorded by a Gen III MCP intensified CCD system and recorded in digital format. Note that in the image the even and odd scan lines have been accurately recorded by the digital recorder (in NTSC video every other line is first scanned, and then the intervening lines, so that a video field is recorded every 1/60 s, with a complete video frame every 1/30 s). An optimized Gen III system can record stars to an apparent limit in sensitivity

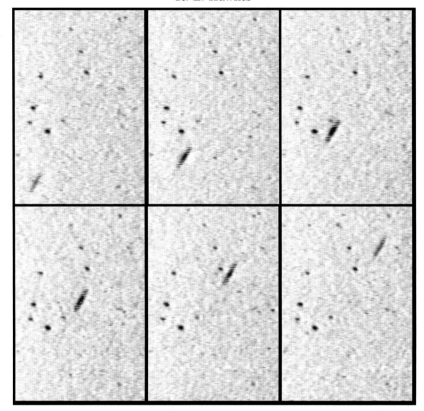

FIGURE 5.4. A typical Gen III image-intensified CCD image of a stellar field and meteor. The faintest stars are about +9 astronomical magnitude in this 3.4×4.9 degree portion of the field of view. The sequence shows every second video frame (i.e. 1/15 s between images). The even and odd scan lines of the interlaced NTSC video frame are clearly evident in this digitally recorded image.

of about +8.5 to +9.0 over a field of view of typically 10 times 14 degrees (with a 50 mm objective lens). Hawkes & Jones (1986) and Hawkes *et al.* (1993) have reviewed the potential performance limitations on the sensitivity of an image intensified video-based meteor detection system. Limitations can be conveniently divided into three categories: quantum limitations due to an insufficient number of photons falling on the detector during the frame integration period, the sky background limitation in which unresolved stellar and nonstellar components of the night sky mask the detection of faint meteors, and an instrumental background limitation in which electronic noise of various types overwhelm the meteor signal. The quantum limitation can be expressed in the form

$$m_\varepsilon = 49.9 - 2.5 \log \left(\frac{nR^2}{\gamma \varepsilon D^2 \tau} \right) \tag{5.4.3}$$

where R is the range to the meteor, n is the number of photons per frame integration time which must be detected, γ is the fraction of incident light transmitted by the lens to the photocathode, ϵ is the quantum efficiency, D is the effective diameter of the lens, τ is the frame integration time and m_ϵ is the effective limiting sensitivity due to the quantum limitation expressed according to the astronomical magnitude scale. All quantities are to be expressed in SI units. If we adopt $n = 2$, $R = 100,000$, $\gamma = 0.9$, $\epsilon = 0.4$, $D = 0.05$, $\tau = 1/30$ we obtain a quantum limitation of +12.9 magnitude.

If we adopt the figure for sky brightness from Allen (1973) as adapted by Hawkes *et al.* (1993) then we obtain the following expression for the background sensitivity limitation

$$m_b = 3.8 - 5.0 \log(\alpha) \tag{5.4.4}$$

where α is the angle (expressed in degrees) subtended by one effective pixel of resolution. Note than in this expression α is the effective resolution, and not simply the number of pixels in the CCD. If one assumed a system with a field of view of 15 degrees which had an effective resolution of 300 pixels, a background limitation of +10.3 astronomical magnitudes is obtained. Clearly, the precise value will depend on the elevation angle of the field of view and the atmospheric characteristics of the observing location. Since most meteor detection systems are background limited, or near background limited, attaining the highest possible resolution is important. Generally speaking, Gen III systems have improved resolution over earlier generations of image intensifiers. Pioneering work in the application of high definition video equipment to meteor research has been made by Watanabe *et al.* (1999) and Abe (2000). By coupling a high resolution image intensifier to higher resolution CCDs equipped with direct digital output, a powerful meteor detection system is possible. The resolution is 1920×1035 pixels, with 10 bit intensity output per pixel.

Computation of the instrumental background noise limitation is complex, since the characteristics of both the image intensifier and the CCD need to be considered. A particularly important figure when comparing image intensifier specifications is the EBI, which is the equivalent background illumination, of the image intensifier. Even with no luminous input the image intensifier will produce an output according to this EBI level applied at the input photocathode. In a poorly designed image intensifier this EBI will become the sensitivity-limiting factor.

The actual limiting magnitude for meteor detection is typically several magnitudes fainter than the apparent stellar limiting magnitude. This is because the meteor moves across a number of pixels during a single frame integration time, so stellar sources can integrate this light whereas the moving meteor spreads it among various integration cells. A simple approximate expression is the following relating the meteor apparent magnitude m_m and the stellar apparent magnitude m_s as a function of the number (d) of pixels moved by the meteor during the frame integration time. As we will see later, there is a photometric procedure which corrects for this effect, but it is important to keep it in mind when one is determining the actual limiting magnitude for meteor observation by intensified video systems.

$$m_m = m_s - 2.5 \log(d) \tag{5.4.5}$$

Up until about 1996 meteor occurrences on video records needed to be located by human observers. Gural (1995) described an automated system for meteor detection, and the initial implementation was reported in Gural (1997). In the last few years, there has been rapid advances in the performance and features of this software, MeteorScan, and it was used with great success for automated real time detection and shower association of meteors in the 1999 Leonid campaign by Brown *et al.* (2000). Essentially the software first computes mean intensity and standard deviation statistics on a pixel by pixel basis for the intensified CCD imagery. It then in real time flags pixels which are brighter than their mean value by a statistically significant amount. A Hough transform is used to search for linear segments of statistically significant pixels. This is currently implemented on a Macintosh G3 or G4 PowerPC RISC system utilizing Scion LG3 frame grabber cards. It operates in real time (30 frames per second with 640×480 pixels per frame). Recent improvements in the code have permitted real time shower association, after positional calibration within the software. Performance measures achieved during the 1999 Leonid campaign in Israel are given in Brown *et al.* (2000). It is estimated that MeteorScan, under optimal operation, detects about 80% of the meteors which would be detected by a careful human observer, with most of the missed meteors being near the limiting magnitude of the system.

Over the last few years an alternative software package (MetRec) for automated detection and shower association has also been developed [Molau (1998a), Molau (1998b)]. This software, which operates under PC/Windows OS, uses the Matrox Meteor line of video digitizing cards. MetRec operates on half resolution PAL format video frames (384×288 pixels), performing differentiation (current frame is subtracted from the previous frame). Both MetRec and MeteorScan have user defined variables which can be used to set the tradeoff between false detections and recovery rate. Also, both allow certain areas of the screen to be masked out (e.g. areas on which a time signal has been superimposed).

5.5. Positional Analysis of Optical Data

A complete treatment of the photometric and positional analysis procedures for photographic and video-based multi-station meteor data is beyond the scope of this review, although we will outline concisely the procedures to be employed. For more detail, the reader is referred to a number of works, including Wray (1967), Whipple & Jacchia (1957), Ceplecha (1987) and Hawkes *et al.* (1993).

The first task is to obtain equatorial coordinates (right ascension, α, and declination, δ) of the meteor head on each digitized frame. We will follow the plate constants approach of Wray (1967) and Marsden (1982). This can be done by measuring the pixel coordinates (x, y) of a number of reference stars, and then applying a least squares fit between these coordinates and what an ideal (no distortion) camera would see – this is the plate constants approach described in detail by Wray (1967), Smart (1977) and Marsden (1982). The image measurement system provides a planar coordinate pair (x, y) for each point of interest (i.e. reference stars and meteor points). We will map (x, y) to a second planar coordinate system (ξ, η) which corresponds to what an ideal zero distortion system would observe. The η axis corresponds to motion along a great circle pointing in the direction of the north celestial pole, while the ξ axis points in the direction of increasing right ascension. The origin of the (ξ, η) system should be the plate center. Wray (1967) and Hawkes *et al.* (1993) have shown that the ideal plate positional coordinates can be given in the form of the following equations.

$$\xi = \frac{\cos(\delta) \sin(\Delta \alpha)}{D} \tag{5.5.1}$$

$$\eta = \frac{\sin(\Delta \delta) + \cos(\delta) \sin(\delta_0) [1 - \cos(\Delta \alpha)]}{D} \tag{5.5.2}$$

$$D = \cos(\Delta \delta) + \cos(\delta) \cos(\delta_0) [\cos(\Delta \alpha) - 1] \tag{5.5.3}$$

$$\Delta \alpha = \alpha - \alpha_0 \tag{5.5.4}$$

$$\Delta \delta = \delta - \delta_0 \tag{5.5.5}$$

In the above (α, δ) represent the right ascension and declination of the meteor or reference star points, while (α_0, δ_0) represent the equatorial coordinates of the plate center.

Typically in image intensified CCD meteor observations 12 to 18 reference stars are measured to establish positional calibrations. For each reference star (x, y) are measured and (ξ, η) are computed. The positional calibrations between the pixel coordinates (x, y) and the ideal plate coordinates (ξ, η) are usually assumed to be represented by the following functional forms. If the system is not seriously distorted, it may be possible to use only the linear terms, which permit an arbitrary rotation, origin shift and scaling between the pixel and perfect coordinates. A least squares best fit is used to determine the values of the a and b coefficients. Hawkes *et al.* (1993) discusses typical accuracies in meteor observation systems.

$$\xi = a_0 + a_1 x + a_2 x^2 + a_3 y + a_4 y^2 + a_5 xy \tag{5.5.6}$$

$$\eta = b_0 + b_1 x + b_2 x^2 + b_3 y + b_4 y^2 + b_5 xy \tag{5.5.7}$$

After the meteor position pixel coordinates have been measured on each video frame the coefficients are used with the above functions to determine the plate constant coefficients (ξ, η). In order to determine the equatorial coordinates (α, δ) for the meteor positions one uses the following set of relationships. Note that the first three define a vector $\mathbf{P}$, which is made into a unit vector before application of the last two equations to obtain the equatorial coordinates of the meteor point.

$$P_x = \cos(\delta_0) - \eta \sin(\delta_0) \tag{5.5.8}$$

$$P_y = \xi \tag{5.5.9}$$

$$P_z = \sin(\delta_0) + \eta \cos(\delta_0) \tag{5.5.10}$$

$$\alpha = \alpha_0 + 2 \tan^{-1} \left(\frac{\widehat{P_y}}{\widehat{P_x} + \sqrt{1 - \widehat{P_z}^2}} \right) \tag{5.5.11}$$

$$\delta = 2 \tan^{-1} \left(\frac{\widehat{P_z}}{1 + \sqrt{\widehat{P_x}^2 + \widehat{P_y}^2}} \right) \tag{5.5.12}$$

5.6. Photometric Procedures

Photometric procedures for photographic meteor data are discussed in Ceplecha (1987) and Rendtel (1993) and will not be discussed here, although the techniques described below for image intensified CCD observations of meteors can also be applied to photographic observations. The photometric procedures must take into account that for moderate and bright sources both the image intensifier and the CCD introduce blooming, or spillover of signal into nearby pixels (see Figure 5.5. For this reason the peak pixel intensity is not a valid measure of the meteor brightness except for the faintest meteors. As mentioned earlier, account must also be taken of the fact that the meteor will spread its brightness over a number of pixels during the video frame integration time. A variety of approaches have been used to perform image intensified video meteor photometry. For very faint meteors the pixel intensity value is a measure (not necessarily linear) of the brightness, while for brighter images the diameter of the bloomed image becomes an indication of the brightness of the meteor. The approach which we have recently employed is described in more detail by Fleming *et al.* (1993) and has also been employed by Campbell *et al.* (1999) and Murray *et al.* (1999). We have found, through simulations with stellar images

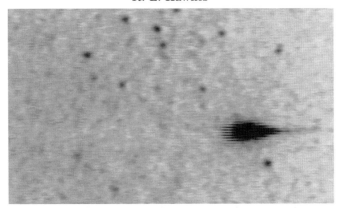

FIGURE 5.5. An image intensified CCD meteor record which illustrates blooming whereby
bright signals spill over from one pixel into nearby pixels.

being slewed at typical meteor angular rates, that the meteor brightness on the astronomical magnitude scale can be related to the integrated pixel intensity values by the
following relationship.

$$mag = a_0 + a_1 \log \left[\sum (p_i - b) \right] + a_2 \log \left[\sum (p_i - b) \right]^2 \qquad (5.6.1)$$

Here the summation is taken over all pixels making up the meteor head and the portion
of the trail since the previous video frame, p_i is the intensity value of each pixel and b
is the mean intensity value of background areas near the meteor. In defining the mean
background intensity value it is important to use areas near the meteor (since most
intensified CCD systems are somewhat to strongly nonuniform in sensitivity) and also to
avoid stellar images to the degree possible. One would only expect the above relationship
to be valid in cases in which the intensifier and CCD response are approximately linear,
and it is surprising that it works as well as it does. All image-intensified systems are more
sensitive in the central portions of the field of view, but surprisingly by subtracting a local
background mean the response becomes almost independent of distance from the center
of the viewing area. A typical photometric calibration curve when using this technique
is indicated in Figure 5.6.

   Corrections for distance can be applied to determine absolute meteor magnitudes
[McKinley (1961)], which are defined to be at a range of 100 km. Atmospheric absorption
corrections will not be applied if the reference stars used in the calibration were observed
at the same elevation angle, which will normally be the case. With care it is possible to
estimate relative points along a single meteor light curve to an accuracy of about $\pm 0.3^m$,
with absolute magnitude uncertainties approximately double this. The accuracy of the
technique worsens both for very bright (because of a lack of reference stars) and very
faint (because of poor signal to noise ratio) meteors. Some correction for apparent motion
is needed, although the effect is not very large for intermediate brightness meteors.

   Photometric masses can be obtained by integrating the light curve (this is based upon
a rearrangement and integration of equation 5.1.1). The most convenient operational
form is the following summation over the light curve.

$$m_{\mathrm{p}} = \frac{2}{\tau_i v^2} \sum_{i=1}^{N} I \Delta t \qquad (5.6.2)$$

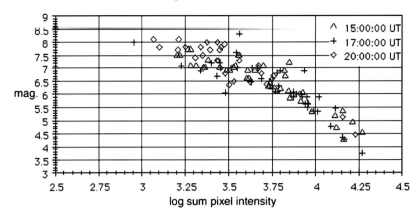

FIGURE 5.6. Plot of the apparent stellar visual magnitude versus the logarithm of the sum of the amount the stellar image is above background for an image intensified CCD system with blooming. Each point represents one reference star. Adapted from Murray *et al.* (1999).

In equation 5.6.2 m_p represents the photometric mass estimate for the meteor (in kg), v is the geocentric speed (assumed constant and expressed in m/s), τ_i is the meteor luminous efficiency factor (expressed as a unit less fraction of the energy which is in luminous form), I is the luminous intensity of the meteor (expressed in W) during the ith video frame, Δt is the time interval for each video frame and the summation is performed over the number N of frames in the trajectory. Since the meteor brightness is usually expressed in terms of the astronomical magnitude scale, the following operational form of the equation is often used.

$$m_\mathrm{p} = \frac{2}{\tau_{i0} v^3} \sum_{i=1}^{N} 10^{-M_i/2.5} \Delta t \tag{5.6.3}$$

In equation 5.6.3 M_i is the absolute magnitude of the ith point in the trajectory and τ_{i0} must be expressed in a form applicable to expressing the meteor luminosity in terms of the number of 0 astronomical magnitude sources. If SI units are used for all other variables, the value of 1.0×10^{-10} is frequently used for τ_{i0} for the case of relatively faint meteors. In this equation we have assumed [following Whipple as quoted in McKinley (1961)] that the luminous efficiency factor varies linearly with velocity (see equation 5.6.4), although, as mentioned in section 5.1, this is not established with any degree of certainty.

$$\tau_i \approx \tau_{i0} v \tag{5.6.4}$$

Many meteors observed with intensified CCD techniques begin or end outside the field of view, and in these cases the photometric mass will underestimate the true mass. These cases should at least be flagged in the results. Also, the above relationships are derived on the basis of individual atomic interactions between the atmosphere and the meteor. Certainly for larger meteoroids air caps are established during atmospheric ablation, and the physics of the light production becomes more complex [see Ceplecha *et al.* (1998) and references therein for discussion of this situation]. The important recent work by Popova *et al.* (2000) indicates that screening by ablation vapor is important even for meteors studied by image intensified video methods.

   In addition to integrating the light curve to determine photometric masses, the shape of the meteor light curve can yield information regarding the mode of atmospheric ablation

of the meteoroid and ultimately its physical structure. The simplest method to specify
the degree of symmetry and the direction of the skew in a light curve is with the F
parameter, applied to photographic work by Jacchia *et al.* (1967). This approach has
been applied to intensified CCD meteor work by Fleming *et al.* (1993) and Murray *et al.*
(1999) among others.

$$F_\delta = \frac{H_{B\delta} - H_{max}}{H_{B\delta} - H_{E\delta}} \tag{5.6.5}$$

Here F_δ represents the meteor light curve F parameter for a difference in magnitude of δ
while $H_{B\delta}$ is the beginning height for a meteor magnitude difference δ (with the ending
point for that difference called $H_{E\delta}$ and H_{max} is the height of maximum luminosity
of the meteor. Once the raw magnitude data have been determined the noise levels
in most intensified video systems dictate that some smoothing be applied before the
light curve parameters are calculated. We have found that a 1–2–1 weighted nearest
neighbors smoothing algorithm works well. One typically defines F values for a series of
δ magnitude differences such as 0.5, 1.0, 1.5, etc. Campbell *et al.* (1999) has developed a
more sophisticated curve fit to the meteor light curve, with one parameter representing
the degree of skew. Campbell *et al.* (1999) and Murray *et al.* (2000) have shown how one
can fit models of the fundamental grain size distribution in a dustball model [e.g. Hawkes
& Jones (1975)] to meteor light curves.

5.7. Multi-Station Observations

Photographic or image intensified video techniques may be applied to simultaneous
observations from two or more stations. Typically, a baseline of 15 to 140 km is used.
A smaller baseline will result in less well determined trajectory and orbital parameters,
but a larger number of meteors in common between the two stations. If cameras with
a relatively small field of view are employed a severe height bias is introduced by the
requirement for overlap between the two stations if the baseline separation is large. Woodworth & Hawkes (1996) has studied the height bias inherent in two station intensified
video observations.

The analysis procedures for multi-station photographic [Wray (1967)], fireball network
[Ceplecha (1987)] and intensified television [Hawkes *et al.* (1993)] observations have already been published, and we will only outline the main steps here. We will place emphasis
on reduction of data from video detectors and also will employ vector analysis notation.
First a least squares best fit line is found for the meteor on each camera, using all points
on the meteor's trajectory. The curvature introduced by gravitational attraction is not
significant for television meteor accuracy. Each raw point on the apparent meteor trail
is transformed to a nearest point on this best fit line. This will tend to eliminate small
errors in the positional measurements.

The geometry of the two station analysis is shown in Figure 5.7. Here $\mathbf{d}$ is a baseline
vector pointing from station 1 to station 2. One must work in a consistent set of units
for the following vector analysis: the baseline is usually initially expressed in a local
Earth-based coordinate system, while the vectors to points on the meteor trail most
conveniently are calculated from the positional analysis using an equatorial astronomical
coordinate system as expressed in the following relationships. Taking the vector cross
product of vectors to the corrected first and last points ($\widehat{\mathbf{b}}$ and $\widehat{\mathbf{e}}$ in Figure 5.7) gives
a vector normal to the plane containing the meteor as viewed from each site, which is
transformed into the unit vector $\mathbf{n}$.

$$\mathbf{n}_1 = \widehat{\mathbf{e}}_1 \times \widehat{\mathbf{b}}_1 \tag{5.7.1}$$

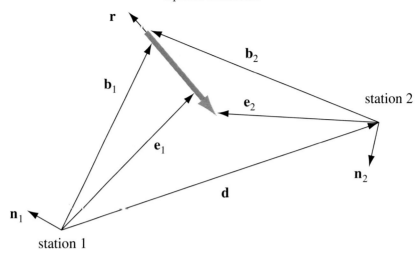

FIGURE 5.7. Geometry of two station meteor analysis. The unit vectors $\widehat{\mathbf{n}}_1$ and $\widehat{\mathbf{n}}_2$ represent normals to the planes containing the meteor as viewed from stations 1 and 2. Vector $\mathbf{d}$ is a baseline from station 1 to station 2, while vector $\mathbf{r}$ points in the direction of the meteor radiant.

$$\mathbf{n}_2 = \widehat{\mathbf{e}}_2 \times \widehat{\mathbf{b}}_2 \tag{5.7.2}$$

The cross product of the normals of these planes must then lie along the trail of the meteor, and the apparent radiant of the meteor can thus be determined. The sign of the vector $\mathbf{r}$, in the direction of the meteor, can easily be obtained by converting to Earth-based coordinates.

$$\widehat{\mathbf{r}} = \pm \frac{\widehat{\mathbf{n}}_1 \times \widehat{\mathbf{n}}_2}{|\widehat{\mathbf{n}}_1 \times \widehat{\mathbf{n}}_2|} \tag{5.7.3}$$

The following vector equations (with reference to the notation of Figure 5.7) constitute systems of equations which can be solved for the range parameters. Each is a vector equation, so the three variables can be solved.

$$c_1 \widehat{\mathbf{b}}_1 - c_2 \widehat{\mathbf{r}} = \mathbf{d} + c_3 \widehat{\mathbf{e}}_2 \tag{5.7.4}$$

$$c_4 \widehat{\mathbf{e}}_1 + c_5 \widehat{\mathbf{r}} = \mathbf{d} + c_6 \widehat{\mathbf{b}}_2 \tag{5.7.5}$$

For example, c_1, c_4 will represent the range from station 1 to the apparent beginning point and ending point of the meteor; c_3, c_6 are the ranges to the beginning and ending points as viewed from station 2. One strength of this approach is that it is not necessary to assume any common point (as viewed from both stations), and indeed the solution yields the offset of the apparent beginning point from one station compared with that as viewed from the second station. Also, it is not necessary for the complete trail to be within the field of view as seen from both stations.

A conversion of the radiant vector to local geocentric coordinates yields the zenith angle. The time information provided by the video frame rate can be used to determine velocities. Heights can be calculated once the range and angle to any specific meteor point is known. While the analysis is done in Cartesian coordinates fixed on station 1, a small correction is applied for the curvature of the Earth in determining the vertical heights of the meteors.

A discussion of computation of orbits is beyond the scope of this chapter. But once the precise time and corrected geocentric velocity vector have been determined, one can readily calculate the orbital parameters (a, e, i, Ω, ω, MA). See Ceplecha (1996) or a

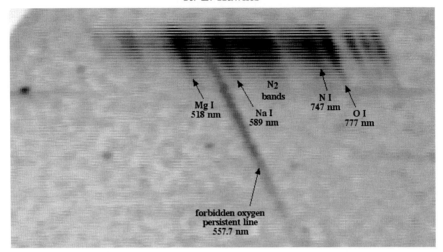

FIGURE 5.8. Low resolution Perseid spectrum obtained with a transmission diffraction grating with an image intensified CCD system. Image courtesy of Murphy (1998).

similar reference for a detailed discussion on the steps involved. One must correct the apparent velocity for atmospheric deceleration, rotation of the Earth and the gravitational attraction of the Earth, and one must correct the apparent radiant for these same effects prior to computation of orbits.

5.8. Meteor Spectroscopy

As mentioned in the introduction, the majority of meteor luminosity in the visible region of the spectrum comes from rapid decay of meteoric atoms and ions which have been excited by atomic collisions with atmospheric constituents. If a transmission diffraction grating is placed in front of the objective lens, meteor spectra can be obtained. Ideally, one uses a blazed diffraction grating, which preferentially directs light into the first order image. Photographic spectra have much higher resolution, but are limited to moderately bright meteors (usually -1 astronomical magnitude and brighter). A recent example of a high resolution photographic spectrum is shown in Airey (1999). Image intensified CCD systems can measure meteor spectra at much fainter magnitudes (typically down to about $+5$ astronomical magnitude), but provide much poorer resolution. We show in Figure 5.8 a typical low resolution spectrum obtained with a wide field image intensified CCD system. Ceplecha et al. (1998) provides an overview of meteor spectroscopy, which is also reviewed by Millman (1980). Extensive lists of identified meteor spectral lines in the visual region from photographic spectra include the work by Halliday (1961), Ceplecha (1971), and Borovička (1994a). Near ultraviolet ($\lambda > 310$ nm) observations have been made by Halliday (1969) and Harvey (1973), while Millman & Halliday (1961) provides near infrared ($\lambda < 900$ nm) meteor spectra. A typical ground-based Gen III microchannel plate image intensifier coupled to a CCD can observe in the spectral region from about 350 nm to 870 nm. Special lenses and observations from high altitudes would permit going deeper into the near ultraviolet.

Atomic (and in some cases ionic) emission lines from Al, Ca, Co, Cr, Fe, H, Li, Mg, Mn, N, Na, Ni, O, Si, Sr and Ti have been definitively identified in meteor spectra, and atomic/ionic lines of Ba, Cu, Sr and V have been tentatively identified. In terms of molecules identified in meteor spectra, AlO, CaO, C_2, CN, FeO, MgO and N_2 have been identified with certainty. In most cases the strongest lines are from the H and K lines of Ca II (396.8 nm and 393.4 nm), the Na I lines at 589.0 nm and the Mg I lines at 383.8 nm and 518.4 nm. In high velocity meteors the forbidden Oxygen auroral line at 557.7 nm is a prominent feature, persisting prominently for typically one second after the passage of the meteor – the identification of this line was first determined by Halliday (1960).

One advantage of video-based meteor spectroscopy is that time resolved information is available. Borovička *et al.* (1999) have spectrally studied the 1998 Leonid shower, which was rich in bright fireballs. They found that the relative concentrations of Mg, Ca, Fe and Na in the Leonid meteoroids were approximately consistent with the concentrations in C I chondritic meteorites. They also found spectral evidence for differential ablation, in which the Na ablated prior to other metallic constituents. They suggest that Na might well be the *glue* in the two component dustball model of Hawkes & Jones (1975).

A full theoretical understanding of the production of meteor spectra is not yet available. Borovička (1994b) developed an interesting model in which two components, at radically different temperatures of about 3500 K and 10 000 K, are required to explain the spectra. The higher temperature spectrum contains relatively few distinct lines, and is confined mainly to high velocity meteors.

5.9. Optical Observations of Meteors

While much of this chapter has been concerned with the techniques and analysis procedures for optical observation of meteors, in this section we will briefly consider some of the main contributions of optical detection methods to meteor studies. Clearly, the space available here does not permit a comprehensive review of the field, and we have placed emphasis on typical papers, particularly those of recent publication.

Visual observations have played the dominant role in characterization of the rate profiles for the major showers on a year to year basis. Indicative of the sort of analysis which can be done on large collections of carefully corrected visual data are the following articles on the Geminid shower by Rendtel & Arlt (1997) and Rendtel (2000); the Perseid shower by Arlt (1998), Arlt (1999a) and Arlt & Händel (2000b); the June Bootid shower by Rendtel *et al.* (1998); the η-Aquarid shower by Rendtel (1997), and the Leonid shower by Arlt & Gyssens (2000a). The precision of the magnitude distribution (and inferred meteoroid mass distribution) provided by visual techniques is perhaps less clearly established.

Image intensified video techniques have played a significant role in determining flux measurements, particularly with respect to the recent Leonid showers [e.g. Brown *et al.* (2000), Gural & Jenniskens (2000), Jenniskens *et al.* (2000a), Brown *et al.* (2001)].

The most important contributions of multi-station image intensified video detectors have probably been in the area of precise atmospheric trajectory information. Major studies in this regard include Hawkes *et al.* (1984), Sarma & Jones (1985), Yoshihiko *et al.* (1997), Campbell *et al.* (2000), and Brown *et al.* (2001). Meteor heights of ablation provide one of the best indicators of physical structure and composition. Recent evidence [Fujiwara *et al.* (1998), Spurný *et al.* (2000a)] suggests that some meteors are at very great heights, higher than can be explained by conventional ablation theory.

Perhaps the most interesting recent optical observations of meteors have been concerned with evidence for the structure and mode of ablation. We can define meteor wake as light production over an extended time period or simultaneously over an extended spatial area (note that some authors use the term wake for only the former case). Wake in meteors could be a product of long duration optical emission (such as the 557.7 nm Oxygen auroral line emission), or due to luminosity from grains which have been dif-

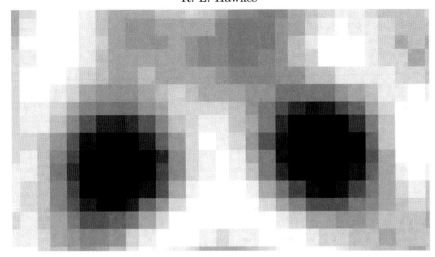

FIGURE 5.9. A short duration image intensified CCD meteor image – note that the light
production region appears almost exactly circular, indicating an absence of wake.

ferentially lagged by aerodynamic forces [Fisher *et al.* (2000)]. If meteors are dustballs
composed of grains with a range of sizes and/or densities, one would expect differential
lag to produce wake. While there is extensive apparent wake in image intensified television records, when the exposure is shortened (see Figure 5.9) there is very little evidence
of instantaneous light emission from an extended area. It appears that wake is relatively
absent in faint meteors [Robertson & Hawkes (1992), Shadbolt & Hawkes (1995) and
Fisher *et al.* (2000)].

Conventional theory suggests that the light production region in optical meteors should
only be a few meters in size, since the interaction with the atmosphere is expected to be
essentially atomic and therefore one would expect the light production to be over only
a few atmospheric mean free path lengths. It came as a surprise when LeBlanc (2000)
found evidence for transverse (to flight of meteor) motion of the order of many hundreds
of meters to kilometers in jet-like short duration features. Independently, Spurný *et al.*
(2000b) found similar evidence, and this has also been confirmed in Mg filtered light by
Taylor *et al.* (2000). The application of higher resolution systems in the future should
allow conclusive determination of the properties of these jet-like features. It appears that
these features are only present in high altitude radiation from high velocity meteors.

One strength of optical detection of meteors is that one can view the entire light curve.
Campbell *et al.* (1999), Murray *et al.* (1999) and Murray *et al.* (2000) provide analyses of
the shapes of meteor light curves as determined from image intensified CCD observations,
and the resultant implications for physical structure of dustball meteors.

5.10. Discussion and Future Trends

We end this chapter with a consideration of some of the trends which we anticipate in
optical detection and analysis of meteors in the coming years. The improvement in automated detection methods in the past few years is impressive, and we anticipate that very
soon there will be several fully automated optical detection systems which will compute
trajectory and orbit information on a daily basis. This will permit collection of large data
sets for purposes such as searching for minor showers, establishing the parameters of the
sporadic complex, and providing information on meteoroids of interstellar origin. In this

way we anticipate that there will be an optical counterpart to automated radar systems such as AMOR. Near-real-time reporting systems [e.g. Brown *et al.* (2000), Jenniskens *et al.* (2000b)] will become more widely employed as one measure of the hazard posed by natural meteoroids to space operations [e.g. Pawlowski & Hebert (2000)].

While there have been some efforts made for simultaneous detection of the same meteors by multiple techniques (radar, intensified CCD, visual or photographic), these have had only limited reported success so far. Such simultaneous observations may help us to calibrate radar and video observations better, since there are uncertainties in both the luminous efficiency factor and the ionization efficiency factor. Also, simultaneous observations with multiple techniques will provide quality assurance on the true errors inherent in the parameters provided by each technique.

There is recent renewed interest in large area fireball network cameras for capturing energetic events. These are of interest not only to establish the mass flux at large masses, but also to determine parameters for potential meteorite-dropping fireballs. The brightest fireballs are also observed by military satellites, providing additional information on the event. While possibly fireball networks will continue to use photographic film, it is more likely that most will now go to video detectors, and incorporate automated detection hardware and software. It is interesting that with the widespread use of personal camcorders that increasingly there will be serendipity observations recorded on video, such as the spectacular Peekskill meteorite fireball [Brown *et al.* (1994)] which was captured by at least 15 different video cameras.

Instrumental improvements in both image intensifiers and CCD detectors will continue to improve, and systems such as the Japanese prototype high definition intensified TV system. These improvements will be particularly important in the field of meteor spectroscopy [Abe (2000)]. A question which bears asking is whether CCD detectors have improved to the point that a useful non-intensified system could be constructed. If so, this would permit much higher resolutions at moderate costs. A related question is whether commercial video rate output, which has been used almost exclusively in scientific meteor observations, is the optimum frame rate. A different approach to CCD observations would be to use longer time exposures, with electronic gating of the CCD (in a non-intensified system) or the microchannel plate image intensifier in order to break the flight of the meteor into a series of segments. In this way one would have only one readout noise contribution for a single frame containing the entire meteor passage. Intermediate approaches are also possible in which higher resolution, lower frame rate digital intensified CCD cameras are used.

Some image intensifiers have electronic gating controls which would permit very short duration exposures. While CCDs themselves can readily be electronically gated, this will not produce truly short exposures unless the phosphor used in the image intensifier has a near instantaneous response. The use of gated image intensifiers coupled to high spatial resolution devices should permit definitive answers to the questions of transverse spread in meteor light production regions and aerodynamically induced lag and apparent wake.

One of the most exciting recent trends has been the use of a large (3 m) liquid mirror telescope coupled to an intensified CCD detector for meteor observations [Pawlowski & Hebert (2000), Pawlowski *et al.* (2001)]. Not only does this extend studies of mass distribution and flux to much smaller meteoroids, but potentially will permit very high resolution studies of fragmentation and ablation processes. Of course such techniques suffer from a very limited field of view with only a small (and unknown) part of the luminous trail observed. Intermediate focal length optics using moderate aperture telescopes connected to intensified CCD detectors would permit study of a larger part of the light curve while still at greatly enhanced spatial resolution.

Optical observations from high flying aircraft and from space will become more common. Not only does this provide ultimate dark skies and a more transparent optical window, but it also permits observation at lower elevation angles. The 1998 and 1999 Leonid Multi-Instrument Aircraft Campaigns [Jenniskens & Butow (1999), Jenniskens *et al.* (2000b)] have displayed some of the advantages, as well as challenges, of this approach. Perhaps one of the most important advantages is the ability to extend observations further into the near ultraviolet in searching for organic and volatile signatures in meteor spectroscopy.

In the last year a new type of optical observation of meteors has achieved success. Telescopes connected to video rate CCD detectors which viewed the moon during the 1999 Leonid shower found a number of conclusive cases of apparent meteoroid-induced lunar impacts. Typically these are present on only a single video frame. Bellot Rubio *et al.* (2000) have provided a detailed analysis of these events, along with projections for future showers which offer promising opportunities for lunar impact detections.

Acknowledgements: Our meteor observation program at Mount Allison University is supported by a Natural Sciences and Engineering Research Council of Canada grant. Jeremy Murphy provided Figures 5.2 and 5.8.

REFERENCES

ABE S., YANO H., EBIZUKA N. & WATANABE Y.-I. 2000. First results of high-definition TV spectroscopic observations of the 1999 Leonid shower. *Earth Moon Planets*, **82–83**, 369–377.

AIREY D. R. 1999. High resolution spectra and monochromatic images of a flaring 1991 Perseid meteor. *J. British Astron. Assoc.*, **109**, 179–188.

ALLEN C. W. 1973. *Astrophysical Quantities*, Athlone Press, London, UK.

ARLT R. 1998. Global analysis of the 1997 Perseids. *WGN J. Inter. Meteor Org.*, **26**, 61–71.

ARLT R. 1999. Global analysis of the 1998 Perseid Shower. *WGN J. Inter. Meteor Org.*, **27**, 237–249.

ARLT R., BELLOT RUBIO L., BROWN P. & GYSSENS M. 1999. Bulletin 15 of the International Leonid watch: first global analysis of the 1999 Leonid storm. *WGN J. Inter. Meteor Org.*, **27**, 286–295.

ARLT R. & GYSSENS M. 2000. Bulletin 16 of the International Leonid Watch: results of the 2000 Leonid Meteor Shower. *WGN J. Inter. Meteor Org.*, **28**, 195–208.

ARLT R. & HÄNDEL 2000. The *new* peak failed: first analysis of the 2000 Perseids. *WGN J. Inter. Meteor Org.*, **28**, 166–171.

BABADZANOV P. B. & KRAMER E. N. 1968. Some results of investigations of instantaneous meteor photographs. In *Physics and Dynamics of Meteors*, ed. L. Kresak & P. M. Millman, Reidel Publishing, Dordrecht, Holland, 128–142.

BELLOT RUBIO L. R. 1995. Effects of a dependence of meteor brightness on the entry angle. *Astron. Astrophys.*, **301**, 602–608.

BELLOT RUBIO L. R., ORTIZ J. I. & SADA P. V. 2000. Observations and interpretation of meteoroid impact flashes on the moon. *Earth Moon Planets*, **82–83**, 575–598.

BETLEM H., JENNISKENS P., VAN LEVEN J., TER KUILE C., JOHANNINK C., ZHAO H., LEI C., LI G., ZHU J., EVANS S., & SPURNÝ P. 1999. Very precise orbits of 1998 Leonid meteors. *Meteoritics Planet. Sci.*, **34**, 979–986.

BONE N. 1993. *Meteors: Sky & Telescope Observer's Guide*. Sky Publishing: Cambridge MA USA.

BOROVIČKA J. 1994. Line identifications in a fireball spectrum. *Astron. Astrophys. Suppl. Ser.*, **103**, 83–96.

BOROVIČKA J. 1994. Two components in meteor spectra. *Planet. Space Sci.*, **42**, 145–150.

BOROVIČKA J., STORK R. & BOCEK J. 1999. First results from video spectroscopy of 1998 Leonid meteors. *Meteoritics Planet. Sci.*, **34**, 987–994.

BROWN P., CEPLECHA Z., HAWKES R. L., WETHERILL G., BEECH M. & MOSSMAN K. 1994. The orbit and atmospheric trajectory of the Peekskill meteorite from video records. *Nature*, **367**, 624–626.

BROWN P., CAMPBELL M. D., ELLIS K. J., HAWKES R. L., JONES J., GURAL P., BABCOCK D. D., BARNBAUM C., BARTLETT R. K., MEDARD M., BEDIENT J., BEECH M., BROSCH N., CLIFTON S., CONNORS M., COOKE B., GOETZ P., GAINES J. K., GRAMER L., GRAY J., HILDEBRAND A. R., JEWELL D., JONES A., LEAKE M., LeBLANC A. G., LOOPER J. K., McINTOSH B. A., MONTAGUE T., MORROW M. J., MURRAY I. S., NIKOLOVA S., ROBICHAUD J., SPONDOR R., TALARICO J., THEIJSMEIJER C., TILTON B., TREU M., VACHON C., WEBSTER A. R., WERYK R., & WORDEN S. P. 2000. Global ground-based electro-optical and radar observations of the 1999 Leonid shower: first results. *Earth Moon Planets*, **82–83**, 167–190.

BROWN P., CAMPBELL M. D., HAWKES R. L., THEIJSMEIJER C. & JONES J. 2002. Multi-station electro-optical observations of the 1999 Leonid meteor storm. *Planet. Space Sci.*, **50**, 45–55.

BUIL C. 1992. *CCD ASTRONOMY: Construction and Use of an Astronomical CCD Camera*, Willmann-Bell Co, Richmond VA USA.

CAMPBELL M. D., HAWKES R. L. & BABCOCK D. D. 1999. Light curves of faint shower meteors: implications for physical structure. In *Meteoroids 1998*, ed. V. Porubčan and W. J. Baggaley, Slovak Academy of Sciences, Bratislava, 363–366.

CAMPBELL M. D., BROWN P. G., LeBLANC A. G., HAWKES R. L., JONES J., WORDEN S. P. & CORRELL R. R. 2000. Electro-optical results from the 1998 Leonid shower: I. Atmospheric trajectories and physical structure. *Meteoritics Planet. Sci.*, **35**, 1259–1268.

CEPLECHA Z. 1971. Spectral data on terminal flare and wake of double-station meteor No. 38421 (Ondrejov April 21 1963). *Bull. Astr. Insts. Czechosl.*, **22**, 219–304.

CEPLECHA Z. 1986. Photographic fireball networks. In *Asteroids, Comets, Meteors II*, ed. C. I. Lagerkvist & H. Rickman, University of Uppsala, Sweden, 575–582.

CEPLECHA Z. 1987. Geometric, dynamic, orbital and photometric data on meteoroids from photographic fireball networks. *Bull. Astr. Insts. Czechosl.*, **38**, 222–234.

CEPLECHA Z. 1996. Luminous efficiency based on photographic observations of the Lost City fireball and implications for the influx of interplanetary bodies into Earth. *Astron. Astrophys.*, **311**, 329–332.

CEPLECHA Z., BOROVIČKA J., ELFORD W. G., REVELLE D. O., HAWKES R. L., PORUBČAN V., & ŠIMEK M. 1998. Meteor phenomena and bodies. *Space Sci. Rev.*, **84**, 327–471.

FISHER A. A., HAWKES R. L., MURRAY I. S., CAMPBELL M. D., & LeBLANC A. G. 2000. Are most meteoroids really dustballs? *Planet. Space Sci.*, **48**, 911–920.

FLEMING D. E. B., HAWKES R. L. & JONES J. 1993. Light curves of faint television meteors. In *Meteoroids and Their Parent Bodies*, ed. J. Stohl & I. P. W. Williams, Slovak Academy of Sciences, Bratislava, 261–265

FUJIWARA V., UEDA M., SHIBA Y., SUGIMOTO M., KINOSHITA M., SHIMODA C. & NAKAMURA T. 1998. Meteor luminosity at 160 km altitude from TV observations for bright Leonid meteors. *Geophys. Res. Lett.*, **25**, 285–288.

GURAL P. 1995. Applying state-of-the-art video and computer technology to meteor astronomy. *WGN: J. Inter. Meteor Org.*, **23**, 228–235.

GURAL P. 1995. An operational autonomous meteor detector: development issues and early results. *WGN: J. Inter. Meteor Org.*, **25**, 136–140.

GURAL P. & JENNISKENS P. 2000. Leonid storm flux analysis from one Leonid MAC video AL50R. *Earth Moon Planets*, **82–83**, 221–247.

HALLIDAY, I. 1960. Auroral green line in meteor wakes. *Astrophys. J.*, **131**, 25–33.

HALLIDAY I. 1961. A study of spectral line identification in Perseid meteor spectra. *Publ. Dominion Obs. Ottawa*, **25**, 3–16

HALLIDAY I. 1969. A study of ultraviolet meteor spectra. *Publ. Dominion Obs. Ottawa*, **25**, 315–322.

HALLIDAY I., BLACKWELL A. T. & GRIFFIN A. A. 1978. The Innisfree meteorite and the Canadian camera network. *J. Roy. Astr. Soc. Can.*, **72**, 15–39.

HALLIDAY I., BLACKWELL A. T. & GRIFFIN A. A. 1983. Meteorite orbits from observations by camera networks. *Highlights of Astronomy*, **6**, 399–404.

HARVEY I. 1973. Spectral analysis of four meteors. In *Evolutionary and Physical Properties of Meteoroids* NASA SP, **319**, 103–129.

HAWKES R. L. & JONES J. 1975. A quantitative model for the ablation of dustball meteors. *Mon. Not. R. Astr. Soc.*, **173**, 339–356

HAWKES R. L., JONES J. & CEPLECHA Z. 1984 The populations and orbits of double-station TV meteors. *Bull. Astron. Inst. Czechosl.*, **35**, 46–64.

HAWKES, R. L. & JONES, J. 1986. Electro-optical meteor observation techniques and results. *Quart. J. Roy. Astr. Soc.*, **27**, 569–589.

HAWKES, R. L. 1990. Video-based observation procedures. *WGN: J. Inter. Meteor Org.*, **18**, 227–234.

HAWKES R. L. 1993. Television meteors. In *Meteoroids and Their Parent Bodies*, ed. J. Stohl & I. P. W. Williams, Slovakia Academy of Sciences, Bratislava, 227–234.

HAWKES R. L., MASON K. I., FLEMING D. E. G. & STULTZ C. T. 1993. Analysis procedures for two station television meteors. In *International Meteor Conference 1992*, ed. D. Ocenas & D. Zimnikoval, International Meteor Organization, Antwerp, 28–43.

HAWKES R. L. 2001. Meteors. In *Royal Astronomical Society of Canada Observers Handbook*, ed. R. Gupta, RASC and Univ. of Toronto Press: Toronto, pp. 200–202.

INTERNATIONAL METEOR ORGANIZATION 2001. http://www.imo.net/visual/vmdb.html.

JACCHIA L. G., VERNIANI F. & BRIGGS R. E. 1967. An analysis of the atmospheric trajectories of 413 precisely reduced photographic meteors. *Smithson. Contr. Astrophys.*, **10**, 1–139.

JENNISKENS P. & BUTOW S. J. 1999. The 1998 Leonid multi-instrument aircraft campaign – an early review. *Meteoritics Planet. Sci.*, **34**, 933–943

JENNISKENS P., CRAWFORD C., BUTOW S. J., NUGENT D., KOOP M., HOLMAN D., HOUSTON J., JOBSE K., KRONK G. & BEATTY K. 2000. Lorentz shaped comet dust trail cross section from new hybrid visual and video meteor counting technique – implications for future Leonid storm encounters. *Earth Moon Planets*, **82–83**, 191–208.

JENNISKENS P. & BUTOW S. J. 2000. The 1999 Leonid multi-instrument aircraft campaign – an early review. *Earth Moon Planets*, **82–83**, 1–26.

KLEIBER J. 1892. The displacement of radiant points due to the attraction, rotation and orbital motion of the Earth. *Mon. Not. Roy. Astron. Soc.*, **52**, 341–354.

KOSCHAK R. & RENDTEL J. 1990. Determination of spatial number density and mass index from visual meteor observations (I). *WGN: J. Inter. Meteor Org.*, **18**, 44–58.

KOSCHAK R. & RENDTEL J. 1990. Determination of spatial number density and mass index from visual meteor observations (II). *WGN: J. Inter. Meteor Org.*, **18**, 119–140.

KRESAKOVA M. 1978. The performance of telescopes for the observation of meteors. *Bull. Astron. Inst. Czechosl.*, **29**, 50–56.

KVIZ Z. 1958. Probability of the perceptibility of a meteor and the independent counting method. *Bull. Astron. Inst. Czechosl.*, **9**, 70–76.

LEBLANC A. G., MURRAY I. S., HAWKES R. L., CAMPBELL M. D., BROWN P., WORDEN S. P., JENNISKENS P., CORRELL R. R., MONTAGUE T. & BABCOCK D. D. 2000. Detection of jet-like features in Leonid meteors. *Mon. Not. Roy. Astr. Soc.*, **313**, L9–L13.

MARSDEN B. G. 1982. How to reduce plate measurements. *Sky & Telescope*, **64**, 284.

MCKINLEY D. W. R. 1961. *Meteor Science and Engineering*. McGraw-Hill, New York.

MCCROSKY R. E. & CEPLECHA Z. 1968. Photographic networks for fireballs. In *Physics and Dynamics of Meteors*, ed. L. Kresak & P. M. Millman, Reidel Publishing, Dordrecht, Holland, 600–612.

MILLMAN P. M. & HALLIDAY I. 1961. The near-infrared spectrum of meteors. *Planet. Space Sci.*, **5**, 137–140

MILLMAN P. M. 1980. One hundred and fifteen years of meteor spectroscopy. In *Solid Particles of the Solar System*, ed. I. Halliday & B. A. McIntosh, D. Reidel, Dordrecht, The Netherlands, 121–128.

MOLAU S., NITSCHKE M., DE LIGNIE M., HAWKES R. L., & RENDTEL J. 1997. Video observation of meteors: history, current status and future prospects. *WGN: J. Inter. Meteor Org.*, **25**, 15–20.

MOLAU S. 1997. The meteor detection software MetRec. In *Proceedings of the 1998 International Meteor Conference*, ed. by R. Arlt & A. Knöfel, International Meteor Organization, Antwerp, Belgium, 9–15.

MOLAU J. 1998. The meteor detection software MetRec. In *Meteoroids 1998*, ed. V. Porubčan and W. J. Baggaley, Bratislava, Slovak Academy of Sciences, 17–21.

MURPHY J. C. 1998. Investigation of perseid meteor composition using low light level television assisted spectroscopy, B.Sc. Honours Thesis, Mount Allison University, Sackville, Canada.

MURRAY I. S., HAWKES R. L. & JENNISKENS P. 1999. Airborne intensified charge-coupled device observations of the 1998 Leonid shower. *Meteoritics Planet. Sci.*, **34**, 949–958.

MURRAY I. S., BEECH M., TAYLOR M. J., JENNISKENS P., & HAWKES R. L. 2000. Comparison of 1998 and 1999 Leonid light curve morphology and meteoroid structure. *Earth Moon Planets*, **82–83**, 351–367.

ÖPIK E. J. 1958. *Meteor Flight in the Atmosphere*, Interscience, New York.

PAWLOWSKI J. F. & HEBERT T. T. 2000. The Leonid meteors and space shuttle risk assessment. *Earth Moon Planets*, **82–83**, 249–256

PAWLOWSKI J. F., HEBERT T. T., HAWKES R. L., MATNEY M. J. & STANSBERY E. G. 2000. Flux of very faint Leonid meteors observed with a 3 m liquid mirror telescope intensified CCD system. *Meteoritics Planet. Sci.*, **36**, 1467–1477.

POPOVA O. P., SIDNEVA S. N., SHUVALOV V. V. & STRELKOV A. S. 2000. Screening of meteoroids by ablation vapor in high-velocity meteors. *Earth Moon Planets*, **82–83**, 109–128.

RENDTEL J. 1993. *Handbook for Photographic Meteor Observations*. International Meteor Organization, Antwerp, Belgium.

RENDTEL J., ARLT R. & McBEATH A. 1995. *Handbook for Visual Meteor Observers*. International Meteor Organization: Antwerp, Belgium.

RENDTEL J. 1997. The η-Aquarid meteor shower in 1997. *WGN J. Inter. Meteor Org.*, **25**, 153–156.

RENDTEL J. & ARLT R. 1997. Activity analysis of the 1996 Geminids. *WGN J. Inter. Meteor Org.*, **25**, 75–78.

RENDTEL J., ARLT R. & VELKOV V. 1998. Surprising activity of the 1998 June Bootids. *WGN J. Inter. Meteor Org.*, **26**, 165–172.

RENDTEL J. 2000. First analysis of global data of the 1999 Geminids. *WGN J. Inter. Meteor Org.*, **28**, 19–21.

REVELLE D. O. & RAJAN R. S. 1979. On the luminous efficiency of meteoritic fireballs. *J. Geophys. Res.*, **84**, 6255–6262.

ROBERTSON M. C. & HAWKES R. L. 1992. Wake in faint television meteors. *Asteroids, Comets, Meteors 1991*, ed. A. Harris & E. Bowell, Lunar and Planetary Institute, Houston, TX, 517–520.

SHADBOLT L. & HAWKES R. L. 1995. Absence of wake in faint television meteors. *Earth Moon Planets*, **68**, 493–502.

SARMA T. & JONES J. 1985. Double station observations of 454 TV meteors: I. Trajectories. *Bull. Astron. Inst. Czechosl.*, **36**, 9–24.

SIMONENKO A. N. 1968. Separation of small particles from meteor bodies. In *Physics and Dynamics of Meteors*, ed. L. Kresak & P. M. Millman, Reidel Publishing, Dordrecht–Holland, 207–216.

SMART W. M. 1977. *Textbook on Spherical Astronomy*. Cambridge University Press, Cambridge UK.

SPURNÝ P., BETLAM H., VAN'T LEVEN J. & JENNISKENS P. 2000. Atmospheric behavior and extreme beginning heights of the 13 brightest photographic Leonids from the ground-based expedition to China, *Meteoritics Planet. Sci.*, **35**, 243–249.

SPURNÝ P., BETLEM H., JOBSE K., KOTEN P. & VAN'T LEVEN J. 2000. New type of radiation of bright Leonid meteors above 130 km. *Meteoritics Planet. Sci.*, **35**, 1109–1115.

TAYLOR M. J., GARDNER L. C., MURRAY I. S. & JENNISKENS P. 2000. Jet-like structures and wake in MgI (518 nm) images of 1999 Leonid storm meteors. *Earth Moon Planets*, **82–83**, 379–389.

VERNIANI F. 1965. On luminous efficiencies of meteors. *Smithson. Contr. Astrophys.*, **8**, 141–171.

WATANABE J.-I., ABE S., TAKANASHI M., HASHIMOTO T., IIYAMA O., MORISHIGE K. & YOKOGAW W. 1999. HD TV observation of the strong activity of the Giacobinid meteor shower in 1998. *Geophys. Res. Let.*, **26**, 1117–1120.

WHIPPLE F. & JACCHIA L. G. 1957. Reduction methods for photographic meteor trails. *Smithson. Contr. Astrophys.* **1**, 183–206.

WOODWORTH S. C. & HAWKES R. L. 1996. Optical search for high meteors in hyperbolic orbits. In *Physics, Chemistry and Dynamics of Interplanetary Dust*, ed. B. A. S. Gustafson and M. S. Hanner, ASP Conf. Ser., **104**, 83–86.

WRAY J. D. 1967. *The Computation of Orbits of Doubly Photographed Meteors*. University of New Mexico Press: Albuquerque, USA.

YOSHIHIKO S., HIROYUKI S. & SHOICHI T. 1997. Doube station TV meteor observations in 1996. *WGN: J. Inter. Meteor Org.*, **25**, 161–165.

ZNOJIL V. 1966. Meteor sighting probability and the problem of actual number of meteors. *Bull. Astron. Inst. Czechosl.*, **17**, 287–293.

ZVOLANKOVA J. 1983. Dependence of the observed rate of meteors on the zenith distance of the radiant. *Bull. Astron. Inst. Czechosl.*, **34**, 122–128.

6
RADAR OBSERVATIONS

By W. JACK BAGGALEY[†]

Physics and Astronomy Department, University of Canterbury, Christchurch, New Zealand

The technique of active probing via radar (rather than via passive detection at optical wavelengths) provides information about meteoroids, their atmospheric trajectories and their orbits in space prior to entry into the Earth's atmosphere. Here we discuss the generation of the ionization that serves as a radar target, the processes that dissipate that ionization and how radio waves interact with the meteoric plasma. Knowledge of these reflection processes enables us to understand the characteristics of meteor radar echoes so that those radar signatures permit the measurement of meteoroid masses, velocities, ablation heights and atmospheric trajectories, and provide a description of the dynamics in space of the parent meteoroid. The fate in the atmosphere of meteoric products subsequent to ablation is described. Finally we summarize the current radar meteor facilities in operation.

6.1. Radar Probing

Radar is a particularly valuable technique providing information on meteoroids for several reasons: immunity to weather and cloud cover (although proximity to electrical storms can interrupt operations); full diurnal sampling; it is feasible to operate programmes on a routine basis; radar has enhanced mass sensitivity compared with surveys at optical wavelengths using cameras and video recording. It is a powerful technique able to monitor meteoroid influx, altitudes, atmospheric trajectories and (with multiple stations) fix heliocentric orbits. The dust distribution in the inner solar system is readily probed. Though possessing greater sensitivity, a radar samples on each radio reflection a much larger volume of a meteor train and therefore cannot probe the fine detail achieved by operations at optical wavelengths: e.g. meteoroid fragmentation data available from video light curve structure have no direct equivalent in radar work. Information about both astronomical and atmospheric aspects is available: parameters describing the meteoroid itself are fixed and with the meteoric plasma acting as a probe in the Earth's atmosphere, a valuable tracer yields data on the atmosphere.

The meteoric plasma generated by a dissipating meteor contains free electrons and positive ions and it is the electrons that are probed by a radar, the ions having too large a mass to have any role in the scattering of radio energy. It is important to be able to relate the observed radar signature to the meteor parameters – mass, physical make-up, speed, altitude, trajectory – and therefore it is necessary to know how the ionization is distributed in space, its formation and dissipation mechanisms and the effect of this ionization assembly on radio waves.

6.2. Production of Meteoric Plasma

The ablation process at meteoroid surface temperatures of about 1850 K releases meteoric atoms and molecules from the meteoroid matrix (see Chapter 11). Retaining the cosmic speed of the meteoroid, the ablated particles impact atmospheric species (principally N_2 and O_2 since ablation occurs between about 70 and 130 km) with the more

[†] Email:j.baggaley@phys.canterbury.ac.nz

volatile species like sodium being preferentially released earlier. Collisional ionization at speeds between about 11 and 72 $km\,s^{-1}$ (the range covered by the Earth-impacting meteoroids) produce predominantly singly ionized species. Since the ionization potentials of the principal atomic constituents (Ca, Fe, Mg, Si and Na) are less than the potentials of O_2 and N_2, ion pairs resulting from impacting meteoric atoms are meteoric Ca^+, etc. (as confirmed by impact emission spectra for large meteoroids). As meteor ablation progresses downward, electrons are released forming a column some kilometres in length and a few atmospheric mean free paths wide. An important quantity controlling radar sampling is the number of free electrons deposited per metre length of path – a deposition which is proportional to the rate of loss of meteoroid mass and the effective ionization coefficient β (which may depend on the atomic species). The coefficient is the number of electrons produced per (vaporized) meteor atom, which for single collisions is less than unity. Since secondary collisions may be important, the effective value of β can exceed unity. If the meteoroid speed is V then

$$q = -\frac{\beta}{\mu V}(\mathrm{d}m/\mathrm{d}t) \qquad (6.2.1)$$

with m the meteoroid mass and μ the atomic mass of the ablated meteoroid atom. Fixing β is important since this affects both the determination of meteoroid influx and limiting mass sensitivity. β increases markedly with speed and the modelling work of Jones (1997) suggests the form $\beta = 9.4 \times 10^{-6}(V - 10)^2 V^{0.8}$ (with V in $km\,s^{-1}$) assuming an ionization threshold near 10 $km\,s^{-1}$ (a value that happens by coincidence to be close to the lower limit of Earth impact speed in the Earth's gravitational field). The question of the exact value of this threshold is clearly important for the detection of low speed material (e.g. in-falling space debris), while for large speeds Jones' expression is close to the dependence found earlier [see the summary given in Bronshten (1983)]. The ionization curve – the variation of plasma density along the meteor trajectory – has a maximum value which is proportional to the pre-entry meteor mass and also to $\cos Z$ (with Z the zenith angle of the trajectory) and has a form which is (in a rudimentary derivation) a parabolic function of atmospheric density. This simple picture can be complicated by the effects of thermal gradients within the meteor body, fragmentation and meteoroid rotation. The altitude regime of the plasma produced by the ablating meteor is governed principally by meteroid mass and entry speed: the altitude of maximum plasma density deposition for a given meteor train varies from $\sim$ 80 km for slow large mass meteoroids to $\sim$ 130 km for high speed small bodies. The plasma generated along the meteor trajectory is distributed over a vertical altitude interval of about 1.5 atmospheric scale heights ($\sim$ 12 km).

For radar-detected meteoroids, the size regime is such that the body size is generally much less than the atmospheric mean free path so that free molecular flow applies. For very large meteoroids (size $\sim$ cm) the development of an aircap and the formation of shock-fronts needs consideration. For very small impacting meteoroids, the ratio of the surface area to the mass becomes large enough that the energy acquired by the meteoroid is radiated away faster than the body temperature can increase: such grains do not ablate and are termed micrometeoroids. Radar systems are able routinely to sample meteoroids in the mass range approximately 10^{-10} to 10^{-6} kg with the micrometeoroid limit being highly velocity dependent and $\sim 10^{-12}$ kg for $V \sim 30$ $km\,s^{-1}$.

The transverse dimensions of the meteor plasma column at formation can have an important effect on radar detection. The collisional history of the ablated atoms (of energy several hundred electron volts) has been studied by Jones (1995) and the thermalization time of the plasma by Baggaley & Webb (1977), the meteoric plasma achieving thermal equilibrium with the ambient gas in < 0.1 s. There is some evidence that some meteoroids have a composition that is dustball in nature and also some that possess rotation: such objects will produce trains of larger radius at formation.

In addition to the plasma deposited along the track of the ablating meteoroid, a roughly spherical ionization volume is generated in the immediate vicinity (few metres) of the meteoroid itself. It is known that the lifetime of this ionization is very short and a full understanding of its formation mechanism has yet to be formulated [see Jones *et al.* 1999].

6.3. Ionization Dissipation

Within a fraction of a second of the deposition of the plasma column the original ionization is dissipated by various processes which affect the electron train content and therefore modification of the radar target. Diffusion, ionic reactions and transport in the neutral wind field are mechanisms that dictate the life of the meteoric plasma and so govern the lifetime of radar echoes.

6.3.1. *Diffusion*

The process of particle flow along a density gradient in the presence of collisions is called diffusion. The radial diffusion of the plasma into the surrounding atmosphere occurs predominantly transverse to the train axis along the major density gradient. Since the free diffusion coefficient for electrons D_e exceeds that for the ions D_i by a factor of ~ 200 a space charge electric field is created which retards the electron flux, maintaining charge neutrally so that the ions and electrons diffuse at the same rate governed by the common ambipolar coefficient $D_a = D_i(1 + T_e/T_i)$ with T_e and T_i the electron and ion temperatures. For thermalized conditions (see above) $D_a \simeq 2D_i$ and the coefficient depends on the atmospheric pressure p and temperature T through $D_a \propto T^2/p$ increasing from about 0.4 m^2 s^{-1} at 80 km to about 140 m^2 s^{-1} at 110 km. The value is dependent on season and latitude and can be estimated from standard atmosphere models. In consequence of the radial diffusion, the number density $n(r)$ in the plasma train decreases with time. If r is the radial coordinate from the train axis, the solution of the continuity equation

$$\frac{\partial n}{\partial t} = D_a \nabla^2 n = D_a \left[\frac{\partial^2 n}{\partial r^2} + \frac{1}{r} \left(\frac{\partial n}{\partial r} \right) \right] \tag{6.3.1}$$

where ∇ is the Laplacian Operator for cylindrical coordinates yields

$$n(r,t) = \left(\frac{q}{\pi a^2} \right) \exp \left(-\frac{r^2}{a^2} \right) \tag{6.3.2}$$

where $a^2 = r_0^2 + 4D_a t$, a is the $1/e$ train radius at any time and q is the linear density of electrons with q time invariant in the absence of ionic reactions.

The plasma column maintains a pseudo-gaussian cross-section [Jones & Jones 1990a], diffusion being isotropic in the absence of the Earth's magnetic field. The controlling factor is the comparison between the collision frequency with atmospheric neutrals ν and the angular gyrofrequency $\omega_B = eB/m$ for the charged particle in the geomagnetic field. At all altitudes the diffusion of any species parallel to the field $D_\parallel$ is unchanged from the free diffusion case. If $\nu \gg \omega_B$ the free diffusion is collision dominated, a situation holding at altitudes < 80 km where the diffusion of meteoric plasma is therefore controlled by the ions with the nett diffusion coefficient being $D_a \simeq 2D_i$. At greater altitudes the diffusion transverse to the geomagnetic field is inhibited with

$$D_\perp = D \left(\frac{\nu^2}{\nu^2 + \omega_B^2} \right) \qquad (6.3.3)$$

applying for both ions and electrons. For electrons $\nu_e \simeq \omega_{Be}$ at ~ 80 km whereas ions remain collision dominated throughout the meteor region with $\nu_i \simeq \omega_{Bi}$ at ~ 140 km. With increasing altitude above 95 km it is possible for $D_{\perp e}$ to become less than $D_{\parallel i}$ so that the nett plasma diffusion across the geomagnetic field is controlled by the electrons instead of ions. The aeronomical result of this is that meteor trains closely aligned (within a few degrees) with the magnetic field at altitudes > 105 km assume an elliptical cross-section with inhibited expansion in a direction orthogonal to the plane containing the train axis and the geomagnetic field. The effect can be large for some geometries and aspect-angles.

When the diffusing plasma has reached the dimensions of the scale of turbulent cells (~ 300 m) subsequent expansion occurs by eddy diffusion (coefficient $\sim 3 \times 10^2$ m^2 s^{-1} in the meteor region). Such an expansion would be attained in a few minutes – longer than the lifetime of most meteor echoes (see section 6.4.2.3).

6.3.2. Ionic Reactions

Radar meteor echo durations predicted on the presence of diffusion only are expected rarely to exceed a few minutes (see section 6.4.2.3), so here we are interested in chemistry that can proceed with reaction time-constants less than this. Electron recombination with meteoric atomic ions directly via a two body reaction is unimportant in meteor trains because of the very low rate coefficients for such radiative processes: even for dense ionization trains where the meteor radar magnitude $\mathbf{M} \simeq 0$ (see definition in section 6.4.2.2), the electron loss time-constant within the generated column exceeds one minute. Permanent loss of electrons by negative ion formation producing O^- and O_2^- cannot be important because in the presence of atomic oxygen the detachment rate greatly exceeds the attachment rate. Routes providing charge exchange of neutral metals with ionospheric ions or photoionization are too slow to be relevant in meteor train chemistry but do play a role in the ultimate fate of meteoric species (section 6.7). The controlling neutral atmospheric gas is ozone: O_3 provides a route for oxide ion formation from the initial (mainly atomic) meteoric ions M^+ present at plasma train formation ($M = $ Mg, Si, etc.).

$$M^+ + O_3 \longrightarrow MO^+ + O_2. \qquad (R1)$$

Subsequent dissociative recombination

$$MO^+ + e \longrightarrow M + O \qquad (R2)$$

leads to loss of the free electrons. For large meteors with dense ionization (which are responsible for persistent echoes), the rate of (R2) much exceeds that of (R1). In addition the rate of the reduction reaction

$$MO^+ + O \longrightarrow M^+ + O_2 \qquad (R3)$$

proceeds more slowly than reaction (R1) early in the life of a meteor train. Taking reaction rate coefficients (cm^3 s^{-1}) [McNeil *et al.* 1998] $k_1 \simeq 2\times10^{-10}$, $k_2 \simeq 5\times10^{-7}$, $k_3 \simeq 1\times10^{-10}$, and for an altitude of 90 km taking $[O_3] \simeq 2\times10^8$ cm^{-3}, $[O] \simeq 10^{11}$ cm^{-3} then for a meteor of magnitude $\mathbf{M} \simeq 0$ so $q \sim 10^{14}$ cm^{-1} with train radius $r_0 \sim 1$ m, the initial ion lifetimes τ for each of the three processes are $\tau_1 \sim 25$ s, $\tau_2 \sim 10^{-3}$ s, $\tau_3 \sim 10^{-1}$ s. At altitudes below about 90 km an additional reaction chain is operative initiated by a three-body charge transfer process

$$M^+ + O_2 + X \longrightarrow MO_2^+ + X \qquad (R4)$$

with dissociative recombination

$$\text{MO}_2^+ + e \longrightarrow M + O_2 \tag{R5}$$

generally proceeding much faster than reduction

$$\text{MO}_2^+ + O \longrightarrow MO^+ + O_2. \tag{R6}$$

Meteoric atomic ions can also react with nitrogen molecules producing clusters,

$$M^+ + N_2 + X \longrightarrow M.N_2^+ + X \tag{R7}$$

with subsequent recombination

$$M.N_2^+ + e \longrightarrow M + N_2 \tag{R8}$$

proceeding much faster than collisional break-up.

For most of the meteor region, the loss of ionization is governed by reaction (R1) (since the dissociative recombination is so rapid) with the electron lifetime against chemical loss increasing from about 3 s at 80 km to about 200 s at 100 km. Meteoric plasma is dissipated by rapid diffusion high in the atmosphere and by the effects of chemistry in the lower regions.

The time variation of the electron number density within the expanding meteor train can be modelled by solving the set of coupled diffusion equations that describe the rate of change of the concentrations of the various ion species in the radially diffusing column with the rate being governed by ambipolar diffusion early in the life of the train. The coupled set of equations, similar to equation 6.3.1 above, will include appropriate production and loss terms governed by reactions (R1)–(R8) with reaction rate coefficients well established from laboratory studies [see the compilation of McNeil *et al.* (1998)]. The coefficients depend on the different meteoric species M (Fe, Mg, Na, etc.) with some reactions being endothermic. Because the atmospheric gases O_3 and O have diurnal variations they have a marked effect on meteor train de-ionization (see section 6.4.2.3).

For the simple cases (e.g. a recombination process only) analytical solutions can be formulated but the full coupled equation requires numerical procedures. The full treatment has been given in Baggaley & Cummack (1974), Baggaley (1978) and Baggaley (1979). Chemical processes lead to severe depletion of meteoric plasma at altitudes below ~ 90 km.

6.4. The Scattering of Radio Waves from Meteoric Plasma

6.4.1. *The Radar Equation*

A single electron has a scattering cross-section to an incident electric field $\sigma_e = 4\pi r_e^2 \sin^2 \gamma$ where r_e is the classical scattering radius and γ the angle between the **E** field vector of the incident wave and the scattering direction. For the radar backscatter case $\gamma = 90°$ and $\sigma_e \simeq 10^{-28}$ m^2. For an assembly of electrons the effective total scattering cross-section results from the summation of phase contributions from individual electron scatter, on the overall geometry and on whether secondary radiative effects are present (and therefore on electron number density). At meteoric altitudes electron-neutral collisions can be neglected in the radiowave scattering process. It is the coherent reflection from the meteor plasma that provides the radar target. There are two particular geometries involved for the case of meteor scatter: reflection from the linear column of ionization so that scattering takes place from a long ($\sim$ kms) cylinder of plasma (providing the *body echo*); and reflection from the spheroidal plasma generated around the ablating meteoroid providing the *head echo*. In either case the radar equation gives the received power from a target of scattering cross-section σ (which here we specify as a single electron) at range R using transmitter and receiver antennas of gains G_T and G_R as: the power flux at the target is $P_T G_T / 4\pi R^2$ so that the flux received at the radar due to scattering from a single electron is $P_T G_T \sigma_e / 16\pi^2 R^4$. A radar receiving antenna of gain G_R has an

W. J. Baggaley

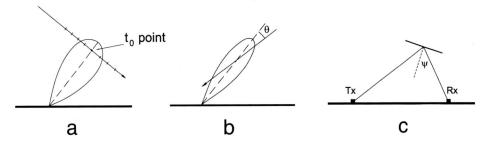

a   b   c

FIGURE 6.1. Radar geometries: (a) Transverse geometry providing backscatter echoes with Fresnel intervals indicated for illustration (see section 6.4.2.2); (b) Radial geometry; (c) Forward scatter.

effective collecting area of $G_R\lambda^2/4\pi$, with λ the radar operating wavelength, so that the signal power available at the antenna due to scattering from a single electron becomes

$$\Delta P_R = \frac{P_T G_t G_R \lambda^2 \sigma_e}{64\pi^3 R^4}. \tag{6.4.1}$$

6.4.2. *The Body Echo*

6.4.2.1. *Plasma Conditions*

For ease of geometrical illustration, discussion is confined to the case where $\gamma = 90°$ – the backscattering situation where the radar transmitter is co-located with the receiver (transverse geometry, Figure 6.1a). The problem reduces to determining an expression for the total scattering cross-section for the particular geometry and meteor train parameters. The large range of masses of meteoroids means a substantial range in ionization densities is generated. For a small meteoroid generating a weak plasma, the radio wave passes through the plasma unaffected, the scattering of the incident radio waves from electrons takes place independently without any secondary or interference effects: the total scattering is then the summation over all electrons in a cross-section of the meteor train. Such a situation is described as an *underdense* train. At the instant of formation of such an underdense train, if the train radius at formation (the *initial radius* r_0) is very small compared to the radar wavelength λ, then for evaluating the reflection, all electrons can be considered as situated on the train axis. For the opposite extreme of a dense plasma, secondary scattering is an important feature, and the incident wave is grossly disturbed: the dielectric constant of the medium is sufficiently negative for total reflection to occur at some critical boundary. As diffusion increases the dimensions of the boundary, the reflection coefficient increases. Train expansion reduces the number density on the axis so the dielectric constant becomes less negative, the critical radius collapses to the train axis and subsequently individual electron scattering operates as the incident wave penetrates the plasma column. Such a dense expanding train can be regarded in its reflection behaviour as a metallic cylinder whose radius expands for a period before collapsing. This regime is terms an *overdense* train.

The transition between the two regimes can be defined in terms of the attenuation of the incident wave: since the refractive index of the plasma is imaginary for the overdense case, the incident wave is not progressive and is termed an *evanescent* wave. If the plasma is of such a density that the amplitude of the evanescent wave is reduced by $1/e$ on the axis (in analogy with the skin depth for a metal), the train plasma is described as being transitional. This definition means the attenuation distance, i.e. the radius of the cylinder, is $\lambda/2\pi$. The corresponding critical electron density, $n_c = q/\pi a^2$, can be related to the radio wave frequency and plasma conditions by $n_c = \omega^2 m \epsilon_0/e^2$ with m

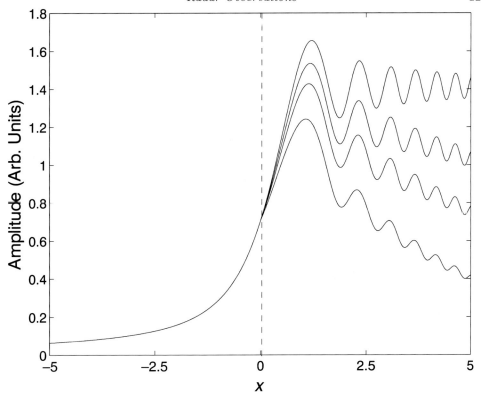

FIGURE 6.2. Echo amplitude as a function of the Fresnel parameter, x. The top trace is for the case of no diffusion and the others show the effect of increasing rate of diffusion.

and e the electron mass and charge respectively and ϵ_0 the free space permittivity: the implied critical electron line density is $q \sim 10^{14}$ m^{-1} – a value independent of the probing radio frequency. Such a value of q corresponds to a meteor magnitude of approximately +5 (near the limit of unaided eye visibility).

6.4.2.2. *Echo Formation Process*

Of fundamental importance in radar is a clear interpretation of the early life of an echo and it is valuable to examine the time behaviour of the meteor reflection coefficient, especially during the formation stage, using a simple model assuming that the free electrons can be regarded as essentially confined to the axis (so $r_0 \ll \lambda$ and diffusion is absent). This approach permits a ready description in terms of diffraction where the relative phases of contributions of segments of the forming plasma sum to provide the scattered field. The time behaviour of the body echo (Figure 6.2) arises as the meteoroid deposits ionization over successive Fresnel intervals. The concept of 'zone' (rather than 'interval') often used in describing meteor diffraction arises from its use in the two-dimensional case in optical diffraction and derives from the zone plate device. An optical zone plate is described by annuli having successive (usually π) increments in phase. Such a geometry and its use of a zone or area is not appropriate in the present meteor case where the contributions are distributed in one dimension and the term 'Fresnel interval' is more appropriate (see Figure 6.1a). It is useful to analyse first the time characteristics of the scattered radio electric field and its phase – and hence echo signal – and then consider the numerical value of the field. If an element of train of length ds contributes scattered waves of uniform phase, then as the range changes over time a phase modulation of the total scattered field will be introduced. The resultant electric field at distance R is proportional to

$$E_{ds} \propto q \exp j \left(\omega t - \frac{4\pi R}{\lambda} \right) ds, \qquad (6.4.2)$$

where ω is the angular radio frequency, and the term $4\pi R/\lambda$ arises from phase changes due to range R (i.e. wave distance $2R$). The analysis is clearer if trigonometrical functions rather than the complex form is employed. The total electric field produced by all electrons in the train follows

$$E \propto q \int_0^s \sin \left(\omega t - \frac{4\pi R}{\lambda} \right) ds, \qquad (6.4.3)$$

where integration is taken over the total train length and q is assumed to be constant along the train. Since the range changes slowly near to the orthogonality position (the t_0 point), changes in trail position s can be related to changes in range by $R \simeq R_0 + s^2/2R_0$. Introducing a change of variable $\chi = \omega t - 4\pi R_0/\lambda$ and $2s = x(R_0\lambda)^{\frac{1}{2}}$ with x the Fresnel parameter. The total scattered field is of the form

$$E \propto q[(C \sin \chi - S \cos \chi)], \qquad (6.4.4)$$

with

$$C = \int_{-\infty}^{x} \cos \left(\frac{\pi x^2}{2} \right) dx \quad \text{and} \quad S = \int_{-\infty}^{x} \sin \left(\frac{\pi x^2}{2} \right) dx \qquad (6.4.5)$$

the Fresnel integrals. The maximum value of the electric field can be regarded as due to an effective train scattering length of one Fresnel interval $\sim (R_0\lambda/2)^{1/2}$ either side of the t_0 point and so the resultant maximum scattering cross-section is given by the square of the number of contributing electrons and so (equation 6.4.1) the echo power available at the radar antenna is

$$P_R = \frac{P_T G_t G_R \lambda^3 \sigma_e}{128\pi^3 R^3} q^2 \left(\frac{C^2 + S^2}{2} \right), \qquad (6.4.6)$$

and numerically

$$P_R = 2.5 \times 10^{-32} \left(\frac{P_T G_t G_R \lambda^3 \sigma_e}{R^3} \right) q^2 \qquad (6.4.7)$$

watts. The maximum echo signal power is $\propto q^2$ and in an analogous way to optical registering of meteor light intensity, it is conventional to use a radio magnitude related to q on a logarithmic scale. Since q also depends on meteoroid speed (via the ionization coefficient) a meteor magnitude $\mathbf{M}$ is commonly used so that $\mathbf{M} = 36 - 2.5 \log_{10} q + 2.5 \log_{10} V$ with q in m^{-1} and V in $km\,s^{-1}$.

The time behaviour of the received signal has a close optical analogue in the angular changes in light intensity in the case of diffraction at a straight edge. The echo amplitude diffraction varies as $V_r = (C^2 + S^2)^{\frac{1}{2}}$ and the phase described by $\tan \phi = S/C$ – the complex diffraction behaving in a way described very usefully by the Cornu spiral (Figure 6.3). The relative echo amplitude at any time (and therefore value of x) is given by the length of a vector from the point $C = -0.5$, $S = -0.5$ ($x = -\infty$) to a point having a particular value of x on the curve. As the meteor approaches the orthogonality position t_0 there is a rapid rise of the signal as an increasing number of half-period Fresnel intervals contribute: the phase changes monotonically reaching a maximum (of $-\pi/6$) at $x = 0.6$ (defining the phase as $-\pi/4$ when $x = 0$).

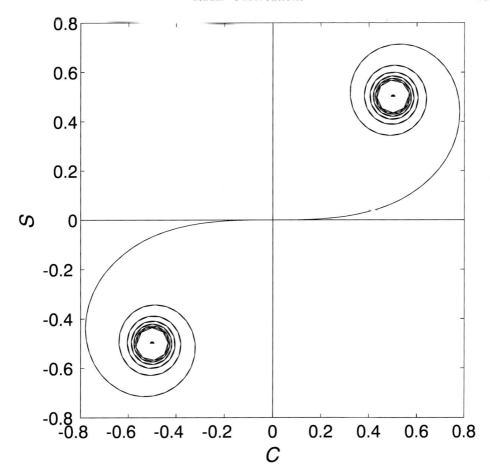

FIGURE 6.3. The Cornu spiral: the Fresnel functions C and S plotted with the Fresnel parameter, x, along the curve. The magnitude of the scattered is field is provided by the length of the vector anchored at the $x = -\infty$ point and extending to any position x on the curve: the maximum value occurs at $x = 1.217$. The phase is given by the angle between the vector and the abscissa.

As the meteoroid moves through successive Fresnel intervals both the signal and its phase oscillate. At the instant of maximum echo power ($x = 1.217$) about 90% of the energy is contributed by the central zone extending therefore over a distance along the meteor ionization train of about 2 km for an HF radar. This represents the sampling size and governs the features that may be discerned with a radar.

The important governing feature is that the early stage of a meteor echo formed with orthogonal geometry is dictated by diffraction – independent of the q value and therefore whether an under or overdense plasma. Indeed the rise-time of an echo can yield a reasonable estimate of meteoroid speed V. Neglecting diffusion effects the echo rise-time defined as the time from the amplitude rising edge $1/e$ point to amplitude maximum is $t_r \simeq 1.35(R_0\lambda)^{\frac{1}{2}}/2V$ s. This method has been shown to be an easily applied technique for routine measurements of V accurate to $\sim 10\%$ [Baggaley *et al.* 1997].

6.4.2.3. *Echo Behaviour after the Formation Stage*

The derivation of train scattering from a general plasma requires the use of a full wave treatment employing wave matching techniques. In a comprehensive treatment for

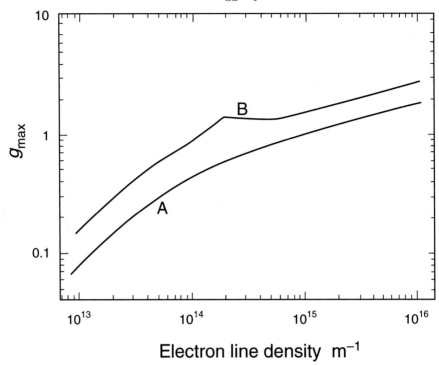

FIGURE 6.4. Reflection coefficient: Full Wave treatment [Poulter & Baggaley 1978] $g_{\parallel}$(A) and $g_{\perp}$(B) as functions of the line density q. The transition region $\sim 10^{14}$ m^{-1}.

the case of no electron loss and general line density q and plasma radius r, Poulter & Baggaley (1977) and Poulter & Baggaley (1978), show how the complex reflection coefficient changes during a plasma lifetime. When the incident electric field is transverse to the column axis the forced motion of electrons across the ionization gradient can lead to enhanced scattering. This short-lived resonance phenomenon is restricted to times early in the train life when $r < \lambda/2\pi$. Figure 6.4 shows the reflection coefficient g for parallel and transverse cases as a function of q over a range of line densities showing the echo transition region while Figures 6.5 and 6.6 show the time behaviour of g. The transition region (see 6.4.2.1. above) actually extends over a factor of $\sim 10^{2}$ in q but for the limiting cases models of the echo characteristics are straightforward.

(*a*) *Underdense Echoes*

The later stages of an underdense echo are dictated by diffusion. As the plasma expands, phase summation of the individual electron contributions in a train cross-section results in a reduction in the echo reflection coefficient, g. If a gaussian train cross-section is maintained during expansion [see Jones & Jones 1990a] the change in g will be an exponential function of time and the echo decay is described by $\exp(-t/t_{\rm d})$ with $t_{\rm d} = \lambda^{2}/32\pi^{2}\,D_{\rm a}$. The theoretical behaviour of an underdense echo is depicted in Figure 6.2 showing the effect of diffusion (an example is given in Figure 6.9). For a more general cross-section function, analytical solutions using Fourier Transforms have been used [Jones 1995].

In section 6.3.1, the discussion of the form of the ionization distribution emphasized the role of the geomagnetic field in producing anisotropic diffusion. The resulting distortion of the plasma column produces a look-angle dependency on the decay of the radar echo. For simplicity of treatment assuming a gaussian profile and circular symmetry at formation, the plasma column distorts into a generally elliptical cross-section.

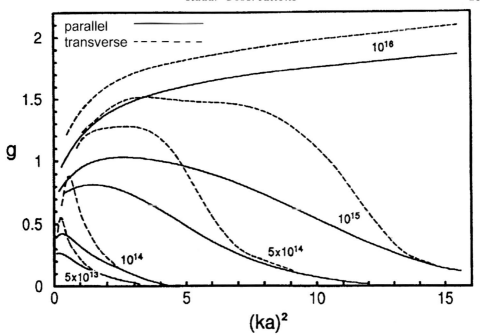

FIGURE 6.5. Reflection coefficient: Full Wave treatment. Reflection coefficient as a function of $(ka)^2$: For $(ka) > 1$, the abscissa is proportional to time. $k = 2\pi/\lambda$ and a is the train radius.

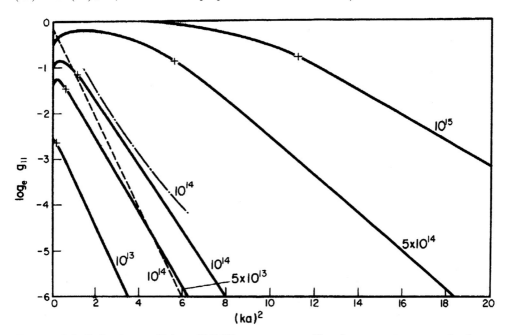

FIGURE 6.6. Reflection coefficient: Full Wave treatment. Showing a greater range of values.

With inhibited diffusion the echo decay rate can be very much reduce and enhanced echo lifetimes can result under favourable geometry [Jones 1991, provides convenient tables].

(b) *Overdense Echoes*

A full treatment of the reflection from a gaussian dense plasma is complex because of the

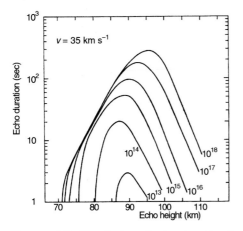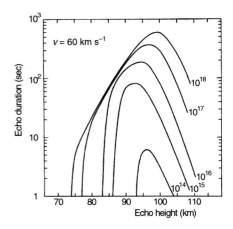

FIGURE 6.7. Simulation of echo duration behaviour for different electron line densities q for a radar frequency of 30 MHz and two values of meteor velocity. The effect of chemistry causes the height of maximum duration to increase.

effects of the weaker ionization in the outer regions on the wave. Figures 6.5 and 6.6 [after Poulter & Baggaley 1978] show results from wave-matching solutions. However, a simple picture enables an estimate to be made of the echo characteristics of an overdense plasma. For a general overdense plasma there will be a critical radius r_c at which the plasma has a critical electron number density n_c (for which the refractive index is imaginary) which will occur when

$$n_c(r_c, t) = \left(\frac{q}{\pi a^2}\right) \exp\left(-\frac{r_c^2}{a^2}\right) \tag{6.4.8}$$

If we define the duration of an overdense echo as the time for the axial electron number density to fall to n_c then $n_c(0, t) = q/\pi a^2 = \omega^2 m \epsilon_0 / e^2$ with $a^2 = r_0^2 + 4 D_a t$, then the overdense duration is

$$T_{ov} = \left(\frac{e^2}{4\pi m \epsilon_0}\right)\left(\frac{q}{\omega^2 D_a}\right) - \left(\frac{r_0^2}{4 D_a}\right) \tag{6.4.9}$$

where the second term represents the 'age' of the train at the start (the time to diffuse to a radius of r_0) under normal diffusion and is only comparable with the first term in the upper meteoric region > 105 km. This diffusion-limited echo duration will be modified by the action of ionic processes removing electrons from the train plasma. Models can be made of how echo durations (chemistry influence is generally negligible for underdense echoes for operating frequencies > 15 MHz) will be curtailed by the effective decrease in the value of the line density, q. Ozone is expected to be the dominant gas initiating the chain of reactions responsible for the de-ionization of trains and therefore, because of the solar control of lower thermosphere O_3 densities, the durations of echoes under day-time conditions will be markedly reduced compared with nighttime. The modelled electron loss characteristics can then be inserted in the echo durations to model the echo lifetimes as a function of altitude. Examples are given in Figure 6.7 of how T_{ov} values are expected to vary with altitude. The approximately exponential downwards increase in T_{ov} due to both q increase (for a given meteor) and atmospheric density increase are countered by the increasingly short chemical lifetime of electrons below about 95 km. A specific illustration of the effect is to be found in the altitude variations of long enduring Perseid meteor echoes presented by McKinely (1961). In that study triple station ranging could fix the echo altitude so that echo duration could be followed as a function of altitude

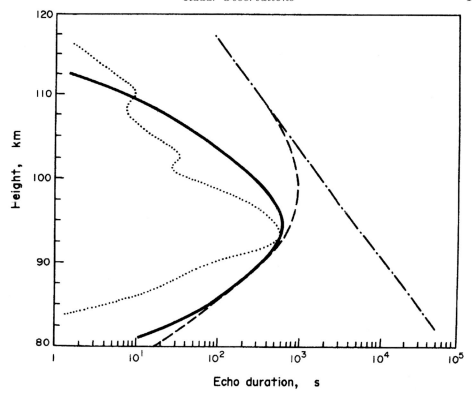

FIGURE 6.8. Example of an enduring (549 s) radar echo from a Perseid meteoroid as recorded by McKinley (1961).

------- using expression T_{ov} no chemistry with $q = 6 \times 10^{18}$ m^{-1} at all heights;

- - - - - including chemistry with $q = 6 \times 10^{18}$ m^{-1} at all heights;

——— chemistry using ionization curve with maximum $q = 6 \times 10^{18}$ m^{-1} at 88 km;

.... observed by McKinley.

during the persisting ($\sim$ 9 mins) echo (Figure 6.8). As predicted [Baggaley 1978], the most enduring section of the diffusing plasma occurred at an altitude close to 93 km. This sensitivity of echo behaviour to ambient gases presents a method of monitoring the atmosphere: measurements of radar echo lifetimes characteristics can enable estimates to be made of the altitude profiles of the important atmospheric species O_3 [Hajduk *et al.* 1999] in a altitude regime which is largely inassessible by other techniques.

6.4.3. *Forward Scatter Geometry*

Full wave solutions of this reflection geometry (Figure 6.1c) for a given forward scatter angle and polarization angle have been given in the comprehensive treatment of Jones & Jones (1990b). The principal characteristics of the body echo for the case of oblique reflection can be regarded crudely as an extension of the backscatter case. The principal features can be described simply by increasing the Fresnel interval by a factor $\cos^{-2}\psi$, with ψ the forward scatter angle and the wavelength increased by a factor $\cos^{-1}\psi$. This *effective* increase in the probing wavelength means that the attenuating effect of initial radius is reduced enabling an upwards extension of the 'height-ceiling' (see below section 6.5).

6.4.4. *The Head Echo*

In contrast to the time characteristics of a body echo, the head echo, resulting from the continuous creation and rapid loss of ionization, is produced by a spherical plasma concentration formed around the meteoroid body and travelling with its cosmic velocity. Estimates of the properties of the head echoing region plasma are of an overdense spheroid of radius $\sim 1\,\mathrm{m}$ [see Jones *et al.* 1999]. A full description of the scattering from such a target of arbitrary shape and radar wavelength can be described in terms of Mie Theory. The form of the scattering behaviour is dependent on the ratio of the circumference of the spheroidal (radius a) target to the radar wavelength, $\alpha = (2\pi\,a/\lambda)$. For radars operating at high frequencies in the UHF band the target scattering cross-section can be regarded as close to the geometrical cross-section and $\sigma \simeq \pi\,a^2$. Conversely, using a long wavelength radar the Mie treatment assumes a form close to the Rayleigh scattering (strictly when $\alpha \ll 1$). The scattering cross-section is then given by the scattering efficiency $4(2\pi\,a/\lambda)^4$ multiplied by the geometrical area of the target. The value of the scattering cross-section is then $\sigma = (64\pi^5 a^6)/(\lambda^4)$ adequate for $\alpha < 0.6$. For a 1 m radius overdense plasma at 20 MHz $\sigma \sim 0.3\ \mathrm{m}^2$ increasing to $\sim 3\ \mathrm{m}^2$ at 300 MHz.

The target reflection cross-section yielding the head echo is some orders of magnitude less than that responsible for the body echo so that for those conditions (specular reflection and near orthogonality) favourable to provide a body echo, the head echo is weak. However, if the meteoroid's trajectory is at a small angle to the radar line of sight so the body echo is absent then the head echo dominates. Radars usually employ antennas with near-vertical radiation patterns to probe this radial geometry situation (Figure 6.1b) [see Wannberg *et al.* 1996 and references therein]. The plasma target's range change and phase behaviour (see section 6.4.2.2) provide a diagnostic for the meteoroid's trajectory and deceleration [e.g. Janches *et al.* 2000].

6.4.5. *Off-Specular Echoes*

It is the nature of wave diffraction and the associated phase addition that is responsible for the restrictive orthogonal geometry governing the transverse reflection mode. There are two conditions necessary for an echo to occur for a given meteor train: for the particular aspect angle the central Fresnel interval should be within the antenna radiation pattern and the scattered power should be above the detection threshold. Complete echoes as discussed above can be expected in cases where a substantial fraction of the central interval (length ~ 2 km) lies in the radiation pattern. This constraint is very strict so that many meteors having radiants above the horizon but which are formed with unfavourable geometry are undetectable. If the t_0 point is formed near the high elevation edge of the radiation beam then contributions from pre-t_0 zones ($x < 0$) are attenuated and the resulting echo is due to $x > 0$ and the large phase growth is absent. Conversely, ionization formed near the low elevation edge of the antenna beam can only produce contributions from the $x < 0$ region. Radars employing very narrow antenna radiation patterns – so that the size of the Fresnel length is comparable with the physical distance intercepted by the antenna at the echoing altitude – will record a significant proportion of such echoes.

6.5. Biases in Radar Sampling

It is useful to summarize some of the effects (some of which are wavelength dependent) that are important both for the design of meteor radar facilities and for a realistic unbiased analysis of meteor echo signal data.

6.5.1. *Initial Train Radius*

For the body echo case, if the ionization column at formation has a gaussian radius r_0, which is comparable to the operating wavelength λ then contributions from electrons in the cross-section will not be in phase (it is this mechanism that is responsible for reduction in reflection coefficient arising from diffusion). Indeed, the initial radius effect can be viewed as a reduction in the scattering cross-section due to a meteor train whose *effective* age is $r_0^2/4D_a$. The target cross-section decreases by $\exp(8\pi^2 r_0^2/\lambda^2)$. The effective value of r_0 for a realistic non-gaussian train has been discussed by Jones (1995). Ablated meteoric atoms suffer about 10 collisions in establishing the ionization column. However, both modelling and experiment suggest that r_0 changes with atmospheric density ρ (and therefore altitude) and meteoroid speed, V, as $r_0 \approx \rho^{-0.3}V^{0.6}$ with $r_0 \simeq 0.7$ m at 90 km.

This attenuation effect providing a wavelength-dependent 'height ceiling' is important in limiting the transverse geometry radar detection of small high-speed meteoroids ablating above about 100 km and imposes a serious limitation to flux measurements using radars. Steel & Elford (1991) contrast the echo height distributions obtained at multiple frequencies to illustrate well this important effect.

6.5.2. *Diffusion*

The body echo maximum target cross-section is produced from a length of train extending a few Fresnel intervals either side of the specular, t_0, point. While the later intervals are being formed earlier sections of the train are diffusing. An instantaneously formed train would behave as previously outlined (in section 6.4.2.2) but for slow meteors some of the pre-t_0 sections of the plasma column may have diffused to a radius comparable to λ thus reducing their contribution. A simple picture illustrates the effect. Of the energy contributing to the echo maximum, 90% is produced by the section of train extending one Fresnel interval either side of the t_0 point (section 6.4.2 and equation 6.4.6). Ignoring the presence of any initial train radius (dealt with separately in 6.5.2 above), the instant when the meteoroid is one interval past the t_0 point the radius of the expanding cone of ionization extending along the train is $r = (4D_a t)^{\frac{1}{2}}$ and specifically at the t_0 point the train has been expanding for a time interval equal to the time taken by the meteoroid to traverse one Fresnel interval $t_{\mathrm{fres}} = (R_0\lambda/2)^{\frac{1}{2}}/V$. With the train wider above and narrower below, the train radius at position t_0 can be taken as a representative average of the expanding section above and the smaller width section below – approximately a cylinder – taking the effective column width due to diffusion only as the width at the t_0 point. Phase summation is applied across the column as above (section 6.5.1). The reflection coefficient is reduced by a factor controlled by the ratio of the time taken by the meteor to transverse one Fresnel interval to the time for the echo to decay, Δ, and is $\exp(-t_{\mathrm{fres}}/t_d)$. The approximations made are not realistic when $\Delta = t_{\mathrm{fres}}/t_d > 1.5$. A more rigorous derivation [Peregudov 1958] shows that the attenuation factor is better described as $[1 - \exp(-\Delta)]/\Delta$.

6.5.3. *Finite Echo Lifetime*

For high altitude underdense trains, rapid diffusion (coefficient D_a in the absence of geomagnetic field effects) will cause the scattering cross-section to decrease exponentially described by the factor $\exp(-t/t_d)$ with $t_d = \lambda^2/32\pi^2 D_a$. For illustration: at an altitude of 120 km and $\lambda = 6$ m (operating frequency 50 MHz) $t_d \sim 10^{-3}$ s. If the (pulsed) radar sampling rate is such that the interpulse interval is comparable to this, the finite echo duration will result in reduced response or even lack of detection. In consequence those meteoroids that produce high altitude echoes (small mass, high speed, small zenith angle) will be under-sampled.

6.5.4. *Ionization Coefficient*

Ablated meteor atoms yield a greater number of ion pairs if their speeds are high. This dependence of ionization density produced in a train (section 6.1) results in the under-sampling of low velocity meteoroids because their low densities may result in their echoes being below detection threshold. The under-sampling effect depends on how the proportion of low mass meteroids increases with decreasing mass, i.e. on the relevant population mass distribution index.

6.5.5. *Fragmentation*

The idealized form of meteor ionization as described by a smoothly varying line density q along its length is not realized for some meteors. Continuous or gross fragmentation of the ablating meteoroid produces irregularities of various scales including structures in the form of separate plasma concentrations yielding complex echoes. Such echoes exhibit rise-time characteristics quite different from the diffraction behaviour produced by non-fragmenting meteoroids.

6.5.6. *Effects of the Local Wind Field*

At meteor ablation altitudes the dynamics of the ambient atmosphere is described by a range of time and spatial scales and that neutral gas motion can be transferred to the embedded meteoric ionization. Above about 100 km where the electron gyro frequency well exceeds the electron-neutral collision frequency, the imparted motion may be affected by the geomagnetic field but below this altitude the local wind effect is direct momentum transfer to the plasma. There are three effects.

Large (several kilometres) scale vertical shears of the horizontal wind can rotate the ionization train: trains initially formed with a non-orthogonal geometry for a particular radar can be sheared so that the train becomes substantially specular – the echo rise is slow (seconds) in contrast to the diffraction governed echo leading edge. Alternatively, trains which have an initially specular geometry can be rotated, leading to change in altitude of the orthogonality point and hence echo altitude. Irregular shear on a scale of a few kilometres can distort the initially linear train developing multiple reflection points producing interference effects and echo fluctuations and curtailing echo durations from overdense trains. The time constant of such inhomogeneity in the wind are several seconds so that underdense echoes recorded using HF and VHF radars are not usually so affected.

6.5.7. *The Radar Propagation Path*

There are circumstances for which the propagation of radio waves to and from the meteor target cannot be safely assumed to be under free-space conditions. Ionization in the lower ionospheric E region and the D region through which the radar path passes can produce two important effects. For a realistic interpretation of meteor echo parameters it is necessary to be aware that absorption of wave energy in the D region can occur if the electron collisional frequency in the medium is not negligible compared with the radar operating frequency. Such non-deviative absorption can occur under solar disturbed conditions for operating frequencies in the low HF band. More serious is likely to be the effect of rotation of the plane of polarization of the radio waves – caused by the anisotropy of the medium in the presence of the geomagnetic field – the medium is bi-refringent and any wave can be thought of as two modes with differing phase speeds. The rotation (termed Faraday rotation) occurs on the up and down propagation paths (the rotation does not 'unwind') increases as λ^2 and will have different values for meteor echoes at different altitudes: the rapid increase with height of the ambient lower E region ionization 90–120 km can cause height-dependent selection effect during day time. The result is that radars using linear polarized antennas can be subject to ionospheric-source signal attenuation. The effect will normally not be major for operating frequencies greater than ~ 30 MHz. Numerical values are given by Elford & Taylor (1997).

6.6. Meteoroid Parameters Available

6.6.1. *Velocity*

The speed of the meteor in the atmosphere, V, is an important quantity and can be measured employing several possible techniques. The measured speed is the scalar speed at the radar echo point: the entry speed before encountering the atmosphere V_{inf} (and therefore astronomically significant) can be inferred either from a directly measured meteoroid deceleration or modelled using assumed ablation coefficients (which depend on the shape and structure of the body): the change $(V_\infty - V)$ is ~ 1–5% depending on mass and coefficient. In radar programmes that require the vector velocity for determining a meteoroid trajectory three speed components need fixing.

6.6.1.1. *Velocity using Radar Transverse Geometry*

(*a*) *Head Echoes*

Large meteoroids – producing meteor radar magnitudes **M** < 0 – sometimes produce an echo (the head echo) in addition to a body echo. An analysis of the hyperbolic range-time function of the head echo can yield a speed. Indeed this was the first radar method [Hey & Stewart 1947] to be employed for speed determination. The small flux of meteoroids producing such echoes renders this technique unsuitable on a routine basis.

(*b*) *Diffraction Oscillations in Echo Power*

The theoretical echo power time characteristics (Figure 6.2) shows the signal oscillation caused by Fresnel diffraction after the meteor has passed the orthogonal t_0 point (example Figure 6.9). Various signal analysis schemes [e.g. Baggaley *et al.* 1994] can be used to extract the meteor speed V. The rate at which a meteoroid travels through successive Fresnel lengths is related to the Fresnel parameter x by $V = 2(R_0\lambda)^{\frac{1}{2}}(x/t)$. The exact form of the oscillations depends on the echo decay t_{d} (section 6.4.2.3) and deceleration [see e.g. Šimek 1966]. The uncertainty in such measured speeds is about 2%. Fragmentation effects, inadequate signal-to-noise ratio and rapid decay results in only a minor fraction of echoes ($\sim 15\%$) rendering usable diffraction patterns. In particular, high altitude rapidly decaying echoes suffer from severe suppression of the diffraction oscillations.

(*c*) *Phase Changes*

The post-t_0 phase fluctuations are weak (extreme deviation $\sim 20°$) whereas those pre-t_0 are substantial. A meteor detected prior to the pre-t_0 stage produces a build up of approaching ionization sensed as a Doppler approach and therefore as an increasing phase. This formative stage of the echo, while the train is developing, is especially valuable because the effects of meteoroid fragmentation, wind shear and echo decay caused by diffusion which may disrupt the diffraction signal later are not operative at this early stage. A phase-coherent radar is necessary – one able to sense the phase of the returned signal (this phase is arbitrary, but comparison with a phase reference, e.g. the radar transmitter will sense changes in phase as the meteor progresses through the t_0 point). An example of such an echo record is shown in Figure 6.9 where the triple antenna

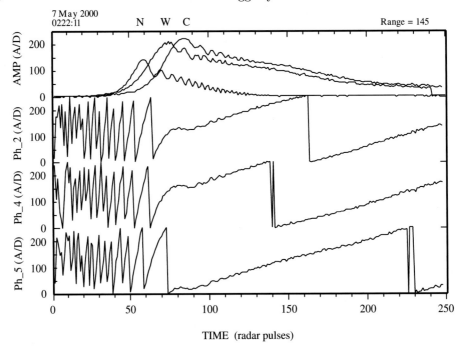

7 May 2000
0222:11   N   W   C   Range = 145

TIME (radar pulses)

FIGURE 6.9. Triple echoes from the multi-antenna AMOR system showing the post-t_0 signal response and the pre-t_0 phase behaviour. The top traces are the three spaced station amplitude profiles. The bottom three traces are the phase records for the three antennas of the elevation-finding interferometer. All three show unwrapped phases over $\sim 10\pi$ up to maximum phases at radar pulse 80. The phases show good coherence for signal-to-noise ratio levels above 6 dB.

system of the AMOR radar interferometer system shows the phase oscillations prior to the t_0 point and phase drift at later times due to transport of the meteoric plasma by the atmospheric wind.

When the meteoroid is located many Fresnel intervals prior to the t_0 point the phase is large and negative (see the Cornu spiral of Figure 6.3) increasing to a maximum value of $-(\pi/6)$ at the t_0 point. The recorded phase will be ambiguous to multiples of 2π and the true phase can be recovered ('unwrapped') by suitable signal processing algorithms. Polynomial expansions exist for the Fresnel integrals for general Fresnel parameter x: when x is outside the region $-1 < x < +1$ taking the first term in the expansion provides sufficient accuracy for many purposes. For $x < -1$

$$C \simeq -0.5 - \left(\frac{1}{\pi x}\right) \sin\left(\frac{\pi x^2}{2}\right) \qquad S \simeq -0.5 + \left(\frac{1}{\pi x}\right) \cos\left(\frac{\pi x^2}{2}\right) \qquad (6.6.1)$$

so that on the approach to the t_0 point and $x < -1$ (and the phase region $\phi < -\pi$) the phase change is well described by $(\phi - \pi/2) = \pi x^2/2 = t^2(2V^2/R_0\lambda)$ yielding a quadratic variation in time and so simplifying the reduction to velocity V. The phase pattern retains good coherence even at low signal to noise, a feature that makes this pre-t_0 phase method a valuable speed determiner (e.g. Figure 6.9). For times after the t_0 point slow phase trends result from the radial motion of the ionization train due to a background wind: such records provide a probe of atmospheric dynamics.

A variation of this phase-sensitive method is to use a continuous wave (CW) radar employing separated transmitter and receiver. Because of the changing path as the meteor approaches the t_0 point the (weak) ground transmission can be used as a reference phase with the combined signal appearing as amplitude fluctuations [McKinley 1951].

(*d*) *Time of Flight*

Using spaced receiving antennas the time delays between echoes will yield meteor speeds. An array of three such sites will provide velocity components and therefore uniquely determine the atmospheric trajectory. Using measured and modelled decelerations this timing technique [Baggaley *et al.* 1994] yields heliocentric velocity components and hence solar system orbits.

6.6.1.2. *Velocity from Radar Radial Geometry*

Radar reflection obtained from the plasma concentration moving with the meteoroid velocity will register a direct radial speed component. With a phase sensitive radar very rapid linear phase changes will be registered at a rate of a factor ~ 100 faster than for the transverse case. From such rapid phase change an unambiguous (i.e. to within 2π) value will not be extractable unless simultaneous accurate range changes can be provided by the radar. Such techniques have enabled the routine determination of meteoroid velocities and decelerations [see Figure 6.10 from Elford 1999].

The observed radial speed is not the true velocity V along the trajectory. However, V can be deduced because during the time the meteoroid is within the antenna beam the angle between the line from the radar to the plasma target and the direction of travel is continuously changing. For illustration: using an antenna beam $3°$ wide a meteoroid crossing the antenna beam axis at an angle $\theta = 20°$ would produce a decrease in radial speed of 1.7%. Such a change is easily measured and can be readily distinguished from the atmospheric deceleration near the end point of the meteoroid's trajectory. The crossing angle θ can also be estimated from the echo duration. Notice that the aspect angle is not known: the true meteoroid radiant lies on a circle of radius θ centred on the beam pointing direction.

6.6.2. *Mass*

Equation 6.4.7 together with the consequences of equation 6.2.1 permit an estimate of the pre-ablation meteoroid mass. Modifications are necessary for departures from solid body behaviour or if fragmentation occurs. Additionally, the factors detailed in section 6.5 above need inclusion. The radar equation also assumes (i) that reflection (the t_0 point) is located at the maximum ionization line density q position on the meteor train and (ii) that the antenna gains G_R G_T correctly represent the geometry, i.e. that the meteor echo direction within the radar antennas is known (by, for example, interferometric methods). On a statistical basis minor uncertainty is introduced by (i) and a factor ~ 3 (in inferred q) will pertain for individual echoes. The antenna pattern might be expected to introduce a similar uncertainty, so that the overall uncertainties in mass of a factor ~ 10 might be expected for individual meteors. Some of the factors summarized in section 6.5 above can be quite severe. In particular, transverse geometry radars operating at frequencies in excess of ~ 50 MHz will suffer from signal attenuation caused by the initial radius effect hence overestimating the meteoroid mass. Small meteoroids ablating at high altitude may be undetectable. The effective 'height ceiling' for a meteor speed of 60 km s^{-1} is $\sim$ 105 km at 50 MHz compared with ~ 125 km at 10 MHz.

6.6.3. *Altitude*

Echo altitude (the t_0 point for transverse geometry or the (changing) height of the spherical plasma concentration for radial geometry) is a parameter central to much of meteor science. Additionally, a good altitude estimate needs to be factored into the biases of sec-

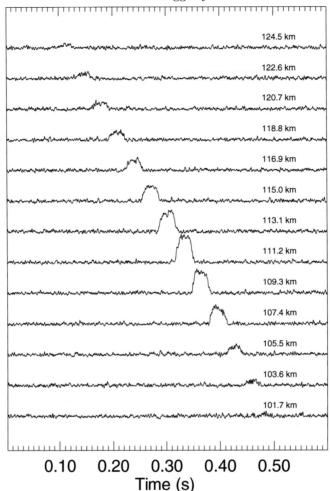

FIGURE 6.10. An example of an approaching head echo obtained with a 54 MHz radar of pulsing rate 1024 Hz. As the head plasma passes through eight successive range bins (each of width 2 km) the echo (amplitude) occurs at progressively later times. [From Elford 1999].

tion 6.5. There are three principal methods of measuring echo altitude, the first two being trigonometric – requiring knowledge of the echo range and elevation or zenith angle: employing radar receiving antennas which have different radiation patterns in the elevation plane and experimentally comparing echo signal amplitudes; employing an interferometer system of spaced (a few wavelengths) antennas with phase comparison providing angular information. Notice that although each antenna will sense the echo diffraction phase behaviour, it is the phase difference registered at the spaced antennas that requires recording. Both these techniques require measurement of echo range, the uncertainty of which is governed by the pulse duration (unless pulse coding techniques or pulse shape analyses are used). A method independent of angle and range measurement is to infer the echo altitude from the signal decay rate t_d (section 6.4.2.3). While the ultimate altitude uncertainty of trigonometrical methods is governed by the Fresnel interval length and is therefore ~ 1 km for HF radars, the echo decay constant provides only a crude estimate with uncertainties of ~ 10 km for individual meteors. The reason for the large experimental spread in decay constants (and therefore in diffusion coefficient, D_a) at a fixed altitude is not clear but may lie in the dependence of effective D_a on meteoroid chemical composition or meteoroid structure and in local small-scale variations of atmospheric pressure p and temperature T at a fixed height. In addition severe vertical shear in the horizontal wind will cause changes in the altitude of the t_0 point during the echo. Radial geometry radars employing narrow antenna beams can straightforwardly derive altitudes from range and echo point direction.

6.6.4. *Atmospheric Trajectories*

The emphasis in section 6.6.1 above was to measure the scalar speed of the meteor at the radar echo point. In order to determine the path of a meteor through the atmosphere three velocity components need to be fixed requiring spaced recording so that time-of-flight in two directions can be determined together with the third component which is fixed by the value of the echo elevation.

6.6.5. *Radar Sensitivity*

For optimum recording conditions – the absence of broadcast interference, static, electrical storm activity or radar scatter via the ionosphere – the radar system sensitivity is limited by the receiving system noise and cosmic noise. For operating frequencies less than about 200 MHz, cosmic noise exceeds internal receiver noise and with the cosmic noise effective temperature (K) increasing with wavelength approximately described by $T_{cos} = 150\,\lambda^{2.3}$ (λ metre) and the noise power delivered by a matched receiving antenna $P_n = kT_{cos}\Delta f$ (Watt) with k the Boltzmann constant and Δf the bandwidth of the radar receiver then $P_n = 2 \times 10^{-21}\lambda^{2.3}\Delta f$ watt. The bandwidth depends on the radar system, the use of coding or the operation of swept frequency (Chirp techniques). For a mono-frequency radar using pulse modulation the ultimate sensitivity for single pulse detection can be assessed by equating P_n to the echo power received for an underdense echo (equation 6.4.7) giving the limiting detectable ionization line density q_n as

$$q_n = \frac{3 \times 10^5 (\Delta f)^{\frac{1}{2}} R^{\frac{3}{2}}}{P^{\frac{1}{2}}(G_R\,G_T)^{\frac{1}{2}}\lambda^{0.35}} \tag{6.6.2}$$

If there is no integration or signal processing to enhance the radar signal then a realistic limiting q_{lim} will be larger than this – depending on the detection specifications for radar echo recording. For numerical illustration the AMOR facility (see section 6.8) with $P_T = 100$ kW, $\Delta f = 30$ kHz, $\lambda = 11.54$ m, $G_T = 600$, $G_R = 300$ and taking a representative range of 150 km then $q_n = 10^{10}$ m^{-1}. If the limited sensitivity is set by a signal-noise ratio of e.g. 9 dB then the sensitivity corresponds to a radio magnitude of $\mathbf{M} = +14$ at a meteoroid speed of 40 km s^{-1}. A realistic sensitivity can be assessed by taking into account the biases detailed in section 6.5.

6.7. Metal Ion Layers

Individual meteor trains having diffused initially by ambipolar expansion and during later stages by turbulent diffusion into the ambient atmosphere deposit metal neutrals and ions. Ablation initially deposits a mixture of M and M$^+$ with the M$^+$ contribution being a minor fraction (taking an average ionization coefficient over a range of meteoroid speeds).

The large range of meteor masses, speeds and entry zenith angles results in meteoric deposition over the altitude regime 80–115 km. The sequence of reactions ending with dissociative recombination [(R5) and (R8) of section 6.3.2 above] is only operative below ~ 95 km so that taking the whole meteor regime and noting that the mass distribution index will result in less material being deposited at the lower heights, most of the initially released M^+ ions will survive. The lower E region daytime ionization composed of O_2^+ and NO^+ (concentration $\sim 10^5$ cm^{-3} at ~ 100 km) can provide a route yielding metal ions via charge exchange

$$M + O_2^+ \longrightarrow M^+ + O_2 \tag{R9}$$

and

$$M + NO^+ \longrightarrow M^+ + NO \tag{R10}$$

with photoionization

$$M + h\nu \longrightarrow M^+ \tag{R11}$$

being much slower.

The nett result of ablation and aeronomy is the production of a broad layer (90–115 km) of metal ions, electrons, and neutrals having a global distribution and maximum ion density that varies between 10^2 and 10^4 cm^{-3}. The subsequent fate of these metallic species can be summarized: M atoms are stable since oxidation to MO via ozone is much slower than reduction in the presence of atomic oxygen. Similarly, since dissociative recombination is so slow in the ambient lower E region compared with reduction, M^+ is maintained against the creation of MO^+ or MO_2^+. The result is that metal ions have long residence times.

There are various dynamical processes that act on the broad ion layer, some mechanisms transporting ions to high altitude while other action leads to the compression of the meteoric plasma into thin sheets. Shears in the horizontal wind originating in tides and gravity waves can force meteoric plasma into thin sheets – ions descending from above and ascending from below. At high geomagnetic latitudes a similar physical feature has been modelled [Bristow & Watkins 1991] with the necessary compression provided by the convective electric field associated with magnetospheric plasma convection. Electric fields of optimum orientation with respect to the geomagnetic field can force meteoric ions into thin (1–2 km) layers. Such ion layers at mid and high latitudes are a well known feature monitored by vertical sounding radars (ionosondes) as sporadic-E (E_s). However, the efficient probing of such layers has been possible with incoherent scatter radars particularly high northern latitude radar facilities: the EISCAT system in Tromsø, Norway [Kirkwood 1997] and at Sondrestrom in Greenland [Bedey & Watkins 1998]. The layers appear of limited horizontal scale (~ 100 km) and duration (few hours), appear at altitudes of 100–130 km and have peak ionization densities 10^4 to 10^6 cm^{-3}. The layer meteoric ion composition has been probed using tunable laser beams (e.g. Alpers *et al.* 1993).

6.8. Facilities Currently in Operation

There are several programmes currently or recently in operation actively involved in radar meteoroid research: an outline is presented for each facility of the radar technique employed, the main thrust of the study and the status. The mode of operation follows the terminology used above. Some radar facilities carried out foundation work but are no longer in operation: Springhill, Canada; Jodrell Bank, Manchester University, UK; Bradfield, Sheffield University, UK: Onsala, Sweden.

6.8.1. *Transverse Geometry*

6.8.1.1. *Orbits*

Atmospheric trajectories and velocities are measured to produce heliocentric orbits.

(*a*) The Advanced Meteor Orbit Radar (AMOR) situated geographic 172°30′ E 43°35′ S near Christchurch, New Zealand, 26.2 MHz 100 kW pulse power fan beam orientated in the N–S and E–W meridians. Spaced stations for time-of-flight velocity components [Baggaley *et al.* 1994, 2001]. Routine operation 1990–present. Speeds from pre-t_0 phases and post-t_0 echo amplitudes. Atmospheric winds.

(*b*) Middle and Upper Atmosphere Radar (MU) radar Shigaraki, Japan, located geographic 136°06′ E, 34°51′ N. 45.5 MHz, 1 MW pulse power 475 crossed dipoles driven by individual transmitters. Rapid beam swinging; speeds from post-t_0 amplitude oscillations. Meteor programme time shared with atmospheric studies [Sato *et al.* 2000].

6.8.1.2. *Echo Directions, not Individual Orbits*

(*a*) Chung Li radar Taiwan geographic 120°55′ E, 25°0′ N 52 MHz; interferometer for echo directions meteor stream activity.

(*b*) CLOVAR, London Ontario, University of Western Ontario. geographic 81°20′ W, 42°55′ N 40.68 MHz.

(*c*) SKiYMET Various locations [see Hocking 1997].

(*d*) Buckland Park geographic 138°30′, 34°30′ S near Adelaide, Australia. 54 MHZ beam-width 3°.8 tilted from zenith by selected fixed angle. Pioneered the method of pre-t_0 speeds; stream parameters especially velocity distributions.

(*e*) Englehardt Observatory, Kazan, Russia. Geographic 48°25′ E, 55°50′ N.

6.8.1.3. *Limited Direction*

(*a*) Ondřejov Observatory, Czech Republic. Geographic 14°47′ E, 49°55′ N. 37.6 MHz common TX/RX antenna 36° beam-width in azimuth, steerable. Operations devoted to epochs of major streams providing an extensive survey for $\sim$ 40 years.

(*b*) Grahamstown, South Africa, geographic 26°30′ E, 33°25′ S.

6.8.2. *Radial Geometry*

(*a*) Buckland Park, Adelaide (as above). Radial velocities and decelerations. Individual orbits.

(*b*) Arecibo Observatory, Puerto Rico, geographic 66°45′ W, 18°21′ N 430 MHz 0°.16 beam tiltable to 18° from zenith [Janches *et al.* 2000]. Use of triple pulses for enhanced sensitivity. Employs radial velocities. Leonid influx. Shared access. Individual orbits.

(*c*) EISCAT, Scandinavia geographic Kiruna 20°27′ E, 67°52′ N 930 MHz steerable beam. Shared access. Leonid epoch times. Head echo studies.

6.8.3. *Forward Scatter*

6.8.3.1. *Dedicated Systems*

Scatter path Bologna–Lecce (Italy) and Bologna–Modra (Slovakia). Central transmitter at Budrio near Bologna, geographic 11°30′ E, 44°36′ N, transmitting 700 km south to Lecce or 800 km to the north to Modra. 99 MHz Operational periods at shower epochs [e.g. Cevolani *et al.* 1995].

6.8.3.2. *Passive Links*

There exist several recording programmes operated by astronomical societies or radio groups where forward scatter is monitored from a suitable distant transmitter involved in commercial traffic or broadcast TV. Networks exist in USA (American Meteor Society), Finland, Germany and Japan. These operations provide monitoring with wide geographical coverage of meteor fluxes which provides valuable data on, for example, small scale structure in shower influx.

6.9. Current Activity

There are several recent areas of radar research where new initiatives are proving of particular value, including: the pioneering work by the Adelaide University group in establishing the phase techniques both in transverse and radial geometry [Cervera et al. 1997; Elford 1999]; the employment of the narrow beam Arecibo radar in the detailed study of meteoroid deceleration, fragmentation, and orbits [Janches et al. 2000]; the use of meteoric plasma as an atmospheric probe [Hocking 1997; Marsh et al. 2000] and the mapping of the inflow of extra-solar-system grains [Taylor et al. 1996; Baggaley 2000].

REFERENCES

ALPERS M., BLIX T., KIRKWOOD S., KRANKOWSKY D., LÜBKEN F. J., LUTZ S. & VON ZAHN U. 1993. First simultaneous measurements of neutral and ionized iron densities in the upper mesosphere, *J. Geophys. Res.*, **98**, 275–284.

BAGGALEY W. J. & CUMMACK C. H. 1974. Meteor train ion chemistry, *J. Atmos. Terr. Phys.*, **36**, 1759–1773.

BAGGALEY W. J. & WEBB T. H. 1977. The thermalization of meteoric ionization, *J. Atmos. Terr. Phys.*, **39**, 1399–1403.

BAGGALEY W. J. 1978. The de-ionization of dense meteor trains, *Planet. Space Sci.*, **26**, 979–981.

BAGGALEY W. J. 1979. The interpretation of overdense radio meteor echo duration characteristics, *Bull. Astron. Inst. Czechosl.*, **30**, 184–189.

BAGGALEY W. J., BENNETT R. G. T., STEEL D. I. & TAYLOR A. D. 1994. The advanced meteor orbit radar facility: AMOR, *Q. J. Roy. Astron. Soc.*, **35**, 293–320.

BAGGALEY W. J., BENNETT R. G. T. & TAYLOR A. D. 1997. Radar meteor atmospheric speeds determined from echo profile measurements, *Planet. Space Sci.*, **45**, 577–583.

BAGGALEY W. J. 2000. Advanced meteor orbit radar observations of interstellar meteoroids, *J. Geophys. Res.*, **105**, 10353–10361.

BAGGALEY W. J., BENNET R. G. T., MARSH S. H. & GALLIGAN D. P. 2001. Features of the enhanced AMOR facility: Advanced Meteor Orbit Radar, *Meteoroids 2001*, ed. B. Wambein, ESA Pub. SP-495, 387–391.

BEDEY D. F. & WATKINS B. J. 1998. Large-scale transport of metallic ions and the occurrence of thin ion layers in the polar ionosphere, *Geophys. Res. Lett.*, **25**, 3767–3770.

BRISTOW W. A. & WATKINS B. J. 1991. Numerical simulation of the formation of thin ionization layers at high latitudes, *Geophys. Res. Lett.*, **18**, 404–407.

BRONSHTEN V. A. 1983. *Physics of Meteoric Phenomena*, D. Reidel Publishing Co., Dordrecht, The Netherlands.

CERVERA M. A., ELFORD W. G. & STEEL D. I. 1997. A new method for the measurement of meteor speeds: the pre-t_0 phase technique, *Radio Sci.*, **32**, 805–816.

CEVOLANI G., BORTOLOTTI L., FOSCHINI L., FRANCESCHI G., GRASSI G., TRIVELLONE A., HAJDUK A. & PORUBČAN V. 1995. Radar observations of the Geminid meteoroid stream, *Earth Moon Planets*, **68**, 247–255.

ELFORD W. G. 1999. New radar techniques: precission measurements of meteoroid velocities, decelerations and fragmentation, in *Meteoroids 1998* pp. 21–28 Astronom. Instit. Slovak Acad. Sci. Bratislava.

ELFORD W. G. & TAYLOR A. T. 1997. Measurement of Faraday rotation of radar meteor echoes for modelling electron densities in the lower ionosphere, *J. Atmos. Terr. Phys.*, **59**, 1021–1024.

HAJDUK A., HAJDUVOVA M., PORUBČAN V., CEVOLANI G. & GRASSI G. 1999. The ozone concentration in the meteor zone, in *Meteoroids 1998* pp. 91–97 Astronom Instit. Slovak Acad Sci. Bratislava.

HEY J. S. & STEWART G. S. 1947. Radar observations of meteors, *Proc. Phys. Soc.*, **59**, 858–883.

HOCKING W. K. 1997. System design, signal processing procedures, and preliminary results for the canadian (London, Ontario) VHF atmospheric radar, *Radio Sci.*, **32**, 687–706.

JANCHES D., MATHEWS J. D., MEISEL D. D. & ZHOU Q. H. 2000. Micrometeor observations using the Arccibo 430 MHz Radar. 1. Determination of the ballistic parameters from measured doppler velocity and deceleration results, *Icarus*, **145**, 53–63.

JONES J., JONES W. & HALLIDAY I. 1999. The head echo problem – a solution at last? in *Meteoroids 1998* pp. 29–36 Astronom Instit. Slovak Acad Sci. Bratislava.

JONES W. 1991. Theory of diffusion of meteor trains in the geomagnetic field, *Planet. Space Sci.*, **39**, 1283–1288.

JONES W. 1995. Theory of the initial radius of meteor trains, *Mon. Not. R. Astron. Soc.*, **275**, 812–818.

JONES W. 1997. Theoretical and Observational determinations of the ionization coefficient of meteors, *Mon. Not. R. Astron. Soc.*, **288**, 995–1003.

JONES W. & JONES J. 1990a. Ionic diffusion in meteor trains, *J. Atmos. Terr. Phys.*, **52**, 185–191.

JONES W. & JONES J. 1990b. Oblique scattering of radio waves from meteor trains: theory, *Planet. Space Sci.*, **38**, 55–66.

KIRKWOOD S. 1997. Thin layers in the high-latitude lower ionosphere, *Adv. Space Res.*, **19**, 149–158.

MARSH S. H., BENNETT R. G. T., BAGGALEY W. J., FRASER G. J. & PLANK G. E. 2000. Measuring meridional mesospheric winds with the AMOR meteor radar, *J. Atm. Solar Terr. Phys.*, **62**, 1129–1133.

MCKINLEY D. W. R. 1951. Meteor observations determined by radio observations, *Ap. J.*, **113**, 225–267.

MCKINLEY D. W. R. 1961. *Meteor Science and Engineering*, p. 227 McGraw-Hill, New York.

MCNEIL W. J., LAI S. T. & MURAD E. 1998. Differential ablation of dust and implications for the relative abundances of atmospheric metals, *J. Geophys. Res.*, **103**, D9 10899–10911.

PEREGUDOV F. I. 1958. On the effect of meteor velocities on the hour number in radio echo detection of meteors, *Soviet Astron.*, **2**, 833–838.

POULTER E. M. & BAGGALEY W. J. 1977. Radiowave scattering from meteoric ionization, *J. Atmos. Terr. Phys.*, **39**, 757–768.

POULTER E. M. & BAGGALEY W. J. 1978. The applications of radiowave scattering theory to radio-meteor observations, *Planet. Space Sci.*, **26**, 969–977.

SATO T., NAKAMURA T. & NISHIMURA K. 2000. Orbit determination of Meteors using the MU Radar, *IECE Trans. Commun.*, **E83-B**, 1990–1995.

ŠIMEK M. 1966. The influence of diffraction on the radio determination of meteor velocities, *Bull. Astron. Inst. Czechosl.*, **17**, 354–360.

STEEL D. I. & ELFORD W. G. 1991. The height distribution of radio meteors: comparison of observations at different frequencies on the basis of standard echo theory, *J. Atmos. Terr. Phys.*, **53**, 409–417.

TAYLOR A. D., BAGGALEY W. J. & STEEL D. I. 1996. Discovery of interstellar dust entering the Earth's atmosphere, *Nature*, **380**, 323–325.

WANNBERG G., PELLINEN-WANNBERG A. & WESTMAN A. 1996. An ambiguity-function-based method for analysis of Doppler decompressed radar signals applied to EISCAT measurements of oblique UHF–VHF meteor echoes, *Radio Sci.*, **31**, 497–518.

7
METEOR TRAILS AS OBSERVED BY LIDAR

By U. von ZAHN[1†], J. HÖFFNER[1‡], AND WILLIAM J. McNEIL[2¶]

[1]Leibniz-Institute of Atmospheric Physics, Schloss-Str. 6, 18225 Kühlungsborn, Germany

[2]Radex Inc., Three Preston Court, Bedford, MA 01730, USA

We report on an extensive set of new observations of meteor trails by ground-based lidars and proceed to a general review of the field. The observations are performed with metal resonance lidars which sound the atom densities of Na, K, Fe, Ca, and Ca^+ in the altitude range between 80 and 105 km. At the Leibniz-Institute of Atmospheric Physics (IAP) we have clustered up to three such lidars at one site for simultaneous common-volume observations of meteor trails. On an experimental basis, we have also enhanced our observations of meteor trails by lidar with simultaneous observations of meteors by an image-intensified video camera.

The total number of lidar-observed meteor trails stands now at more than 1300, of which only 43 are two-element trails and 6 three-element trails. For the IAP lidars, the average rate of trail detections is 0.8 meteor trails per hour of lidar observations. We show that the capability of any lidar to detect meteor trails is strongly altitude dependent and limited to the altitude range of 80 to 105 km. We estimate that the lidar-observed trails were produced by meteoroids ranging from 10 μg to several grams.

The most important result of our research is the discovery that lidar-observed meteoroids ablate almost exclusively differentially and not homogeneously. Differential ablation shows up in (a) the number of single-element trails being very large in comparison with that of two- and three-element trails and (b) the lidar-measured ratios of metal abundances in meteor trails. The mean observed ratios K/Ca, K/Fe, and Fe/Ca deviate significantly from CI carbonaceous chondrite composition. A preponderance of differential ablation processes could indicate that during their atmospheric entry a large percentage of all meteoroids desintegrate into many tiny particles.

We also review the frequent searches for a response of the metal layers to short-term changes of the influx of meteoric matter into the upper atmosphere.

7.1. Introduction

7.1.1. *Motivations for Lidar Studies of Meteor Trails*

In this chapter we will use the nomenclature as given in Kerridge & Mathews (1988): A meteor is "the light phenomenon produced by a meteoroid experiencing frictional heating when entering a planetary atmosphere" and a meteoroid is "a natural small (sub-km) object in an independent orbit in the solar system". We will deal with meteoroids in the size range of 0.2 mm to about 10 mm (approximate mass range of 10 μg to several grams). An important attribute of this size range is that it contributes considerably to the total influx of meteoric mass into the Earth's atmosphere (see chapters 1 and 4). When meteoroids enter the Earth's atmosphere, they do so with entry speeds between 11 and 72 km/s. Due to these high entry speeds, most of the meteoroids in the subject

---

[†] Email: vonzahn@iap-kborn.de

[‡] Email: Hoeffner@iap-kborn.de

[¶] Email: William.McNeil@hanscom.af.mil

size range experience, through aerodynamic friction, strong enough heating to ablate entirely and/or vaporize in the upper atmosphere at altitudes between 140 and 70 km. The ablating meteoroids produce spatially well-defined trails of meteoroid debris in the atmosphere, called meteor trails, which last from seconds to tens of minutes before they are dissipated into the ambient atmosphere by chemical and diffusive processes. The composition and abundances of metal atoms set free in the debris trail depend in yet to be explored ways on the initial composition of the meteoroid and on the heating and ablation processes which act on the meteoroid during its atmospheric flight (see chapter 11). It is at this point that ground-based lidars come into play because they give us the capability to quantify the abundances of a number of elements in these meteor trails and to allow a characterization of the ablation processes. Lidars have demonstrated the capability of measuring quantitatively the absolute number densities of metal atoms in volumes that are separated from the lidar by 100 km and with detection thresholds of better than 1 atom cm^{-3}. The suite of metals which can be investigated by this technique extends from the moderately volatile Na and K to the refractory Ca. In the trails left behind by ablating meteoroids, all these elements must be present in amounts substantially higher than the detection thresholds of these lidars. To study them by lidar "only" requires sampling such trails by the lidar right after their formation and before they are dissipated.

Another motivation for lidar-studies of meteor trails is to seek further support for our belief, that the ablation of meteoroids is the dominant source process for producing the free metal atoms which we know to exist with densities up to 15 000 atoms cm^{-3} in the altitude range of 80 to 100 km. If this is indeed so, one would expect that the abundances of the metals in this altitude region and their temporal and spatial variations should reflect, in some fashion, the properties of the incoming meteoroid population. This assumption, though, has been difficult to prove in any detail (see section 7.6.2). The situation that none of the state-of-the-art models of the different metal layers is able to satisfactorily model the source strength of ablating meteoroids for the metal layers is also of concern. We develop this argument by first denoting the measurement of the mass flux of meteoroids into the Earth's atmosphere by Love & Brownlee (1993) to be the most direct and accurate one available, clearly deserving a much higher level of confidence than the estimated value given by Hughes (1978). Secondly, we presume for the sake of argument that this infalling matter is of CI chondritic composition. Then all models of metal layers predict much too high atom densities in the metal layers, if not constrained by arbitrary assumptions. In order to bring the model results into agreement with the observations, the modelers have to introduce scaling factors for the mass flux called "ablation efficiencies" which are much smaller than unity. Plane et al. (1999) need for their model of the Na layer a scaling factor of 0.38; Helmer et al. (1998) need for the Fe layer a factor of 0.13; Eska et al. (1999) need for the K layer a factor of 0.05; Gerding et al. (2000) need for the Ca layer a factor of 0.01. In other words, the mass flux of meteoric metals into the metal layers has to be "arbitrarily" scaled down by 1 to 2 orders of magnitude in the cases of the Fe, K, and Ca layers. This situation raises the question as to which portions of this deficiency in our current modeling capabilities might be caused by unknowns of (a) the mean composition of the meteoroids, (b) the ablation process or (c) the chemical and dynamical sink processes for the metal atoms. The recently initiated studies of meteor trails by ground-based lidar have opened up the possibility of quantitative investigation of the transfer of metals from the meteoroid minerals through the ablation process into the steady state metal atom layers. Recent results of these studies will be reported below.

PLATES

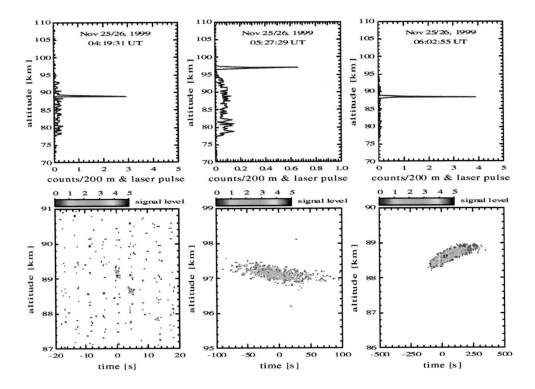

Plate 7.1. Three K meteor trails with quite different temporal developments and probed by the IAP K lidar on 26 Nov. 1999 (from Höffner *et al*., 2000; their figure 1). The upper panels give count rate profiles at the time of maximum K densities in the trails. Integration times are 0.5 s, 4 s, and 4 s for the left, middle, and right profiles, respectively. Altitude resolution is 200 m and 15 m in the upper and lower panels, respectively. The lower panels show the passage of the wind-driven meteor trail through the laser beam on much longer time scales, but more focused altitude scales. Here, times "zero" are identical to the times given in the upper panels. Data for the lower panels are smoothed with a running mean of 0.5 s, 4 s, and 4 s for the left, middle, and right panels, respectively.

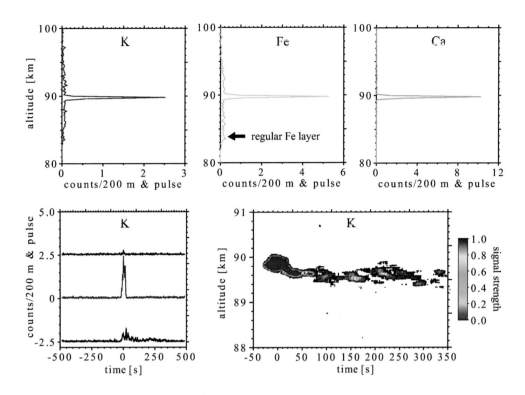

Plate 7.2. A meteor trail and subsequent train simultaneously observed by three metal lidars at Kühlungsborn 04:54 UT on October 27, 1997. The upper three panels show the profiles of photon counts recorded per 200 m altitude and laser pulse for K, Fe, and Ca, respectively, at the time of maximum atom density in the trail (with 4 s integration time). The lower left panel shows the temporal development of the K density in the altitude channel of maximum density and one channel higher and lower. Time "0" is that of the maximum density (with 4 s running mean). The lower right panel shows the temporal development of the trail as recorded by the K lidar and at an altitude resolution of 7.5 m and based on the arrival time of each individual photon (with 10 s running mean). After the passage of the dense trail, there followed a low density train which lasted for about 5 min.

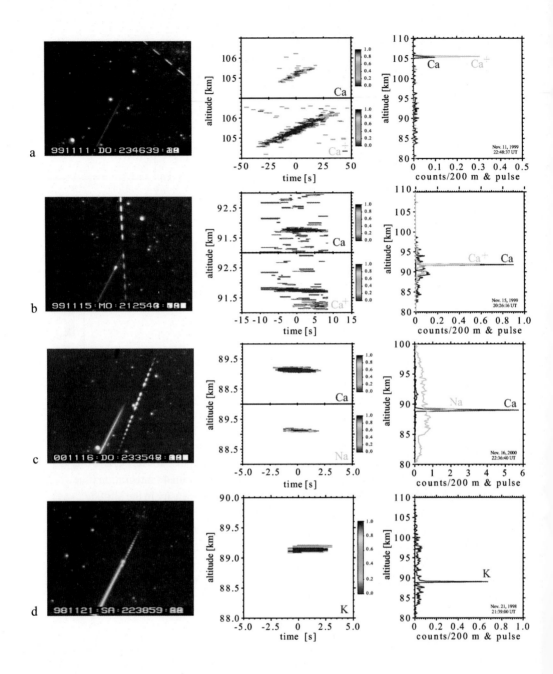

Plate 7.3. Examples of camera images of meteors and the ensuing lidar-observed meteor trails. The meteor images are in fact the sum of a number of video frames (see text for details). Each of the four rows shows in the left panel the camera image of the meteor, in the middle panels the temporal development of the lidar-observed trail(s) at high spatial and temporal resolution and in the right panels the count profile(s) at the time of highest metal density in the trail(s). Row (a) shows a Leonid trail, row (b) a sporadic, row (c) a Southern Taurid, and row (d) a flight of a meteoroid directly through the laser beam. Times are given on the camera images in CET, on the lidar profiles in UT.

TABLE 7.1. Bulk composition of selected classes of chondrites [Kerridge & Mathews (1988)] and percentages in Antarctic meteorite finds [Sears & Dodd (1988)]

	K (mg/g)	Na (mg/g)	Fe (mg/g)	Ca (mg/g)	K/Ca[a]	K/Fe[a]	Fe/Ca[a]	In Antartica (%)
CI	0.53	5.0	182	9.4	0.058	0.0042	13.9	0
H	0.80	6.4	275	12.5	0.066	0.0042	15.8	63
L	0.80	7.0	215	13.1	0.063	0.0053	11.8	21

[a]Ratios of atom number densities.

Concerning our above comparison of the composition of the atmospheric metal layers with that of CI chondrites, we note the following: On the one hand, CI chondrites are rare species among meteorites. Sears & Dodd (1988) gave the fall frequency of CI chondrites as 0.6% of all meteorites and their find frequency as 0% of the more than 1000 Antarctic finds (see Table 7.1). On other hand, more than 70% of all meteorite falls are either H or L chondrites; in the case of the Antarctic finds this number is 84%. Therefore, we cannot expect any direct link between the metal layers and CI chondrites. Fortunately for our further deliberations, the differences in composition between CI, H, and L chondrites do not matter here because we are dealing with only four selected metals (Na, K, Fe, Ca). In the three chondrite classes CI, H, and L, the abundance ratios of these metals vary little. The differences between the three classes are small (less than a factor of 1.4) in comparison with the variations found in meteor trails and the metal layers. Therefore, we take the liberty to use for comparison purposes the CI chondritic composition (as done very often in the literature), still recognizing the fact that CI meteorites are *not* the primary source of metal atoms in the upper atmosphere metal layers. The latter may be even the more true, as the largest influx of extraterrestrial matter into the Earth's upper atmosphere occurs at meteoroid sizes which are many orders of magnitude below those of CI meteorites.

Due to limitations of space we will not review those lidar studies of meteor trails which are aimed at understanding properties of the upper atmosphere, such as upper air winds, temperature, diffusion coefficients, and minor constituent abundances and reactions.

7.1.2. *Brief History of Lidar Observations of Meteoroid–Atmosphere Interactions*

In the following, we will deal with two aspects of the interactions between meteoroids and the upper atmosphere:

(a) One aspect is that it has become possible to study the properties of individual meteor trails by ground-based lidars. Lidar-observations of five short-lived meteor trails were first reported by Beatty *et al.* (1988), who used a Na lidar for their work. Three years later, She *et al.* (1991) reported about a single meteor trail, observed again with a Na lidar. A much more substantial number of meteoroids was studied by Kane & Gardner (1993a). They reported about the occurrence frequency and vertical distribution of 101 sodium and five iron meteor trails. Next came the results of the group at the Leibniz-Institute of Atmospheric Physics (IAP), which studied the 1996 Leonids by a K lidar [Höffner *et al.* (1999)], a March 1997 meteor shower by simultaneous and common-volume observations of a K and a Ca lidar [Gerding *et al.* (1999)] and the 1998 Leonids by a cluster of K, Ca, and Fe lidars [von Zahn *et al.* (1999)]. A first two-laser-beam observation of a meteor trail by a Na lidar was reported by Grime *et al.* (1999). Additional results

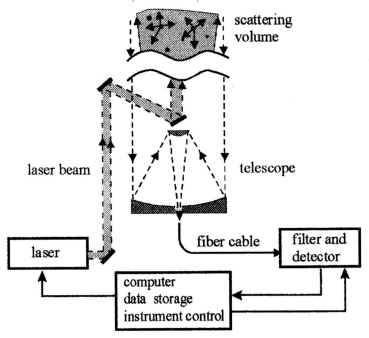

FIGURE 7.1. Schematic diagram of a lidar instrument. Pulses of laser light are emitted into a column of air which is also observed by a telescope. Photons backscattered into the telescope are transferred via a fiber cable into a filter and detector assembly. There the photons are analyzed for their wavelength, time-of-flight (= altitude of scattering volume), and rate of their recordings. Emitting the laser beam along the axis of the telescope field of view provides for certain advantages, but is not a mandatory feature of lidars.

on the 1998 Leonids, obtained by Fe and Na lidars, were reported by Chu *et al.* (2000a), Chu *et al.* (2000b), Kelley *et al.* (2000), and Grime *et al.* (2000). Last, but not least, we note that since late 1996, the IAP group analyzed the properties of more than 1000 meteor trails, which were observed by their Na, K, Fe, Ca, and Ca^+ lidars. The results of these studies pertain, as we will see, mostly to the ablation process of the meteoric material. We will present and discuss them in sections 7.3–7.5.

(*b*) The other aspect of meteoroid-atmosphere interactions has been studied for decades, starting with Hake *et al.* (1972), Aruga *et al.* (1974), Mégie & Blamont (1977): It is the longer term reaction of the metal layers to a time-variable influx of meteoric material. In these studies one is looking for changes in the atom densities of the metal layers on time scales of many hours and longer, as reaction to, e.g., transient meteor showers or seasonal variations in the flux of sporadics. We will deal with this topic in section 7.6.2.

7.2. The Observation Technique

7.2.1. *Metal Resonance Lidars*

Light radars, or in short "lidars", use pulsed laser beams to illuminate intermittently columns of air in our atmosphere (figure 7.1). Atoms, molecules, and particles contained in this column scatter part of the photons of the laser beam, some of which are directly backscattered towards a receiving telescope at the site of the lidar. The photons collected by this telescope are then analyzed for their time-of-flight between their emission from the lidar and reception at the telescope (yielding the altitude of the scattering volume), for the rate at which they are received (yielding information on the density of the scatterers) and for their exact wavelength (yielding information on motions, temperature, and composition of the scattering particle).

Lidars can be classified according to the scattering process which they use for detection of the scatterer: *Rayleigh scatter* requires the scatterer to be of much smaller size than the wavelength of the laser in use. This is clearly the case if the scatterers are air molecules or atoms. Throughout the lower atmosphere and up to 30 km height, there are, however, considerable numbers of particulates (aerosols, cloud particles) which, due to their size, scatter according to the *Mie scatter* process. Due to their size being many orders of magnitude larger than air molecules, the backscatter cross sections for Mie scatterers are many orders of magnitude larger than those of Rayleigh scattering air molecules. That translates into a much higher sensitivity of these Rayleigh/Mie lidars for particulates than for air molecules. To take a numerical example: At a wavelength of 532 nm, the Rayleigh backscatter cross section for a N_2 molecule is 6×10^{-32} m^2 sr^{-1}, the Mie backscatter cross section of a 0.1 and 1 μm diameter aerosol particle is, however, as large as 1×10^{-17} m^2 sr^{-1} and 1×10^{-14} m^2 sr^{-1}, respectively. The sensitivity of lidars is correspondingly many, many orders of magnitudes larger for particulates than for air molecules.

At this point, a brief discussion of the terms "sensitivity" and "detection threshold" of a lidar may be helpful. A definition of the lidar sensitivity would need to include the efficiency of the scattering process. It could be the ratio of the rate of received photons from 1 scatterer per unit volume over the rate of emitted photons. The ratio would be taken for a standard distance between lidar and scattering volume, typically 30 km. A so-defined sensitivity would be directly proportional to the backscatter cross section of the scatterer. Many times one is more interested, however, in the "detection threshold" of a lidar. This threshold is the lowest constituent density, in (cm^{-3}), which the lidar can detect in a given integration time and in the presence of a certain flux of background photons and of multiplier noise.

One can increase the sensitivity of lidars by many additional orders of magnitude by using the process of *resonant scatter*. Here the laser photon excites an optically allowed, resonant transition of an atom. The excited atom falls back to its ground state with a time constant on the order of 10 to 100 ns (the equivalent light path being a few to tens of meters) and by re-emitting the absorbed photon. In first approximation, the photon is emitted isotropically from the atom and thus has a finite chance of being backscattered into the lidar telescope. The great advantage of these lidars is that resonant scatter involves huge backscatter cross sections which are typically in the order of 1×10^{-16} m^2 sr^{-1} (see table 7.2). They are thus more than 15 orders of magnitude larger than the Rayleigh backscatter cross section of a N_2 molecule. Lidars using resonant backscatter achieve impressively low detection thresholds. For integration periods of, say, 5 min and altitude intervals of 200 m, the ground-based metal resonance lidars sound the metal layers at 80 to 100 km altitude at nighttime with detection thresholds between 0.1 and 1 atoms cm^{-3} These can be compared with the densities of the metal atoms near the peaks of the metal layers which range from 30 cm^{-3} in the case of the Ca layer to 20 000 cm^{-3} in the case of the Fe layer [see, e.g., Gerding *et al.* (2000)]. Obviously, ground-based lidars are able to study the properties of these metal layers with large signal-over-noise ratios.

In using resonant scatter processes for sounding the metal layers one has to cope, though, with the fact that the resonance lines of the metals are very narrowband features (see table 7.2 for their full width at half maximum (= FWHM)). Therefore, the

TABLE 7.2. Metals and their resonant transitions used in lidar soundings

Metal	Wavelength (in air) [nm][a]	FWHM of line [pm][a]	Line/transition	Backscatter cross section [m² sr⁻¹][b]
^{56}Fe	371.993	0.5	Fe I / $a^5D_4 - z^5F_5^0$	8.15×10^{-18}
^{40}Ca$^+$	393.366	0.63	Ca II/$^2S_{\frac{1}{2}} - {}^2P_{\frac{3}{2}}^o$	1.12×10^{-16}
^{40}Ca	422.673	0.68	Ca I/$^1S_0 - {}^1P_1^o$	4.17×10^{-16}
^{7}Li	670.776	4	Li(D$_2$)/$^2S_{\frac{1}{2}} - {}^2P_{\frac{3}{2}}^o$	1.12×10^{-16}
^{23}Na	588.995	3 for D$_{2a}$ + D$_{2b}$	Na(D$_2$)/$^2S_{\frac{1}{2}} - {}^2P_{\frac{3}{2}}^o$	7.78×10^{-17} at peak of D$_{2a}$
^{39}K	769.897	1.85	K(D$_1$)/$^2S_{\frac{1}{2}} - {}^2P_{\frac{1}{2}}^o$	7.51×10^{-17}

[a]References: All wavelengths from Morton (1991). Other values as follows: Fe from Alpers et al. (1990); Ca and Ca$^+$ from Mégie (1988) and Alpers et al. (1996); Li from Mégie (1988); Na from Fricke & von Zahn (1985); K from von Zahn & Höffner (1996).
[b]At peak of fine structure line and 200 K.

light source of the lidar, the laser, must be tunable to the metal-specific wavelength and it must be very narrowband to allow full use of the peak resonant cross section. Further requirements for the lasers are short pulse length (< 300 ns) at high pulse energy. For more than two decades, these requirements could only be achieved by dye lasers which were pumped by flash lamps or excimer lasers. More recently, however, all solid-state lasers have been developed which allow sounding of the K layer by means of an alexandrite laser [von Zahn & Höffner (1996)] and of the Na layer by use of a twin Nd:YAG laser system [Jeys et al. (1989); She & Krueger (2001)]. Further details of these lidars can be found in a review by She (1990) and the literature cited therein. Nevertheless, we will discuss in the next section some special features of lidars used in meteor work.

7.2.2. Lidars Used in Studies of Meteor Trails

To make meteor trails detectable for ground-based lidars, the debris of ablated meteoroids must be illuminated at least for a few seconds by the laser beam of the lidar. In reality, however, meteoroids pass very, very rarely right through the very narrow laser beam of the lidar. An overwhelming percentage of all lidar-observed trails are caused by meteoroids entering the upper atmosphere well to the side of the laser beam (figure 7.2).

The lifetime of the metal atoms in the trail and the trails's coherence are long enough, however, that the upper atmosphere winds can blow the trail through the laser beam tens of minutes after the meteoroid entry and while the trail is still recognizable by the lidar. Thus, there is no strict temporal correlation between an entering meteoroid ("shooting star") and the lidar-observed passage of the meteor trail through the laser beam.

Lidar studies have shown that meteor trails may have diameters from a few meters to many 100 meters, depending on their age and altitude. At meteor heights, the laser beams have diameters of typically 50 m. In addition, wind velocities at meteor heights

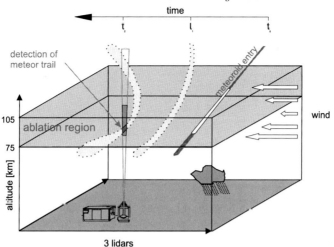

FIGURE 7.2. Illustration of the scenario in which meteoroids enter the upper atmosphere well to the side of the laser beam. Shortly thereafter, the meteor trail is blown through the laser beam by the upper atmosphere winds (obviously depending on their direction). Figure adapted from von Zahn *et al.* (1999).

are typically of the order of 20 m/s [Fleming *et al.* (1990)]. These conditions, considered together, support our experience which shows that most of the meteor trails pass through the laser beams in between 2 and 10 seconds. A few can be monitored by the lidar for up to minutes. The latter statement explains in part why "normal" lidars have not noticed meteor trails for decades: They employed integration times that were much too long thereby swamping the normally feeble signals of the meteors by those from the metal layers and by the background.

A lidar, dedicated to meteor work, therefore has to have a data acquisition system that allows it to collect altitude profiles of the backscattered photons with an integration period as short as possible. An ideal system stores the data of each individual laser pulse. After data acquisition, post-integration of these data with a running mean can always be used to reduce the time resolution to the desired value. Use of such short integration periods drives the currently used lidars close (or below, depending on altitude) to their useful limits of photon count statistics. In the individual profiles obtained from one laser pulse, most of the altitude channel contain either none or one count. Only strong meteors produce a few counts per altitude channel and laser pulse and can be recognized readily in the raw data stream. The average meteor can be recognized only after integrating the data over periods of a few seconds.

The IAP lidars store the count profiles from each laser pulse resulting typically in 2 million profiles for an observation time of 10 hours and using three simultaneously operating lidars. It is obvious that this amount of data cannot be processed visually. An automated computer code is absolutely required to search and identify meteor trails versus background counts in each of the sampled altitude profiles. For this search we use running means of typically 4 s. Höffner *et al.* (1999) describe their automatic computer code which takes into account, among other peculiarities, Poisson statistics of received counts.

At this point, two figures illustrate the type of data which meteor lidars acquire. Figure 7.3 shows the profile of a Fe meteor trail probed by an airborne lidar on 17 Nov. 1998 [from Chu *et al.* (2000a)]. The altitude resolution is 24 m and the temporal resolution

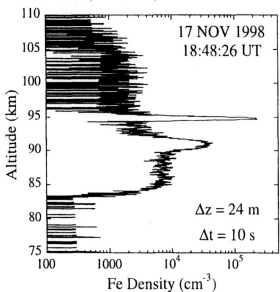

FIGURE 7.3. Profile of a Fe meteor trail close to 95 km altitude and probed by an airborne lidar (from Chu *et al.* (2000a); their figure 1). The temporal resolution is 10 s, the altitude resolution 24 m. This was one of the most dense meteor trails probed during that night of a Leonid shower. The detection threshold for meteor trails above 95 km is about 8000 Fe atoms cm^{-3}.

10 s. This was one of the most dense meteor trails probed during that night of a Leonid shower. The detection threshold above 95 km is about 8000 Fe atoms cm^{-3}. Plate 7.1 shows three K meteor trails, all probed by a ground-based lidar on 26 Nov. 1999 [from Höffner *et al.* (2000)] and selected for this figure on account of their different time scales. The upper panels give the count rate profiles at the time of maximum K densities in the trails. Integration times are 0.5 s, 4 s, and 4 s for the left, middle, and right profiles, respectively. The lower panels show the passage of the wind-driven meteor trail through the laser beam on much longer time scales, but more focused altitude scales. Here, times "zero" are identical to the times given in the upper panels. Altitude resolution is 200 m and 7.5 m in the upper and lower row of panels, respectively. The leftmost trail is detectable only during the period at which the laser wavelength is tuned closely to the maximum backscatter cross section while the laser is scanning across the $K(D_1)$ fine structure line once each 4 s (the wavelength scan is performed to determine the ambient temperature). At any given altitude channel, the trail is recognizable for less than 2 s. The rightmost trail, which is quite dense, behaves quite differently and can be followed for 400 s. Its maximum vertical extent is a little less than 400 m. The apparent vertical motions of the three trails are due to the inclined entry trajectory of the meteoroid being nearly parallel or antiparallel to the wind direction in the altitude range where the lidar detects the trails.

7.2.3. *Mie-Scatter Lidars*

Many attempts have been made and described in the literature to detect by means of Rayleigh/Mie lidars the smoke particles suggested by Hunten *et al.* (1980) or any other particulates left in meteor trails after the passage of the meteoroid. To our knowledge, none was ever successful and therefore we will not dwell on these experiments here.

7.3. Single-Element Trails: Observations and Modeling

7.3.1. *The IAP Database*

In October 1996, the IAP lidar group started its studies of meteor trails at Kühlungsborn with a K lidar. This lidar is tuned to the $K(D_1)$ resonance line and is used for studies of the K layer and the thermal structure of the mesopause region [von Zahn & Höffner (1996)] as well as of K meteor trails. In January 1997, an additional twin lidar system for studies of the mesospheric Na, Fe, Ca, and Ca^+ layers [Alpers *et al.* (1996)] became operational at the Kühlungsborn site. Since that time, we have operated these three lidars during more than 450 instrument-nights (if during one night we operated three lidars, we will call this three "instrument-nights"). As of today, this effort has given us a data base of about 1300 meteor trail observations.

In the remainder of this section, we will discuss these observations as if they were collected entirely independently. This should disclose the similarities and differences of the Na, K, Fe, Ca, and Ca^+ trails. In the next section we will discuss those meteor trails which were observed simultaneously by two or three lidars.

7.3.2. *Observed Rate of Trail Detections*

In table 7.3 we have listed for each of the elements under study the total number of trails that we have detected and analyzed since October 1996, the number of nights and hours that our lidars have observed the specific metal layer, and the rates per night and per hour of lidar sounding. The total number of nights in which we operated our lidars is 467, the total number of observing hours is 2072 h. Hence, the average "night of lidar observation" lasted 4.4 h, being limited mostly by variable cloudiness and rather short summer nights (at 54° N latitude). We recognize that (a) the mean rate of detection (= trails h^{-1}) is similar for each of the four metals and that (b) the overall mean rate of trail detections has been 0.8 trails h^{-1}. At first sight, one may have expected to see that rate to scale somehow with the absolute densities of metals in the trails. This is clearly not the case. For the rather abundant Fe, one of the reasons for this is the small oscillator strength of the Fe line that is used by the lidar. Our rates can be compared only with those of Kane & Gardner (1993a) as this is the only other publication containing a sizable number of lidar-observed meteor trails. These authors detected 106 meteor trails in 1084 hours of lidar observations or 0.1 trails h^{-1}. We attribute our eight times higher rate mostly to our considerably shorter integration times.

The rate at which our lidars detect meteor trails is clearly variable from night to night. Somewhat, but not dramatically, higher rates were reached during the Leonid showers (see below). Yet, significantly higher rates have been observed in sporadic enhancements of meteor activity. Gerding *et al.* (1999) have described one case in which over a period of slightly more than 5 h our Ca lidar detected (at least) 19 meteor trails. Thus the rate on this night approached 4 h^{-1}. The situation with respect to Ca^+ meteor trails in table 7.3 is somewhat disappointing even though our computer software had identified about 200 such trails. A closer inspection of the computer-identified trails revealed, however, that in the case of Ca^+ trails there is a continuous transition from short-lived, narrow meteor trails into longer-lived, broader sporadic layers. It appears almost impossible to define objective criteria which would enable us to distinguish between the two phenomena. This topic is certainly worthy of analysis and reporting in a future paper. But until then, we prefer to delete the Ca^+ trails in the statistics of table 7.3. Still, we will give examples of Ca^+ meteor trails later on. Those we have identified as meteor trails by the fact that another element shows at the same time and altitude a meteor trail.

TABLE 7.3. Statistics of meteor trail observations by IAP lidars (Kühlungsborn and Tenerife; Oct. 1996–Dec. 2000)

Element	Number of trails	Number of nights	Number of hours	Trails/ night	Trails/ hour	Mean altitude (km)
Na	83	30	111	2.8	0.75	92.4
K	519	199	1060	2.6	0.49	88
Fe	107	28	122	3.8	0.88	89.3
Ca	570	137	530	4.2	1.08	89.9
Ca$^+$	(?)	73	249	(?)	(?)	
Sum or mean	1279	394[a]	1823[a]	3.3	0.8	89.9

[a]Sums exclude Ca$^+$ numbers.

TABLE 7.4. Detection thresholds of IAP lidars at 4 s integration time and the highest recorded trail for each of the four metals

Element	Detection threshold (atoms cm^{-3})	Maximum altitude of trails (km)
Na	(500)	107.8
K	20	106.4
Fe	15000	105.2
Ca	80	106.0

7.3.3. Altitude Distributions

We begin the discussion of the altitude distributions of Na, K, Fe, and Ca meteor trails by first introducing the altitude distribution of these trails versus the measured metal density in the trails. In the four panels of figure 7.4 each of the observed trails is shown as one dot according to its altitude and its maximum absolute atom density. We notice the following:

(a) From the panels one can read for our lidars the approximate detection thresholds for metal atoms in meteor trails. They are given in the middle column of table 7.4 and are based on a common 4 s integration time for all trail analyses.

(b) For at least K and Fe, the detection thresholds of the lidars are low enough, even at the very short integration times used here, to show the "shadow" of the regular metal layers, in which altitude range thin meteor trails cannot be detected. In the case of the Ca lidar, our detection threshold of about 80 cm^{-3} is higher than the peak density of the regular Ca layer, which is only 22 cm^{-3} [Gerding et al. (2000)]. Therefore, the "shadow" of the regular Ca layer does not show up in our altitude distribution.

(c) The K, Fe, and Ca trails appear to be most abundant near the 90 km level. This initial impression will be corrected slightly, however, with figure 7.5.

(d) The panels show well the maximum altitude at which meteors where observed for each of the four elements. The numbers are given in the right-hand column of table 7.4. All four lidars lose their detection sensitivity for meteor trails above 105 km. This is an

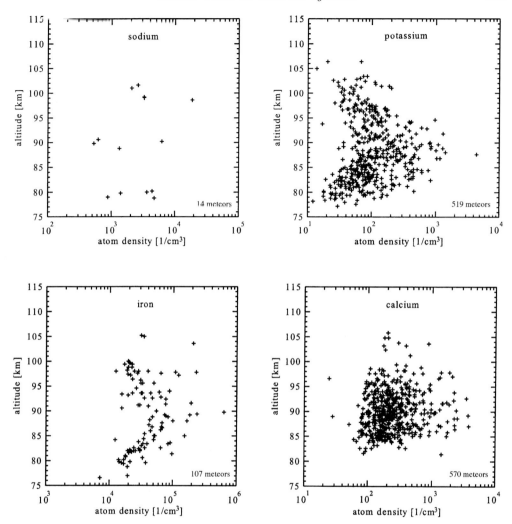

FIGURE 7.4. The altitude and maximum atom density of meteor trails observed by the Na, K, Fe, and Ca lidars of the IAP during the period Oct. 1996 through 2000. For a note on the number of Na trails, see text.

important detail for lidar-studies of meteor trails and was first recognized by Höffner *et al.* (1999).

(*e*) Finally, we draw attention to the unexpected result that for each of the four different metals, the lidar-observed trail parameters are very similar with respect to (1) the mean occurrence rate of trails, (2) the altitude of maximum trail rate, and (3) the maximum trail altitudes.

Now we turn our attention to the altitude dependence of the rate at which our lidars detect meteor trails. In figure 7.5 we express this function as "meteor trails detected per (km and h)". We have summed up for each kilometer the meteor trails shown in figure 7.4 and have smoothed the resulting altitude profile with a five-point Hanning filter (total width 5 km). The smoothed curve thus extends 2 km lower and higher than the lowest and highest data points. This effect is most conspicuous for the case of the Fe profile. We offer the following comments:

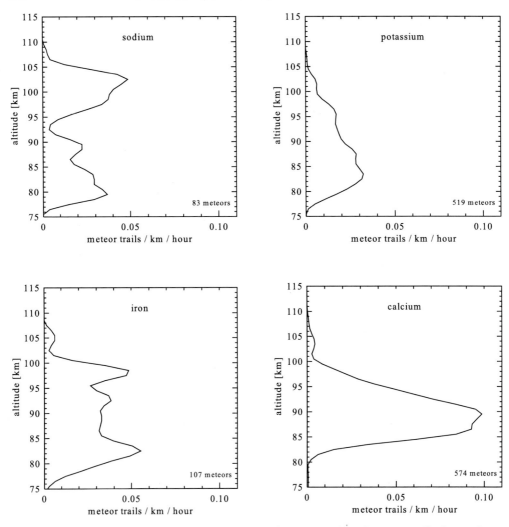

FIGURE 7.5. The rate of lidar-detected meteor trails, expressed as "meteor trails detected per (km and h)", as function of altitude and calculated from the 1279 meteor trails listed in Table 7.2.

(*a*) For a meteor trail to be shown in figure 7.4, the absolute metal density in the trail must have been derived before. This task has not been completed fully for our Na trails. Therefore, in figure 7.4, there are fewer Na trails than in figure 7.5, the latter not being dependent on the absolute Na density in the trail.

(*b*) Only for the cases of K and Ca, the profiles exhibit a shape as we expect it: single peaked and maximizing in the 90 km altitude range.

(*c*) The altitude of the maximum rate of K trails is 83.5 km and falls 6 km lower than that for Ca trails. The altitude of this K maximum is much effected by a cluster of trails at altitudes below, say, 87 km, and densities below 100 cm^{-3} (see figure 7.4). This cluster may be produced by, for example, a particular family of sporadics. In addition, we speculate that due to differential ablation, the Ca atoms of those sporadics are ablated well below 80 km altitude where they become undetectable for the lidars due to their very short lifetime as free atom. Hence, a similar cluster of data points is missing in the Ca profile.

(*d*) The Fe profile shows a deep depression where one would initially expect a maximum to occur. We attribute this to the "shadowing" effect of the regular Fe layer for faint meteor trails which is clearly demonstrated in the distribution of figure 7.4. Following this argument for the Fe profile, one is not surprised to see the same "shadowing" effect in the Na profile as the regular Na layer peaks at densities from 3000 to 5000 cm^{-3}.

(*e*) Our Na profile is strikingly different from the one published by Kane & Gardner (1993a) who showed a single-peaked profile, extending from 82 to 98 km. This difference is not due to poor statistics because either group had assembled in the order of 100 profiles. The fact that for the Kane & Gardner observation the trail maximum falls close in altitude to the regular Na layer maximum indicates to us that they have counted mostly trails, which were dense enough to outweigh the shadowing effect of the atoms in the regular Na layer. Their emphasis on dense trails would also explain a large part of their lower rate of meteor detection (0.1 h^{-1}) in comparison with our rate (0.8 h^{-1}).

We take the data presented above as strong evidence for the fact that the altitude distributions shown in figure 7.5 have little to do with the intrinsic altitude distribution of meteor trails. The lidar-measured altitude distributions resemble basically profiles of lidar sensitivity for trail detection, or in other words, instrument functions. For altitudes above 90 km, this conclusion is opposite to that reached by Kane & Gardner (1993a), who took their lidar-measured profile to resemble the distribution of incoming meteoroids. Our arguments in favor of an instrument function are: (1) the odd shapes of measured Na and Fe profiles; (2) the sharp cut-offs at both the lower boundary (near 80 km) and the upper boundary (near 105 km) which cannot be produced by the incoming meteoroid population; (3) the case of the Leonid meteors which we present in section 7.3.4; (4) results of a numerical model of the dependence of lidar sensitivity and detection threshold on altitude which we present in section 7.3.5.

7.3.4. *The Altitude Distribution of Leonid Meteor Trails*

7.3.4.1. *Why are Leonids of Special Interest for Lidar Studies?*

Single-element lidars allow us to determine the altitude and width of a meteor trail, the absolute metal atom density in the trail and the drift time of the trail through the laser beam. Lidars yield no information, however, on the total speed or the direction of the meteoroid which produced the observed trail. Without supporting external information, the lidar data cannot be linked to a particular meteoroid. If, however, the lidar-observed trail can be assigned uniquely to a particular meteor shower, then the total speed, the entry angle, and the direction of the meteoroid, which formed the trail, become well-known. Only then can the lidar-obtained information be related, for example, to the results of models for the ablation processes of meteoroids or with the results from photometric and camera studies of individual shower meteors.

Among the many known meteor showers, the Leonid shower has two particular properties that facilitate the identification of a lidar-observed meteoroid as Leonid:

(*a*) Due to their extremely high velocity of 72 km/s, the Leonids ablate at high altitudes, thereby helping to distinguish them from the common sporadic meteoroids.

(*b*) Since 1999, the time periods of enhanced Leonid shower activity have become predictable to an amazing accuracy [Asher (1999); McNaught & Asher (1999)]. This again alleviates the task of distinguishing between trails produced by sporadics and those by Leonids.

7.3.4.2. *Altitude Distributions from Electro-Optical and Photographic Studies of Leonid Meteors*

Even though the emphasis of this section is on lidar observations of Leonid trails, it is useful to start with a brief summary of results obtained by other techniques with emphasis on the altitudes of Leonid meteors (the latter can be compared directly with lidar observations). We cite here only some recent results from the rich material collected over the past decades by means of photographic triangulation, image-intensified video cameras and sensitive TV systems. For details of the acquisition of these data, see chapter 5.

A list of 55 Leonid orbits was published by de Lignie *et al.* (2000); these were all obtained through double-station video camera observations during the 1998 Leonid shower and from meteors, the visual magnitudes of which ranged from +6 to −1. In addition, de Lignie studied with the same technique 13 Leonids during the 1995 Leonids (de Lignie, personal communication). The end heights of these 68 Leonid trails range from 120.6 km to 90.1 km. Hence, not a single Leonid was observed to penetrate below 90 km. The mean end heights for the 1998 Leonids were: For the fireball night 97.7 km, for the following night 101.2 km [de Lignie *et al.* (2000)]. The height of maximum brightness for the complete sample of 1998 Leonids was 110 km.

Campbell *et al.* (2000) observed the 1998 Leonid shower by a cluster of 10 image intensified video cameras. In their paper, the authors discuss the results of 79 precisely measured Leonids meteors with maximum brightness from +6 to 0 astronomical magnitude. The mean ending height for their entire event sample was 95.3 km, about 2 km lower than the result of de Lignie *et al.* (2000).

The sensitivity of photographic methods is such that it covers only Leonid fireballs with magnitudes ≤1. Betlem *et al.* (2000) published the orbits of 47 Leonids, recorded by double-station photographic cameras during the night of November 17/18, 1999. The visual magnitude of the entire sample ranges from +1 to −5. The range of their end heights was 108.3 km to 87.3 km with a mean of 100.4 km. Only two of the 47 Leonids penetrated below 90 km, and those only slightly. These findings were extended towards even much brighter meteors by Spurný *et al.* (2000). During the 1998 Leonids, their double-station cameras captured photographically 67 Leonid meteors. The end height of a magnitude −5.7 Leonid (photometric mass 1 g) was measured to be 92 km, Leonids up to magnitude −9 (photometric mass 50 g) did not penetrate below about 88 km.

We conclude from this brief review that the brightest meteors of "normal" Leonid showers have typically a visual magnitude of −5, photometric masses of 1 g and end heights above 90 km.

Another important property of Leonid meteors is the near-constancy their end heights. Jenniskens *et al.* (1999) show in their figure 7a the dependence of observed end heights of Leonid meteors versus their magnitude (for the range +5 to −6 mag). For the case with the Leonid radiant in the zenith, their equation (14) gives the end height H_e (in km) as

$$H_e = 93 + 1.3 \times \text{mag} \qquad (7.3.1)$$

Hence, these end heights increase by only 1.3 km per magnitude, which is a rather weak dependence. Campbell *et al.* (2000) quantify the dependence of observed "last" heights in their video images of Leonid meteors versus their brightness M for an unspecified range of radiant elevations as follows:

$$H_{\text{last}} = 93.6 + 1.6 \times M \qquad (7.3.2)$$

Campbell *et al.* (2000) comment: "The fact that the heights are largely independent of meteor magnitude (and mass) is indicative of total clustering of the meteoroid into fundamental grains prior to the onset of intensive evaporation of the grains themselves." Also, Brown *et al.* (2000) report that according to their optical and radar observations, end heights of Leonid meteors show no significant dependence on meteoroid mass.

The feature of Leonids, to have a rather weak dependence of end heights on meteoroid mass, makes it impractical to derive a Leonid mass distribution from the low altitude part of an altitude distribution of Leonid trails obtained from lidar observations.

7.3.4.3. *Lidar Studies of Leonid Meteor Trails*

So far, only two groups have studied the Leonid meteors and the reaction of the metal layers to the Leonids by use of lidars. That of the IAP at Kühlungsborn, Germany and that of the University of Illinois at Urbana-Champaign, IL, USA (see Table 7.5). The Kühlungsborn group observed the Leonids in 1996 with their K lidar at Kühlungsborn [Höffner *et al.* (1999)], the 1998 Leonids with three lidars (for K, Fe, and Ca) at Kühlungsborn [von Zahn *et al.* (1999)], and both the 1999 and 2000 Leonids with a K lidar at Tenerife [Höffner *et al.* (2000)] and with Ca and Na lidars at Kühlungsborn. The Urbana group observed the 1998 Leonids with an airborne Fe lidar near Okinawa, Japan [Chu *et al.* (2000a)] and with a ground-based Na lidar near Albuquerque, NM, USA [Chu *et al.* (2000b)].

An initial task for lidar studies of the Leonids is to establish an altitude distribution of Leonid trails. This task is complicated by the fact that even during the nights of the Leonid shower, other shower meteors (e.g. Northern and Southern Taurids) and sporadic meteors will be recorded by the lidar. To illustrate this difficulty, we quote the results from our narrow field of view and zenith-oriented video camera for two nights of our Leonid lidar observations at Kühlungsborn. The camera software registered during the night of 16/17 November 1998 four Leonids and four Taurids and during the night 17/18 November 2000 ten Leonids and five Taurids. Clearly, we can not dismiss the occurrence of Taurids in lidar-studies of the Leonids as has been done in a few early lidar-studies of the Leonids. During the morning hours of November 17, 1996, Höffner *et al.* (1999) performed the first lidar observations of Leonids. During 3.4 h close to the peak of that shower, they observed seven meteor trails, all of which the authors assigned to the Leonid shower. The trail altitudes fell between 83.2 km and 96.6 km. Later the authors realized that only Leonids of very large mass penetrate to altitudes below 90 km (see above). Therefore, in a later paper the authors limited the assignment of lidar-observed trails to Leonid meteors for those which were observed at altitudes above 87 km [von Zahn *et al.* (1999); their table 5b]. This left from the original seven meteors only three as Leonids. Their altitudes ranged from 91.0 km to 96.6 km. The other trails are now assumed to be caused by Taurids or sporadics.

During the "fireball night" of November 16/17, 1998, von Zahn *et al.* (1999) obtained 3 hours worth of lidar observations. During this period with a mean Leonid radiant elevation of $15.6°$, their K- and Ca-lidars observed three Leonid trails at altitudes between 97.6 and 103.6 km (table 7.5). During the night of November 17/18, 1998, Chu *et al.* (2000a) obtained 18 meteor trails using an airborne lidar during a 5.5 h period. These trails ranged in altitude from 89.6 km to 102.5 km and all were attributed to Leonids. We suggest that this identification may not be certain for two reasons:

(*a*) Considering the very low elevation of the Leonid radiant during the early part of their observations, it seems improbable that Leonid trails near 90 km altitude and even below could have been observed during this period. In addition, the Taurid meteors contribute significantly to trails seen during the early part of a Leonid shower night. Plate 3c (see colour plate section) shows an example of a bright Taurid during a Leonid shower, recorded by our a meteor camera and lidar. For these reasons it seems advisable

TABLE 7.5. Leonid meteors, observed by lidar

Year	Day & month	Time UT	Altitude (km)	Element	Location[a] (°N)/(°E)	Radiant ALT (°)	Reference
1996	17 Nov.	03:38	95.0	K	54.1 / 11.8	50.5	v. Zahn *et al.* (1999)
1996	17 Nov.	03:52	91.0	K	54.1 / 11.8	52.2	v. Zahn *et al.* (1999)
1996	17 Nov.	04:04	96.6	K	54.1 / 11.8	53.3	v. Zahn *et al.* (1999)
1998	16 Nov.	22:55	97.6	Ca & Fe	54.1 / 11.8	11.2	v. Zahn *et al.* (1999)
1998	16 Nov.	23:07	97.8	Ca & Fe	54.1 / 11.8	12.8	v. Zahn *et al.* (1999)
1998	17 Nov.	00:16	103.6	Ca & Fe	54.1 / 11.8	22.7	v. Zahn *et al.* (1999)
1999	11 Nov.	22:49	105.4	Ca, Ca$^+$ & K	54.1 / 11.8	12	This work
2000	17 Nov.	00:24	106.0	Ca	54.1 / 11.8	24	This work
2000	17 Nov.	04:06	98.2	Ca	54.1 / 11.8	53	This work
2000	17 Nov.	04:36	107.8	Na	54.1 / 11.8	55	This work
2000	17 Nov.	04:47	98.5	Ca	54.1 / 11.8	56	This work
2000	17 Nov.	05:21	98.0	Ca	54.1 / 11.8	57	This work
2000	17 Nov.	05:47	94.2	K	28.3 / −16.5	66	This work
2000	18 Nov.	03:03	99.2	K	28.3 / −16.5	31	This work
2000	18 Nov.	06:09	93.8	K	28.3 / −16.5	71	This work
1998	17 Nov.	17:07	101.1	Fe	25.5 / 127.7	24	Chu *et al.* (2000a)
1998	17 Nov.	17:17	93.6	Fe	24.9 / 128.2	26	Chu *et al.* (2000a)
1998	17 Nov.	17:25	90.1	Fe	24.5 / 127.8	28	Chu *et al.* (2000a)
1998	17 Nov.	18:48	94.8	Fe	26.0 / 126.8	46	Chu *et al.* (2000a)
1998	17 Nov.	19:51	102.5	Fe	23.8 / 126.0	59	Chu *et al.* (2000a)
1998	17 Nov.	20:11	100.7	Fe	24.2 / 125.0	63	Chu *et al.* (2000a)
1998	17 Nov.	20:56	96.2	Fe	26.0 / 127.3	75	Chu *et al.* (2000a)
1998	17 Nov.	21:00	95.4	Fe	25.7 / 127.7	77	Chu *et al.* (2000a)
		mean:	98.1				

[a] Locations for Fe trails taken from Jenniskens & Butow (1999).

to delete the first three trails enumerated by Chu *et al.* (2000a) from their list of Leonid trails.

(*b*) Trails with numbers 4 and 5, 9 through 11, and 14 through 18 each occur as a group in rather short time periods. This is most dramatically demonstrated by the five trails 14 through 18. They all occur within 2 min and 17 s and within the very restricted altitude range of 93.3 to 95.4 km. Our interpretation is that each "group of trails" was caused by just one meteor trail which was blown at different altitudes in various directions. Perhaps a good way to use the data is to retain within each group the highest trail as the most original Leonid trail and to discount later closely timed recordings. Overall, these procedures reduce the number of Leonid trails in the Chu *et al.* (2000a) dataset to eight.

The Leonids 2000 were observed by the IAP lidars at both Kühlungsborn and at Tenerife during the nights of 16/17 and 17/18 November (see table 7.5). During the night 16/17 Nov., observation conditions at Kühlungsborn were excellent between 22 UT till dawn of the next morning (05:50 UT). Throughout this period, we operated a Ca and Na lidar which detected 14 and six meteor trails, respectively. Out of those 20 trails, we designated five to be Leonid trails. Among them is the highest trail ever observed by lidar, a Na trail at 107.8 km. During the same observation period, we operated a zenith-looking video camera. It recorded 62 meteors, of which 10 were automatically identified as Leonids and four as Taurids. The K lidar of the IAP was located at the Observatorio

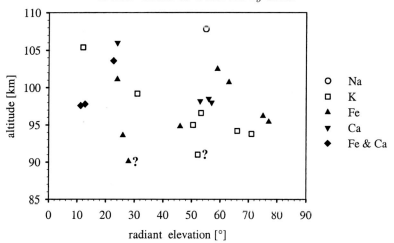

FIGURE 7.6. Altitudes of Leonid meteor trails versus radiant altitude, as observed by ground-based lidars and listed in table 7.4. The two question marks indicate our doubt whether these "low" altitude trails were indeed produced by Leonids.

del Teide of the Instituto de Astrofísica de Canarias. During the period of Leonid shower activity, it was operated from 20:41 UT of November 16 until 12:00 UT November 17 and from 20:24 UT November 17 until 12:00 UT November 18. The Leonid radiant rose about 00:35 UT. The lidar recorded 1 and 2 Leonid trails during the morning hours of 17 and of 18 November, respectively (for details, see table 7.5). We comment on this low rate of detection later in this section.

The altitudes of all 23 lidar-observed Leonids range from 90.1 to 107.8 km and their mean altitude is 98.1 km. This result means that even the highest of the lidar-observed trails is definitely below the mean altitude of maximum brightness of Leonids as measured by video cameras [de Lignie *et al.* (2000); Campbell *et al.* (2000)]. This contributes to the fact that in a given area near zenith the human eye will see more Leonids than a lidar can detect. In figure 7.6, we have collected the various lidar-observed Leonid trails in a plot of trail altitude vs. radiant altitude. Two of the trails are still marked by question marks because we consider their altitude "uncomfortably" low for a Leonid trail. Overall, there is only a slight, if any trend of trail altitudes vs. radiant altitudes.

We evaluate the rate at which Leonid trails are detected by ground-based lidars based on the periods of unobstructed lidar observations during nights of enhanced activity of the Leonid shower. We identify a trail as "Leonid trail" mostly by considering the combination of trail altitude and the elevation of the Leonid radiant. The results are summarized in table 7.6. The mean rate of Leonid trail observations by lidars is slightly more than one trail per hour. This is not even a factor two higher than our rate of trail detections at Kühlungsborn averaged over all sporadics and shower meteors over four years. A factor which certainly contributes to a general lowering of the Leonid detection rate is the fact that many of the Leonid meteors ablate above an altitude at which the lidars can detect these trails.

During the Leonids 2000 campaign at Tenerife, the K lidar was operated continuously throughout the two important nights of 16/17 and 17/18 November. The low number of lidar-observed meteor trails (see table 7.6) is in contrast to a considerable rate of visually-observed Leonids sighted by the lidar operators. This scenario might have been caused by the 2000 Leonid shower having a considerably higher population index (= fewer large-sized particles) than the 1998 Leonid shower [Arlt & Guyssens (2000); Arlt (1998)]. In

TABLE 7.6. Rates at which Leonid trails have been observed by lidars during periods of enhanced shower activity

Year of Leonid shower	Duration of observations (h)	Number of trails	Rate of trail detections (h^{-1})	Reference
1996	3.4	3	0.9	Höffner *et al.* (1999)
1998	3.0	3	1.0	von Zahn *et al.* (1999)
1998	~ 6	8	1.3	Chu *et al.* (2000a)
2000 @ Kühlungsborn	~ 7	5	0.6	This Work

consequence, the average particles of the 2000 Leonids ablated at higher altitudes than those of the 1998 Leonids, putting many more of them above the maximum altitude of lidar detectability.

In summary, the altitude distribution of Leonid meteor trails as we have derived it here from lidar observations is very different from that based on photographically or electro-optically measured Leonid trail heights. We take the latter distribution as being much closer to the "intrinsic" distribution than the lidar-derived one. This difference allows two important conclusions:

(*a*) The lidar-observed Leonid trails provide us information exclusively on a late phase of ablation of the trail-producing Leonid meteoroids,

(*b*) More generally speaking: Lidar-derived altitude distributions of meteor trails cannot be taken to represent truthfully the intrinsic meteor altitude distributions.

7.3.5. *Modeled Lidar Sensitivity vs. Altitude*

We have developed a relatively sophisticated model of the response, or sensitivity, of the lidar observations to trails deposited by particular meteors at particular altitudes. The probability of detecting a trail from a meteor belonging to a particular meteor shower depends on several factors, including (1) the velocity of the meteor and the mass distribution of the shower, (2) the relative volatility of the element under study, (3) the rate of diffusion which dilutes the number density of the trail at higher altitudes, (4) the rates of reactions that remove free metal atoms from the trail at lower altitudes, and (5) the horizontal winds which blow the trail through the lidar beam. The end result of the model is the probability of detection of a trail in a particular element from a particular type of meteor stream at a particular altitude. By keeping the normalization of this probability constant, we are able to compare detection probabilities from various streams and elements. Obviously, there are a great number of unknown and imprecisely known parameters involved in this modeling. Even so, the predictions are valuable in the understanding of several aspects of the measurements.

The first step in the modeling is the calculation of the trail line density for a meteor of chosen mass, or size, as a function of altitude. The deposition rate of metal along the trail is computed following the equations presented in Love & Brownlee (1991) but with allowance for varying volatilities of the various elements. Here we concern ourselves mainly with calcium and potassium, although the model has been extended to iron as well [von Zahn *et al.* (1999)]. The vapor pressure relation employed for calcium is the same as that used in the calculation of the differential ablation of calcium and sodium by McNeil *et al.* (1998). This treatment gave good results for the nearly 100-fold depletion of calcium relative to sodium in the atmosphere, when compared with the nearly equal clemental abundance in CI chondritic material. Differential ablation is discussed in more depth in chapter 11. For potassium, we have assumed a vapor pressure profile comparable to sodium, but lower by one-half an order of magnitude. There is good evidence [Eska *et al.* (1999)] that potassium is depleted relative to sodium in the atmosphere, again compared with CI abundances, by up to a factor of 10, but not nearly to the extent that calcium is depleted. There is substantial uncertainty in the volatility of both of these assumptions, but in any case, these two choices for elemental volatility allow us to examine the effect of differential ablation on the predictions of the model.

The line density of atoms of each species is computed at each altitude from the ablation rate and the instantaneous meteor velocity and this is done for a variety of masses spanning the effective detection threshold of the lidar to very large particles. At each meteor mass division and at each altitude, the trails are allowed to evolve in time according to altitude dependent molecular and eddy diffusion, assuming an initial trail radius of one mean free path. With time, the free neutral atom density is also reduced by an altitude dependent chemical–kinetic removal rate which represents the rate at which neutral metal atoms are converted into complexes at lower altitudes and thereby lost from the system. Details of the model treatment of these processes can be found in Höffner *et al.* (1999). The total number of free atoms within the $50 \times 50 \times 200$ m lidar volume is then computed at each time step until this number falls below the detection threshold (see table 7.4). Doing this as a function of time gives an effective lifetime of the trail, again at each initial meteor mass and altitude, during which the trail could be observed *if* it passed through the lidar beam.

Once the lifetime of each trail at each altitude has been computed, this number is multiplied by the altitude dependent horizontal wind speed at each altitude to account for the increased or decreased probability, as a function of wind speed, of a trail being blown through the lidar beam and hence observed. This gives a relative probability of observation as a function of altitude and meteor size. To compute a total probability of observing a meteor from a particular shower, the mass distribution of the shower must then be folded into the calculation. Details on how these mass distributions are modeled are given in McNeil *et al.* (2001). The critical factor in determining the effect of a shower population on the resulting probability is the slope of the mass distribution, although variation of this slope tends only to scale the probability and move it only slightly in altitude. The end result, after the mass distribution is folded in, is a relative probability as a function of altitude of observing a trail in a particular element from a particular source shower.

The result of this calculation depend most strongly on two factors: the volatility assumed for the element and the velocity of the incoming meteor. Both these factors dramatically change the altitude at which the metal deposition takes place. For purposes of illustration, we have shown in figure 7.7 altitude profiles of the trail detection probability for Ca and K atoms, under the assumed volatility rules outlined above, and for several different velocities. These curves were calculated assuming a 60° zenith angle for the meteor and assuming the mass distribution with a population index of 3.

What is perhaps most striking in comparing the results for the two different elements is the fact that we are *most* likely to observe a Ca trail from a meteor arriving at 45 km/s, a relatively fast meteor, while we are *least* likely to observe a Ca trail from a meteor at 15 km/s, a slow meteor. For K trails, on the other hand, the situation is basically the reverse. We are most likely to observe a trail from a slow meteor, while trails from fast meteors would be observed with a lower probability, and also at a higher altitude. This difference, which is entirely attributable to the differential ablation of the two species,

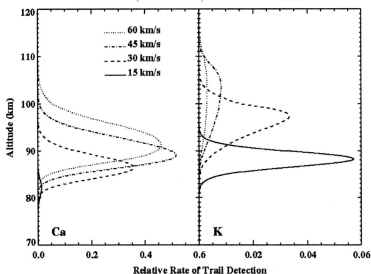

FIGURE 7.7. The computed probability of detecting a trail from a meteor shower with population index 3, as a function of altitude for a Ca trail (left panel) and a K trail (right panel) at various meteor velocities.

may explain why Ca and K trails are very rarely observed together. The most probable trails for observation simply arise from different meteors at different velocities.

Also of interest in figure 7.7 is the fact that the relative detection rate of the most probable Ca meteor is some ten times that of the most probable K meteor (note the different scales on the x-axis of the figure). This is in part due to the fact that Ca is some 20-times more abundant than is K in CI chondritic material [Mason (1971)]. At the same time, the detection threshold of the lidar for Ca is only a factor of four higher than that for K (table 7.4) and Ca trails are observed only about twice as often as K trails (table 7.3). In view of the model results, this would seem to be a result of the velocity distribution of incoming meteors which, if we assume it is skewed toward slower meteors, just by chance results in a nearly equal number of K and Ca trails being observed.

In figure 7.8, we present our model results concerning the dependence of the rate of Ca trail detection of a 30 km/s meteor shower on the entry angle from zenith (left panel) and on the mass of individual meteoroids (right panel). The lack of pronounced dependencies on both entry angle and on meteoroid mass is a clear indication of how strongly the detection rate is determined by the "instrument function" of the lidar and not by the unbiased occurrence profile of meteor trails.

Obviously, the volatility of the respective elements and the nature of their host phases are critical to accurate modeling of the altitude of trail deposition. For those we have, at present, educated guesses at best. Even so, the separation of trails in Ca from trails in K, both predicted and observed, suggests that the differential ablation process does indeed take place for meteors of substantial size. Further discussion of the evidence for differential ablation obtained from the lidar trail results can be found in von Zahn *et al.* (1999).

7.3.6. *Estimation of the Mass of Meteoroids which Produce Lidar-Observable Trails*

In the following, we derive an estimate for the mass of the lightest meteoroids, the trails of which can still be observed by lidars. Here we ask for the *order of magnitude* of a limiting meteoroid mass (not anything like 10% accuracy). To this end, we compare the

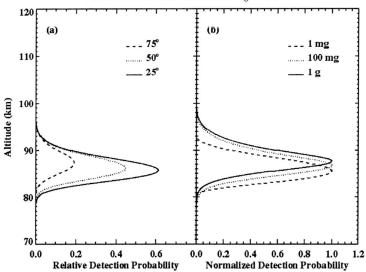

FIGURE 7.8. The computed probability of detecting a Ca trail from a meteor shower with a velocity of 30 km/s and a population index 3 as a function of altitude and entry angle from zenith (left panel) and that on the mass of individual meteoroids (right panel).

rates at which meteors are detected by two independent methods: (a) visual observations of meteors and by video cameras and (b) through detection of their trails by lidars. Methods (a) not only provide total meteor rates (such as the zenithal hourly rate: ZHR), but additional information such as the spectrum of luminosity of the meteors and derived photographic masses for each individual meteor. Availability of the latter parameters allow us to assign a minimum mass to those meteors, the trails of which are just barely observable by lidars.

We define a lidar zenithal hourly rate $LZHR = LHR \times 1/\sin(h_r)$ with LHR being the hourly rate of lidar-observed meteor trails and h_r being the mean altitude of the Leonid radiant during the period of observation. We take the parameters from two specific observations during Leonid showers:

Observation no. 1: Nov. 17, 1998; 19:30–21:30 UT; $h_r = 68.5°$.
mean LHR = 2 [4 Leonids observed by Chu *et al.* (2000a)]
mean LZHR = 2.2
mean ZHR = 127 (Arlt, 1998)
Observation no. 2: Nov. 17, 2000; 04:00–05:30 UT; $h_r = 55.3°$.
mean LHR = 2.7 (4 Leonids of table 7.5, this work)
mean LZHR = 3.2
mean ZHR = 47 [Arlt & Guyssens (2000)]

First we determine the solid angles in the sky, over which the two methods collect their events. For method (a) the the ZHR refers by definition to the full half-sphere $= 2\pi$ sterad. For method (b) that solid angle can only be estimated. It is determined by the speed of the wind which transports the trails from the volume of its formation into the laser beam and by the period of detectability of such trails for the lidar. For the mean wind speeds at these altitudes we have chosen 20 m/s [Fleming *et al.* (1990)]. The ages of lidar-observed meteor trails have been derived by Chu *et al.* (2000a) from their precisely measured trail width. For their work an appropriate value for the period of detectability is 1000 s. These numbers yield for the lidar method a solid angle of detectability at the sky of 0.15 sterad. This is 2.5% of the half-sphere. However, at any one altitude, half of the meteor trails will be statistically in the upwind and half of the trails in the downwind direction of the laser beam. The sensitive sky area is therefore only half the above value, or 1.25% of the half-sphere. Comparing the above quoted values for ZHR and LZHR we get

for observation no. 1: LZHR / ZHR = 1.7% or 1.4 times more than 1.25%.

for observation no. 2: LZHR / ZHR = 6.8% or 5 times more than 1.25%.

From this comparison we deduce that method (b), observation by lidar, can detect slightly weaker meteors than those detected by methods (a). As the ZHR values are given after normalization to a limiting brightness of +6.5 m, we conclude that the lidars can pick out Leonid trails formed by meteors as faint as about +7 m.

Campbell *et al.* (2000) and Brown *et al.* (2000) give conversions of maximum luminosity of Leonid meteors into photometric mass. For a Leonid of +7 m, both figure 12 of Campbell *et al.* (2000) and equation (4) of Brown *et al.* (2000) yield about 10 µg for its photometric mass. We consider this value a reasonable estimate for the limiting mass of lidar-observed Leonids. Furthermore, we obtain for lidar observations the limiting mass of typical sporadics to be 0.7 mg if we apply the Jenniskens *et al.* (1999) dependence of limiting mass on meteoroid velocity.

We believe that this estimation yields the limiting mass with an uncertainty of about one order of magnitude. This we consider a "small" uncertainty if compared with the total mass range of objects impacting the Earth's atmosphere, which is given in chapter 4 to be more than 25 orders of magnitude.

The mass of one of the brightest Leonids observed by the IAP lidars was modeled by von Zahn *et al.* (1999), who arrive at an approximate mass of 10 g for this Leonid. It is for these reasons that we give the mass range of meteoroids, of which the lidars can observe the meteor trails, to fall between 10 µg and a few grams. The upper "limit" is a practical limit and given by the rapidly diminishing probability for the occurrence of ever brighter fireballs.

7.4. Two- and Three-Element Trails: Observations and Modeling

Even though resonant lidars have demonstrated remarkable capabilities for detection of meteor trails and the measurement of absolute metal densities in these trails, studies based on one single-element lidar have a limited potential for extracting new knowledge about the properties of meteoroids and/or their ablation processes. This situation is much improved by employing two or even more lidars for different metals which observe simultaneously the same volume at meteor heights. Then, the *abundance ratios of different metals* in meteor *trails* can be measured. This type of information produces new and interesting knowledge (e.g. about the complex ablation processes of the meteoroids) that is independent from other methods. In plate 7.2 we show an outstanding example of this type of observation, which is the trail of a fireball observed 27 October 1997 above Kühlungsborn. The upper three panels show the profiles of photon counts recorded per 200 m and laser pulse for K, Fe, and Ca, respectively, at the time of maximum atom density in the trail (with 4 s integration time). The metal densities in this trail were so high that the trail was also quite outstanding in our count profiles with 3 min integration period. At the core of this trail, the number density ratios Ca/Fe and K/Fe were 1.6×10^{-2} and 1.3×10^{-3}, respectively. The lower left panel of plate 7.2 shows the temporal development of the photon counts as recorded by the K lidar in the altitude channel of maximum K density (in red) and that in channels 200 m higher and lower. The lower right panel shows the development of the same trail as recorded by the K lidar at an altitude resolution of 7.5 m and time resolution of 10 s. The trail proper

TABLE 7.7. Trail statistics for nights with two simultaneously observing metal lidars of the IAP

Element combination	Number of nights	Number of two-element trails	2-Element trails per night	Mean altitude (km)
Fe & K	18	7	0.39	93.4
Fe & Ca	27	9	0.33	93.2
Ca & Ca$^+$	74	11	0.15	93.2
Ca & K	75	9	0.12	92.9
Ca & Na	27	4	0.15	91.3
K & Na	11	1	0.09	(88.9)
K & Ca$^+$	41	1	0.02	(94.5)
Sum or mean	273	42	0.15	92.9

passed through the laser beam in 50 s. It was connected with a much more extended, but much lower density layer of debris (= train) which was noticeable for additional 7, 5, and 5 min for the Ca, K, and Fe lidars, respectively. Here we emphasize that all our metal lidars have a fixed viewing direction towards the zenith. It is the first time that such data about metal density ratios in the non-luminous, cool part of meteor trails become available in considerable numbers and we will present those in the followings sections.

7.4.1. *Two-Element Trails: Frequency of Occurrence and Mean Altitudes*

Since October 1996, we have operated two metal lidars (with various combinations of metals) during 273 nights. This number includes the nights that we measured with three lidars. Table 7.7 gives in the first column the element combinations which were measured simultaneously and in the fourth column the rate of detection of two-element trails. The average rate is 0.15 trails per night. Here we remind the reader of our rate of trail detections by single lidars, which is 3.3 trails per night (see table 7.3). If the same detection rate would apply for two-element trails, the 273 nights of two-lidar measurements should have given us about 900 trails with two observed elements. In reality, however, we observed during these nights only 42 meteor trails in which both species of metals could be detected and be quantitatively evaluated. This rate is merely 4% of the single-element trail detection rate and indicates that it takes about 6 nights of operations with two lidars to record one two-element trail. It appears significant that the trails involving Fe (combinations nos. 1 and 2) have about three times the occurrence rate of combinations involving Ca (nos. 3–5). This can not be due to an enhanced sensitivity for Fe trails relative to that for Ca trails as consultation of the last column of table 7.3 easily reveals.

It is an open question as to what part of this substantial difference in detection rates for single- and two-element trails might be due to a limited dynamic range of the observable atom densities in the meteor trails for either element involved. It is difficult to give a quantitative answer as this would require performing many hundreds of case studies of the actual detection threshold of the "missing" element and relating them to a somehow predicted atom density for this element. Such an analysis was performed by von Zahn *et al.* (1999) for five strong K-trails observed in two nights of November 1998 which all were missing entirely a Ca signature. They calculated a predicted Ca density in the K-trails for the assumption that the two elements would be present in a CI chondritic

TABLE 7.8. Trail statistics for nights with three simultaneously observing metal lidars of the IAP

Element combination	Number of nights	Number of 3-element trails
K, Ca & Fe	17	5
K, Ca & Ca$^+$	50	1
K, Ca & Na	10	—
K, Fe & Ca$^+$	1	—
Sum	78	6

abundance ratio Ca/K = 17.1. They showed that the Ca detection threshold of the lidar for Ca was between eight and 71 times lower than the predicted Ca trail density. Thus, the observation of the "K-only" trails implied highly significant deviations of the Ca/K ratio in the trails from CI composition. It is undeniable that the detection thresholds may play a role in these arguments. But examination of the data strongly suggests that another process is also shaping the results of the two-lidar observations; this, we suggest, is the differential ablation of the meteoroids.

The mean altitudes of the families of two-element trails scatter surprisingly little about the weighted mean of 92.9 km. One can not overlook, though, the small statistics for some of the trail families which certainly allows for some chance results. Nevertheless, even the strongest statistical sample, the {Ca + Ca$^+$} trails, have a mean altitude which is only 3.3 km higher than that of single Ca trails. We expect that strong Ca$^+$ trails are formed preferentially by fast meteoroids which in turn ablate at higher altitudes than slow meteoroids. Therefore we anticipated the {Ca + Ca$^+$} trails to occur on average considerably higher than the single Ca trails. The fact that our observations show this difference to be only rather moderate may again indicate that the observed altitude distributions are strongly driven by the common instrument function of the lidars, as discussed in section 7.3.5.

7.4.2. Three-Element Trails

After realizing the low rate of detection of two-element trails, we wondered what the probability for detection of three-element trails might be. We have had 78 nights with three metal lidars observing simultaneously a common zenith-directed column of the upper atmosphere. In table 7.8 we have listed the various element combinations which we studied during these nights. We recorded only six three-element trails or, in other words, 0.08 three-element-trails per night. The drop in detection rate is certainly less dramatic in going from two- to three-element trails than going from single- to two-element trails. Yet, the statistics on three-element trails becomes so low that one wonders whether it still is worth the effort which goes into obtaining these data.

As there are comparatively few three-element trails, it is possible to list in table 7.9 the dates, altitudes, and possible origins of these trails. These trails are formed almost solely by fast meteoroids such as the Leonids and Perseids. Independent from the dates of the six three-element trails, the average altitudes point towards a higher entry speed of those meteoroids producing three-element trails than of those producing only single-element trails. The average altitude of the three-element trails is 94.4 km, whereas that of all the single-element trails is 89.9 km.

TABLE 7.9. Dates and altitudes of the three-element trails

Element combination	Date	Altitude (km)	Suggested origin
K, Ca & Fe	May 13, 1997	89.2	sporadic
K, Ca & Fe	Aug. 05, 1997	91.5	Perseid
K, Ca & Fe	Oct. 27, 1997	89.8	Orionid fire ball
K, Ca & Fe	Nov. 16, 1998	97.8	Leonid
K, Ca & Fe	Nov. 17, 1998	103.5	Leonid
K, Ca & Ca^+	Nov. 03, 1997	94.5	Leonid

There is also a tendency for the two-element trails to occur during showers of fast meteoroids. One is attempted to speculate, therefore, that slow meteoroids preferentially produce single-element trails. We have no means to prove this point on a case-by-case basis (lacking camera information to quantify the speed of the meteoroids for many events). We suggest, though, that a clustering of single-element K trails at altitudes below 88 km is a result of this property of meteoroid ablation. This low-altitude cluster of trails, having densities < 150 cm^{-3}, can be recognized in figure 7.4. We add that a K meteor trail with a density of 100 cm^{-3} is a very prominent feature and much stronger than the regular K layer.

7.4.3. *Ratios of Metal Abundances*

In this section we present and discuss the ratios of the number densities $n(X)$, where X can stand for K, Fe, Ca, and Ca^+, and where densities were simultaneously measured in two-element trails by our lidars. These ratios spread out over many orders of magnitude and are thus difficult to comprehend and compare. This wide "dynamic range" can be restricted considerably by normalizing all the ratios to that in CI carbonaceous chondrites (see our remark in section 7.1). In figure 7.9, the centerpiece of this paper, we present measured abundance ratios from our observed two- and three-element trails for K/Ca, K/Fe, Fe/Ca and Ca $^+$/Ca. The former three are normalized by dividing the measured ratios by those in CI chondrites (table 7.1). The normalized ratios are plotted against the number density of one of the constituents, where we take the latter as a proxy for the size of the entering meteoroid. Two features are immediately apparent: (a) The means of the three normalized elemental ratios are significantly different from 1 (= the CI composition); (b) All four scatter plots show a distinct trend of the data coming the closer to unity the higher the absolute density is. We will discuss these two features below:

(*a*) The fact that our observed ratios all come out differently from CI composition can be interpreted in two different ways: Either we suggest that our lidar-measured trail composition indeed reflects that of the meteoroids. Then the composition of the meteoroids must be quite dissimilar to that of CI chondrites. Or we suggest that the meteoroids have a composition like or at least close to that of CI meteorites. Then the differences of composition between meteoroids and trails would be caused by specific processes acting during the trail formation. We prefer to follow the second argument guided by two considerations. Firstly and as pointed out in section 7.1, chondrites contribute by far the largest share to the total influx of meteorites. We consider it unlikely that the population of meteroids, which are sampled by our lidars, has a mean composition *drastically* different from the population of meteorites. Secondly, as pointed out above, two- and

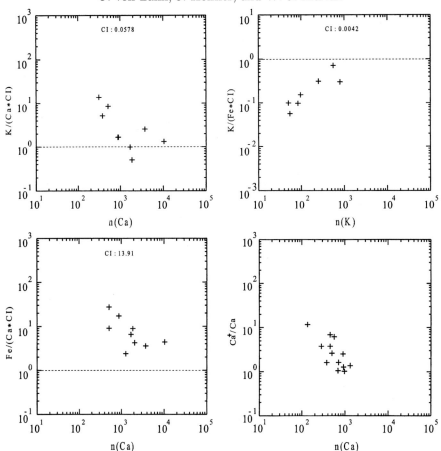

FIGURE 7.9. Ratios of the number densities of two elements (or ion and atom) measured in two-constituent meteor rails versus density of one of the constituents. The ratios are normalized to the ratios in CI chondritic meteorites. At the dashed lines, the measured abundance ratios would equal the CI ratios. The lower right panel shows the ratio of ionized over neutral Ca as measured in {Ca$^+$ plus Ca} trails.

three-element trails are formed preferentially by fast meteoroids. Those are more likely to be of cometary (and not of asteroidal) origin and hence their composition closer to CI carbonaceous chondrites than to any other.

If we average the density ratios obtained in the different types of multi-element trails, we derive the mean ratios (and their full range of measured values) as given in table 7.10. An immediate conclusion drawn from the numbers in table 7.10 is that in the meteor trails Ca is much underabundant with respect to Fe and K. In addition, K is underabundant with respect to Fe. Note that the three ratios of the three constituents are derived from independent sets of trail observations and do not constitute an overdetermined set of values. Causes contributing to the underabundance of Ca could be a lack of complete vaporization and a vaporization at too low altitudes of Ca during the ablation process. The lidar-observed underabundance of Ca in meteor trails goes a long way towards explaining the weakness of the upper atmosphere regular Ca layer in comparison with the regular Fe layer.

(b) The strong trends exhibited by all four scatter plots of figure 7.9 need consideration too. We take them as an indication of differential ablation of the meteoroids. Complete

TABLE 7.10. Means (and the ranges) of metal abundance ratios in meteor trails as measured by IAP ground-based lidars

Element ratio	Number of meteor trails	Mean of lidar-observed ratios	Ratio in CI meteorites	Ratio of lidar mean over CI
K/Ca	9	0.23 (0.03 ... 0.79)	0.0578	4.0
K/Fe	7	0.001 1 (0.000 23 ... 0.003)	0.0042	0.25
Fe/Ca	9	129 (33 ... 380)	13.9	9.3
Ca^+/Ca	11	3.9 (1.0 ... 12)	—	—

differential ablation is the process by which the elements are vaporized from the meteoroid sequentially with increasing temperature of the meteoroid body. The process can be important only over volumes in which the time constants of thermal heat conduction and material diffusion are small in comparison with that of the temperature increase at the volume's surface. This allows differential ablation to be important only for small sized meteoroids. But what is "small" in this connection? We interpret the trends in the scatter plots as indicating that differential ablation becomes stronger and stronger the smaller the meteoroid is. As mentioned before, we take here the measured density as proxy for the size of the meteoroid. In all three cases of ratios of neutrals, the deviations from the CI composition (that we would interpret as representing homogeneous ablation) become stronger the smaller the size of the meteoroid is. The other two clues for the prevalence of differential ablation of lidar-observed meteoroids come from the predominance of single-element trail observations over two-element trail observations and from the fact that we observe hardly any meteoroid with CI composition.

Other gross features of the scatter plots in figure 7.9 are noteworthy. The trends in the K/Ca and Fe/Ca ratios could mean that for the smaller meteoroids the ablation temperature does not become high enough to vaporize Ca fully. This effect works, at least qualitatively, in the expected direction (producing an underabundance of Ca). If we carry this argument over to the measured K/Fe ratios, it would imply, though, that K is the element of the two which ablates (unexpectedly) at the *higher* temperature or for any other reason less completely than Fe. Does it indicate the possibility of an eutectic that depresses the activity of K in the condensed phase, leading to a lower K vapor pressure?

At the end of this section we reiterate that we observed many Ca trails without any detectable concurrent Ca^+ trail. Therefore, the mean ratio Ca^+/Ca = 4 as derived from {Ca^+ plus Ca}-trails, must not be implied for all Ca trails which we have observed.

7.4.4. Comparison with the Literature

A comparison of our lidar-measured metal abundance ratios with other lidar results is not possible as we do not know of any other lidar results. We can only mention the report of Kane & Gardner (1993a) on the observation of a single {Na plus Fe}-trail. The absolute atom densities in this trail remain, however, quite uncertain due to rather long and different integration times applied in the two lidars. On the topic of two-element trails, the authors comment: "Fe meteor trails have been observed at several occasions at Urbana when Na meteor trails were not". While interesting within the context of the previous discussion, the statement does not lend itself to a quantitative comparison with our data.

The optical phenomenon of meteors has been studied in sophisticated spectroscopic experiments, which subject is treated at some length in the review on meteor phenomena by Ceplecha et al. (1998) and in chapter 5. In these experiments one investigates, however, the hot plasma conditions very close to the ablating meteoroid. In addition, the sensitivity of the spectrographic instruments generally limits the observations to fireball-type meteors at altitudes below 70 km. The results of Borovička et al. (1999) obtained by video spectroscopy of Leonid meteors could be better suited for purposes of comparison with our lidar observations. They selected for their study 20 Leonids which they observed on November 17, 1998, and having apparent maximum visual magnitudes between +2.4 and −3.6. Borovička et al. (1999) observe Na to evaporate earlier than other metals. Yet, they give the derived ratios of metal abundances only after having the line intensities integrated along the entire meteor trajectory. By this analysis method, they undoubtedly smooth over the potentially present effects of differential ablation which makes a quantitative comparison with our lidar results impractical.

7.5. Differential Ablation: Review and Consequences

Differential ablation of the meteoric material as it passes through the atmosphere is a process by which different elements are released from the meteor at different points along the meteor's trajectory. It is suggested that the point at which any given element is released is related to its volatility relative to the remainder of the meteoric material at the time of release. The more volatile elements, such as Na, are hypothesized to be released early in meteor flight, while the less volatile elements, such as Fe and Ca, are released later or perhaps retained to some extent in the meteor, never being released at all, especially in slow meteors.

We have seen in the modeling section that an initial assumption as to the relative volatilities of K and Ca leads to a very different prediction for the sensitivity of the lidar to trails in one or the other of these elements. For example, figure 7.7 shows that the K released from a 15 km/s meteor would be detected quite strongly at around 88 km. The same meteor would produce only a very weak Ca trail, if it produced one at all, near 82 km. For a 30 km/s meteor and those at higher velocity, there is predicted a large altitude separation in the detection probabilities of the meteor in K and in Ca. These differences are the direct result of K being released during the meteor flight much sooner than Ca. Of course, since the lidar detects only a short portion of the trail, as it is blown through the laser beam, this altitude separation leads directly to the result that trails are generally detected in one element or another, but not simultaneously in two or three elements. This finding, that more than 95% of the trails detected contain only one element, is therefore strong evidence that differential ablation is taking place in these meteors.

A second feature of these data that argues for differential ablation is the disagreement between the relative composition of those trails detected in two or more elements and the expected ratios based on CI composition. If the meteors ablated homogeneously, one would of course expect the elemental ratio of the trails to be preserved. That this is clearly not the case, and that there are definite trends in the observed ratios away from CI composition, is evident from figure 7.7 and the accompanying discussion. Especially convincing is the trend toward the correct CI composition as the overall density of the trail increases. One would expect larger meteors to undergo differential ablation less completely since the process likely requires uniform melting of the meteor and transport of the element currently being vaporized to the surface. This process would be hindered in larger meteors.

There are additional, independent indications that differential ablation takes place. Heating experiments carried out on meteoric material have shown that the elements are released more or less sequentially according to volatility [Notsu *et al.* (1978)]. Detailed thermodynamic modeling of vaporization of chondritic magma [Fegley & Cameron(1987)] has led to similar theoretical results. Finally, models of differential ablation have been applied to the relative abundance of Na and Ca in the Earth's permanent metal layers with good results in explaining the depletion of Na relative to Ca [McNeil *et al.* (1998)]. Thus, it would seem that the differential ablation hypothesis is supported on several fronts.

The unexpected result of our observations is the fact that much larger sized meteoroids than hitherto considered undergo differential ablation. Differential ablation shows up in the unexpectedly smaller number of two-element trails than single-element trails and in the lidar-measured ratios of metal abundances in meteor trails. Differential ablation has been observed and is expected on theoretical grounds for micrometeorites with masses of, say, below 1 µg (or sizes smaller than 0.1 mm). We argued above, however, that the meteor trails recorded and studied by our lidars were produced by meteoroids with masses in the milligram-to-gram range. Our former expectation that lidar-observed meteoroids undergo homogeneous ablation processes was based on earlier modeling results which assume, among other constraints, the meteoroid to consist of a solid grain or piece of rock. The preponderance of differential ablation processes for meteoroids with masses in the gram range could indicate that during the atmospheric entry a much larger percentage than expected of all meteoroids disintegrate into many tiny particles. The average size of these secondary particles would be so small that they independently ablate differentially. Still, the secondary particles would all travel closely together and thus create a single meteor trail (at least as looked at with our current spatial resolution of a few meters).

We recognize that a possibility exists that this fragmentation process leads to the formation of secondary particles with distinctly different composition and hence ablation processes, as discussed in chapter 4. Our current lidar observations of meteor trails neither specifically support nor refute the existence of this process leading to a chemical differentiation of the secondary particles.

7.6. Meteoroids and the Regular Metal Layers

7.6.1. *Metal Abundances in Meteor Trails and the Regular Metal Layers*

An indirect way to test the (reasonable) assumption that meteoroids give rise to the metals found in the regular metal layers is to compare the density ratios of metals in the metal layers with those in meteoroids. To this end, we compare the annually averaged column densities of Na, K, Fe and Ca (table 7.11) with CI chondritic composition (table 7.1). In figure 7.10 we present a plot of the metal number densities in CI chondrites against the column densities in the regular metal layers (black circles), all in reference to Fe. Of the three layers, Na, K, and Ca, only the column abundance of the K layer scales with that of the Fe layer the same way as the ratio of these two elements in CI chondrites. In the metal layers, Na is significantly overabundant and Ca strongly depleted with respect to Fe. Figure 7.10 indicates clearly that the composition of the upper atmosphere metal layers deviates significantly from that of ordinary chondrites. This conclusion is, no doubt, different from that reached in chapter 9. It has been often been speculated

TABLE 7.11. Annual mean column densities of the regular metal layers at mid-latitudes

Element	Column density (cm^{-2})	Reference
Na	3.5×10^9	She *et al.* (2000)
Na	3.7×10^9	States & Gardner (1999)
K	4.4×10^7	Eska *et al.* (1998)
Fe	10.6×10^9	Kane & Gardner (1993b)
Ca	2.4×10^7	Gerding *et al.* (2000)

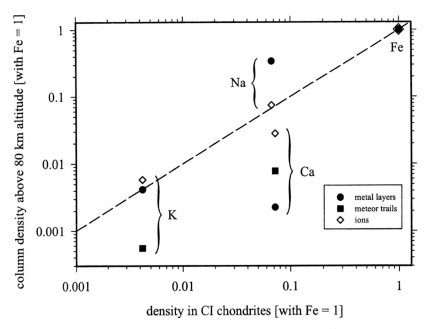

FIGURE 7.10. Comparison of metal atom abundances (normalized to that of iron) in CI chondrites, in the atmospheric metal layers (black circles), and of metal ions in the lower E region (diamonds). Also included are the ratios of the abundance measurements of Ca and K in our lidar-observed two-element trails (black squares).

that the comparatively large abundance of Na atoms in the upper atmosphere is at least in part be due to upward transport of NaCl from below, a process which was defended strongly by Chamberlain (1961). It is interesting to note that the mean column densities of the Na$^+$ and Ca$^+$ ions in the lower ionosphere (open diamonds) fall much closer to CI chondritic conditions than in case of Na and Ca, respectively [Kopp (1997)]. The ion abundances have been studied and interpreted with the help of numerical modeling by McNeil *et al.* (1998); the topic will not be pursued further here.

Figure 7.10 also shows the number density ratios of Ca and K over Fe in meteor trails (black squares) as derived from those lidar observations which allowed the simultaneous measurements of at least two metals. In these type of meteor trails, we find the Ca/Fe ratio to be about one order of magnitude lower than the CI ratio, but still not as low as in the metal layers. The same situation exists for the K/Fe ratio, which we find again to be one order of magnitude lower in the meteor trails than in CI chondrites. A possible rationalization for these findings could invoke our assumption that the majority of the two- and three-element trails are indeed formed by fast meteoroids. These would start ablating the volatiles Na and K at altitudes too high to be accessible for the lidars. Once the meteoroids come in the "lidar-region" below 100 km altitude, much of their volatiles would have been ablated already and the lidar-observed trails would show this as a deficit in K versus Fe. The same argument can not explain, though, the behavior of the refractory metal Ca. Here we invoke, based on our modeling results, (a) a rather incomplete transformation of the Ca contained in the meteoroid minerals into free Ca atoms in the trail and (b) a large percentage of the Ca ending up in the ionized state instead the neutral one (see figure 7.9).

The data on conditions in the meteor trails shown in figure 7.9 are burdened, however, with a definite weakness. So far, they entirely neglect the fact that we have observed numerous meteor trails with either Ca-only (and no detectable Fe) or K-only (and no detectable Fe). It is easy to speculate that the Ca-only trails represent the end phase of fast meteoroids, the K-only trails the beginning phase of slow meteoroids. But quantitative confirmation of such speculation requires a much higher quality of information on the speed of the meteoroids which formed the lidar-observed trails. If such information could be made available, the origin of the single-element trails would very likely become apparent which in turn would set more stringent boundary conditions for numerical models of meteoroid ablation. It is with this goal in mind that we have experimented in recent years with an image-intensified meteor cameras and have combined its operation with that of our lidars. In section 7.7 we will give a summary of our experiences in these experiments.

7.6.2. Reaction of the Regular Metal Layers to Meteor Showers

If we believe in the assumption that the atoms of the regular metal layers derive ultimately from the ablation of meteoroids, then it is natural to expect that variations in the influx of meteoroidal matter are "mirrored" somehow in the properties of the metal layers (e.g. their column density, peak number density, or altitude of layer peak). In fact, if such correlated variations in meteoroid influx and in metal layer properties could be observed, it would provide evidence for the correctness of our assumptions about the origin of the metal atoms. It is with this background that since right after the invention of the metal resonance lidar by Bowman *et al.* (1969), scientists have strived to observe and quantify such correlations. Three years later, Hake *et al.* (1972) made the claim that they observed the transit of the 1971 Geminid meteor stream as a fourfold increase of Na column density. In the following two decades, this paper was followed by many others in which it was claimed that the reaction of one of the metal layers to the passage of a meteor shower had been observed. As part of this review we will propound a skeptical examination of these reports.

We see three reasons which make the early claims unreliable:

(*a*) In the 1970s there was no and in the 1980s very little knowledge about the rather frequent occurrence of sporadic metal layers. These need to be distinguished from transient layers possibly produced by meteor showers or by a single fireball as reasoned by, for example, von Zahn & Hansen (1989). Sporadic layers have an altitude extent of typically 2 km, which is much more than the typical width of a meteor trail and much less than expected for a "layer" caused by a meteor stream. The latter phenomenon has been modeled just recently for Fe and Mg by McNeil *et al.* (2001). Sporadic layers can lead to significant increases in metal column density and these increases have frequently been observed by lidars and wrongly attributed to meteor streams.

(*b*) At least into the mid-1980s, there was little appreciation about the strength of variations in layer number densities and column densities caused by a rich spectrum of gravity and tidal waves passing nearly continuously through the metal layers. In order to prove that meteor showers are responsible for observed layer changes, there must be a verification of such claim by comparison with observations obtained nights before and/or after the shower. No such verification was ever done in the subject papers published before 1992.

(c) No capability existed or was developed to actually monitor the concomitant rate of meteors during the lidar observations, which would have helped greatly to substantiate the claims.

More specifically, our critique applies to the papers by Hake *et al.* (1972), Aruga *et al.* (1974), Mégie & Blamont (1977), Mégie *et al.* (1978), Clemesha *et al.* (1978), Clemeshal *et al.* (1980), Jegou *et al.* (1980), Uchiumi *et al.* (1985), and Mégie (1988). We comment in detail only on one study. Mégie & Blamont (1977) tried, in their chapter 8, to correlate observed anomalous increases of Na column densities with the occurrence of meteor showers. The authors claim "in words" and without showing any data to prove the point: "It is significant that these observations can be correlated one by one with permanent meteoritic showers." The only data which the authors show is the temporal variation of Na column density during two consecutive nights of June 1974 which the authors incorrectly assign to the Perseid shower (the latter in fact peaks on August 12). During the night of June 10/11, the column density increased before midnight by a factor of about three. But during the next night, the column density remained low and stable throughout the night. Our interpretation of these data is that the density increase observed during the night June 10/11 was just one of the many sporadic Na layers. Hence, in our opinion the paper does not present evidence for even a single case of a meteor shower interacting directly with a metal layer.

Matters changed for the better with the paper of Uchiumi *et al.* (1993) because these authors attempted to show, to our knowledge for the first time, repeatability for the observed phenomenon. They present Na lidar data collected during the Perseids of the years 1978, 1979, 1981, and 1983. The authors claim to see a number of effects related to the meteor showers, some of which we have to attribute to sporadic layering, however. Yet, we consider their approach of comparing data collected during a particular meteor shower in different years a real advance for the experiment.

Next we cite Gerding *et al.* (2000) with a remark from their comprehensive paper on the Ca and Ca$^+$ layers: "The overall behavior of the Ca and the Ca$^+$ layers looks as if it is partly driven by the occurrence of high-speed meteor showers. Meteoroids from these high-speed meteor showers should more effectively ablate Ca than slow meteoroids, because Ca is one of the most refractory metallic constituents." The authors do not, however, give a detailed reasoning for this suggestion.

Finally, we come to those papers on the topic which make use not only of lidar observations of metal layer parameters, but also of individual meteor trails. Höffner *et al.* (1999) evaluate the reaction of the K layer to the 1996 Leonid shower. The authors compare the observed nocturnal changes of K column density during the night of the Leonid shower with those observed during six non-shower nights. The authors conclude: "... moderate changes of K column density are frequently observed. The rise during the Leonid shower is, however, one of the strongest observed so far and occurs just at the time one would expect it to happen." The authors do not claim that the increase observed during the Leonids was significantly above the level of increases observed earlier on. Gerding *et al.* (1999) studied the reaction of the Ca and K layers to an intense meteor shower (of unknown radiant). The shower, observed for about 5 h, was identified solely through a significantly increased rate of lidar-observed Ca trails. During this burst of meteor trails, the column densities of the Ca and K layers increased by factors of four and two, respectively. These increases were far stronger than observed in nights with negligible meteor shower activity, similar to the result of Höffner *et al.* (1999). Because of the combined observations of meteor trails and the metal layer responses, the Höffner *et al.* (1999) and Gerding *et al.* (1999) studies may have given the best evidence yet for the assertion that, at least under certain circumstances, metal layer densities can increase due to meteor showers. But after employing the same observational techniques in their study of the 1998 Leonids, Chu *et al.* (2000a) conclude: "Observations suggest that the 1998 Leonid shower did not have a significant impact on the abundance of the background Fe layer." Similarly, Höffner *et al.* (2000) studied K meteor trails and the regular K layer during the 1999 Leonids and summarize in their abstract: "We can conclude that the 1999 Leonid meteor storm has not led to an outstanding enhancement of the upper atmosphere potassium layer."

In conclusion, any reaction of metal layers to a temporarily increased flux of meteoric matter has been very difficult to prove in the past, even with sophisticated experiments. The studies of Höffner *et al.* (1999) and Gerding *et al.* (1999) indicate that the K and Ca layers can react quickly to intense meteor showers. But how intense must a meteor shower be to cause recognizable changes in the metal layers? Evidently, we need much new data before we can identify and quantify the various conditions which make a metal layer change significantly in response to a meteor shower.

7.7. Experiments Including a Meteor Camera

Image-intensified video cameras can be used to observe the same area in the sky that the lidar is sampling. These cameras use the common video rate of 25 images/s and individual meteors appear at sequences of images which may be as long as 20 images. The timing accuracy on the individual images can be made quite high. The position accuracy can also be made high if one initially calibrates the image field against the background star field. The camera observations can provide the urgently needed information about the entry time, the angular velocity across the field of view and flight direction of that meteoroid which produces the meteor trail that is shortly thereafter sampled by the lidar. But there are also hurdles to overcome. Before the upper air winds blow the trail through the laser beam, some time passes, which can be as large as tens of minutes. If during this period additional meteors fly through the field of view of the camera or clouds obscure the vision of the lidar, the task of assigning a specific meteor to a specific lidar-observed trail can become difficult or even impossible to solve. Furthermore, we can expect that a certain percentage of the camera-registered meteors passed the station at altitudes above 105 km, where the lidar is unable to detect a meteor trail. For that reason, even meteors which are bright and pass the camera/lidar station close to the zenith may not produce any lidar-detectable trail. Only a true field experiment will tell us how easy or difficult the task of correlating camera and lidar observations in fact is. It is for that reason that the IAP acquired in late 1996 an image-intensified video camera and the software required to analyze its data. Here we report on some of our experiences gleaned from operating a lidar–camera combination.

What we have briefly called a "camera" actually consists of an image intensifier, transfer optics, a CCD video camera, and analysis software for real-time analysis of meteor properties (flight direction, angular speed, brightness, assignment to specific meteor showers). The camera is for detecting meteors with brightness down to +5 at a rate of $> 75\%$ in a field of view of $9.6° \times 12.9°$. The software used for realtime data processing and analysis is the MetRec package Molau, 2001. The camera looks vertical and is mounted approximately 150 m distant from the cluster of our lidars, which avoids overloading the camera with backscattered light from the laser.

Plate 7.3a–d give four examples of imaged meteors and the ensuing lidar-observed meteor trails. The meteor images are in fact the sum of a number of video frames (here between 4 and 10) which are summed up by the MetRec3 software as long as substantial changes in image content occur from frame to frame. The summed images (shown here) are displayed in real-time to the lidar operator, analyzed for direction, angular velocity, and brightness of the meteor, and stored on the hard disk of a PC. Each of the four rows shows

- In the left panel: The camera image with the time of the last summed video frame at the lower right side. Times are given on the images are in CET (not UT) because the camera clock is wireless synchronized by the German Timing System. The laser beam of the lidar enters the image from the lower left. In the images, North is up and West towards the right. The altitude at which the laser beam becomes invisible is not accurately known.

- In the middle panel: The temporal development of the lidar-observed trail(s) at high spatial and temporal resolution. Time "zero" is that of maximum metal densities in the lidar-observed trail(s). The densities in the trail(s) are color-coded in arbitrary units.

- In the right panel: The count profile(s) at the time of highest metal density in the trail(s) and integrated over 4 s. Near the ordinate scales, the regular metal layers can be noted. They only appear weak due to the short integration time of 4 s. The middle and right panels can also be used to test how close in altitude the twin trails are observed.

Plate 7.3a shows a two-element Leonid trail, one of the highest Leonids that our lidars could detect. The Ca^+ trail becomes detectable for the lidar about 30 s after the meteor; the Ca trail is comparatively weak. From the count rates shown in the right panel, we derive a density ratio of 12 of Ca ions over atoms. Plate 7.3b shows a two-element trail of a sporadics. Plate 7.3c shows another two-element trail with Ca and Na of a Southern Taurid. Without the camera image, we would have probably mis-identified this trail as a Leonid trail. Plate 7.3d shows a meteoroid (dotted trail above laser beam) that flew directly through the laser beam. The time separation between the occurrence of the meteor and of the start of the lidar-observed K trail is less than 1 s. We add that this is our only case in which we could observe such a close hit.

The number of events recorded during a good observation night are illustrated by our data from the night November 16/17, 2000. During this night, we had about 8 h of good observing conditions for our camera plus Ca and Na lidars. During this period, the camera recorded 62 meteors of which 10 were automatically identified as Leonids. The Ca and Na lidars recorded four and one Leonids, respectively. Here, our assignment of lidar-trails to Leonid meteoroids is based "only" on their high altitude: All five occurred at or above 98 km. Camera-recorded Leonids preceded three of those lidar-observed trails by 21 min, 6 min, and 16 min, respectively. Hence, for this night we could correlate only three out of ten camera-Leonids with lidar-Leonids, whereas from five lidar-Leonids we could correlate three with camera-Leonids.

Turning to our more general experiences in combining lidar information with imaging data, we offer the following comments: (1) Foremost, one can collect quite a number of cases, in which the correlation of the camera and the lidar-observed phenomena is easy and solid. These cases make possible, at least, the exact determination of the age of the lidar-observed trail, offering interesting possibilities for doing atmospheric science studies (diffusion coefficients, wind component, etc). Even better are cases in which the meteor is recognized as a shower meteor. Then the total velocity of the meteor and its cntry angle also become known; these parameters greatly constrain any model which is used to analyze the ablation processes of the meteoroid and the fate of the ablated metal atoms. (2) There are many lidar trails and also camera images which cannot be correlated with each other in a convincing way. Reasons for the observation of camera trails without a subsequent lidar trail are manifold: the meteor passes at too high an altitude through the laser beam; the wind direction is such that is does not blow the trail through the laser beam; the trail is already too far dissipated when it is blown through the laser beam; intermittent clouds obscure the searched-for lidar trail. Reasons for the occurrence of strong lidar trails without an earlier camera-recorded meteor are fewer. Again, intermittent clouds may obscure the camera event, but not the lidar observation. This may not look very likely. However, we frequently operate our lidars under conditions of broken or thin clouds and we can, interestingly enough, demonstrate these conditions with video camera sequences clearly showing Leonid fireballs above the clouds. Another cause for lidar-observed trails without obvious camera meteors can be flybys of rather small meteoroids (of which there are many!) close to the laser beam. They may be able to produce a good sized lidar signal, but are not recognized by the computer code MetRec3 which automatically searches for meteors in the camera images.

Considering the large effort and considerable funds required to operate routinely a number of lidars, the added burden of operating a meteor camera seems small in comparison to the gain in information which this instrument is able to give to the entire experiment.

7.8. Conclusions

Had we started to study meteor trails by lidars with the intention of learning something new about the composition of meteoroids, then reaching this goal is still ahead of us. There is a simple, but important reason for this: the lidars do not seem to find meteoroids in the sky which ablate homogeneously. Only with this condition fulfilled can we expect to relate in a straightforward way the composition recorded in a meteor trail with that of the meteoroid. The primary and unexpected result of our study is to discover that meteoroids with masses up to the gram range ablate almost exclusively differentially and not homogeneously. Differential ablation shows up in (a) the number of single-element trails being very large in comparison with that of two- and three-element trails and (b) the lidar-measured ratios of metal abundances in meteor trails. We interpret our observations as indicating that during atmospheric entry most meteoroids in the 10 µg to gram range break up into many smaller, secondary particles which are small enough to undergo independently differential ablation. Whether or not this mechanical fragmentation of meteoroids leads also to a chemical fractionation of the secondary particles remains an open question.

Concerning a better understanding of the capabilities and promises of state-of-the-art lidars for meteor work, we show through modeling and observational results that the capability of any lidar to detect meteor trails is strongly altitude dependent. It is limited to the altitude range from 80 to 105 km. This range appears to be appropriate for studies of many of the sporadics, but not well suited for studies of fast meteoroids of cometary origin such as the Leonids and Perseids. Furthermore, for "solitary" single lidars we see the potential for making significant progress in meteor research as being very limited. We see this potential as very good, however, for work using two lidars for future common-volume studies of two-element trails. Adding a third lidar to this instrument cluster just for reasons of meteor work does not seem to be an efficient use of resources. We would strongly recommend, though, enhancing a twin lidar assembly by a meteor camera and highly integrating the operations and data processing of the three instruments. We believe that this instrument cluster combined with a strong modeling effort has great potential for new and interesting results on meteor trails and on meteoroids. A combination of lidar(s) with a radar (but without camera) will probably encounter considerably more problems in correlating individual radar-observed meteors with lidar-observed trails than a combination of lidars(s) with a camera.

Acknowledgments: U.v.Z. and J.H. acknowledge the highly spirited assistance in collecting the IAP lidar data by M. Alpers, V. Eska, C. Fricke-Begemann, M. Gerding, T. Köpnick, and J. Oldag. Furthermore, our work with the video camera has greatly benefitted from efficient support by S. Molau. U.v.Z. and J.H. were in part funded by the Deutsche Forschungsgemeinschaft, Bonn, Germany under grant no. Al 458/2-1. Also, U.v.Z. wishes to thank EOARD for a Window-on-Science visit to AFRL, helping greatly our collaboration. W.J.McN. was supported in part by Air Force contract F19628-98-C-0010.

REFERENCES

ALPERS M., HÖFFNER J. & VON ZAHN U. 1990. Iron atom densities in the polar mesosphere from lidar observations, *Geophys. Res. Lett.*, **17**, 2345–2348.

ALPERS M., HÖFFNER J. & VON ZAHN U. 1996. Upper atmosphere Ca and Ca$^+$ at midlatitudes: First simultaneous and common-volume lidar observations, *Geophys. Res. Lett.*, **23**, 567–570.

ARLT R. 1998. Bulletin 13 of the International Leonid Watch: The 1998 Leonid meteor shower, *WGN, J. Intern. Meteor Org.*, **26**, 239–248.

ARLT R., & GYSSENS M. 2000. Bulletin 16 of the International Leonid Watch: Results of the 2000 Leonid meteor shower, *WGN, J. Intern. Meteor Org.*, **28**, 195–208.

ARUGA T., KAMIYAMA H., JYUMONJI M., KOBAYASHI T., & INABA H. 1974. Laser radar observation of the sodium layer in the upper atmosphere, *Rept. Ionosphere & Space Res. Japan*, **28**, 65–68.

ASHER D. J. 1999. The Leonid meteor storms of 1833 and 1966, *Mon. Not. R. Astron. Soc.*, **307**, 919–924.

BEATTY T. J., BILLS R. E., KWON K. H. & GARDNER C. S. 1988. CEDAR lidar observations of sporadic Na layers at Urbana, Illinois, *Geophys. Res. Lett.*, **15**, 1137–1140.

BETLEM H., JENNISKENS P., SPURNÝ P., VAN LEEUWEN G. D., MISKOTTE K., TER KUILE C., ZARUBIN P. & ANGELOS C. 2000. Precise trajectories and orbits of meteoroids from the 1999 Leonid meteor storm, *Earth Moon Planets*, **82–83**, 277–284.

BOROVIČKA J., STORK R. & BOČEK J. 1999. First results from video spectroscopy of 1998 Leonid meteors, *Meteorit. Planet. Sci.*, **34**, 987–994.

BOWMAN M. R., GIBSON A. J. & SANDFORD M. C. W. 1969. Atmospheric sodium measured by a tuned laser radar, *Nature*, **221**, 456–457.

BROWN P. *et al.* 2000. Global ground-based electro-optical and radar observations of the 1999 Leonid shower: First results, *Earth Moon Planets*, **82–83**, 167–190.

CAMPBELL M. D., BROWN P. G., LEBLANC A. G., HAWKES R. L., JONES J., WORDEN S. P. & CORRELL R. R. 2000. Image-intensified video results from the 1998 Leonid shower: I. Atmospheric trajectories and physical structure, *Meteorit. Planet. Sci.*, **35**, 1259–1267.

CEPLECHA Z., BOROVIČKA J., ELFORD W. G., REVELLE D. O., HAWKES R. L., PORUBČAN V. & SIMEK M. 1998. Meteor phenomena and bodies, *Space Sci. Rev.*, **84**, 327–471.

CHAMBERLAIN J. W. 1961. *Aurora and Airglow*, Academic Press, New York, NY; reprinted by the American Geophysical Union with notes by the author and errata, p. 469, 1995.

CHU X., PAN W., PAPEN G., GARDNER C. S., SWENSON G. & JENNISKENS P. 2000a. Characteristics of Fe ablation trails observed during the 1998 Leonid meteor shower, *Geophys. Res. Lett.*, **27**, 1807–1810.

CHU X., LIU A. Z., PAPEN G., GARDNER C. S., KELLEY M., DRUMMOND J. & FUGATE

R. 2000b. Lidar observations of elevated temperatures in bright chemiluminescent meteor trails during the 1998 Leonid shower, *Geophys. Res. Lett.*, **27**, 1815–1818.

CLEMESHA B. R., KIRCHHOFF V. W. J. R., SIMONICH D. M. & TAKAHASHI H. 1978. Evidence of an extra-terrestrial source for the mesospheric sodium layer, *Geophys. Res. Lett.*, **5**, 873–876.

CLEMESHA B. R., KIRCHHOFF V. W. J. R., SIMONICH D. M., TAKAHASHI H. & BATISTA P. P. 1980. Spaced lidar and nightglow observations of an atmospheric sodium enhancement, *J. Geophys. Res.*, **85**, 3480–3484.

DE LIGNIE M., LANGBROEK M., BETLEM H. & SPURNÝ P. 2000. Temporal variation in the orbital element distribution of the 1998 Leonid outburst, *Earth Moon Planets*, **82–83**, 295–304.

ESKA V., VON ZAHN U. & PLANE J. M. C. 1999. The terrestrial potassium layer (75–110 km) between 71° S and 54° N: Observations and modeling, *J. Geophys. Res.*, **104**, 17 173–17 186.

ESKA V., HÖFFNER J. & VON ZAHN U. 1998. The upper atmosphere potassium layer and its seasonal variations at 54° N, *J. Geophys. Res.*, **103**, 29 207–29 214.

FEGLEY B., JR. & CAMERON A. G. W. 1987. A vaporization model for iron/silicate fractionation in the Mercury protoplanet, *Earth Planet. Sci. Lett.*, **82**, 207–222 .

FLEMING E. L., CHANDRA S., BARNETT J. J., & CORNEY M. 1990. Zonal mean temperature, pressure zonal wind and geopotential height as functions of latitude, *Adv. Space Res.*, **10**, No. 12, 11–59.

FRICKE K. H., & VON ZAHN U. 1985. Mesopause temperatures derived from probing the hyperfine structure of the D_2 resonance line of sodium by lidar, *J. Atmos. Terr. Phys.*, **47**, 499–512.

GERDING M., ALPERS M., HÖFFNER J. & VON ZAHN U. 1999. Simultaneous K and Ca lidar observations during a meteor shower on March 6–7, 1997, at Kühlungsborn, Germany, *J. Geophys. Res.*, **104**, 24 689–24 698.

GERDING M., ALPERS M., VON ZAHN U., ROLLASON R. J. & PLANE J. M. C. 2000. The atmospheric Ca and Ca^+ layers: Midlatitude observations and modeling, *J. Geophys. Res.*, **105**, 27 131–27 146.

GRIME B. W., KANE T. J., COLLINS S. C., KELLEY M. C., KRUSCHWITZ C. A., FRIEDMAN J. S. & TEPLEY C. A. 1999. Meteor trail advection and dispersion; preliminary lidar observations, *Geophys. Res. Lett.*, **26**, 675–678.

GRIME B. W., KANE T. J., LIU A., PAPEN G., GARDNER C. S., KELLEY M. C., KRUSCHWITZ C. & DRUMMOND J. 2000. Meteor trail advection observed during the 1998 Leonid shower, *Geophys. Res. Lett.*, **27**, 1819–1822.

HAKE R. D., JR., ARNOLD D. E., JACKSON D. W., EVANS W. E., FICKLIN B. P. & LONG R. A. 1972. Dye-laser observations of the nighttime atomic sodium layer, *J. Geophys. Res.*, **77**, 6839–6848.

HELMER M., PLANE J. M. C., QIAN J. & GARDNER C. S. 1998. A model of meteoric iron in the upper atmosphere, *J. Geophys. Res.*, **103**, 10 913–10 925.

HÖFFNER J., FRICKE-BEGEMANN C. & VON ZAHN U. 2000. Note on the reaction of the upper atmosphere potassium layer to the 1999 Leonid meteor storm, *Earth Moon Planets*, **82–83**, 555–564.

HÖFFNER J., VON ZAHN U., McNEIL W. J. & MURAD E. 1999. The 1996 Leonid shower as studied with a potassium lidar: Observations and inferred meteoroid sizes, *J. Geophys. Res.*, **104**, 2633–2643.

HUGHES D. W. 1978. Meteors, in *Cosmic Dust*, edited by J. A. M. McDonnell, Wiley, New York, pp. 123–185.

HUNTEN D. M., TURCO R. P. & TOON O. B. 1980. Smoke and dust particles of meteoric origin in the mesosphere and stratosphere, *J. Atmos. Sci.*, **37**, 1342–1357.

JEGOU J.-P., CHANIN M.-L., MÉGIE G. & BLAMONT J. E. 1980. Lidar measurements of atmospheric lithium, *Geophys. Res. Lett.*, **7**, 995–998.

JENNISKENS P. & BUTOW S. J. 1999. The 1998 Leonid multi-instrument aircraft campaign – an early review, *Meteorit. Planet. Sci.*, **34**, 933–943.

JENNISKENS P., DE LIGNIE M., BETLEM H., BOROVIČKA J., LAUX C. O., PACKAN D., & KRUGER C. H. 1999. Preparing for the 1998/99 Leonid Storms, in *Laboratory Astrophysics and Space Research*, edited by P. Ehrenfreund, Kluwer, Dordrecht, The Netherlands, 425–455.

JEYS T. H., BRAILOVE A. A. & MOORADIAN A. 1989. Sum-frequency generation of sodium resonance radiation, *Appl. Opt.*, **28**, 2588–2591.

KANE T. J. & GARDNER C. S. 1993a. Lidar observations of the meteoric deposition of mesospheric metals, *Science*, **259**, 1297–1300.

KANE T. J. & GARDNER C. S. 1993b. Structure and seasonal variability of the nightime mesospheric Fe layer at midlatitudes, *J. Geophys. Res.*, **98**, 16875–16886.

KELLEY M. C., GARDNER C., DRUMMOND J., ARMSTRONG T., LIU A., CHU X., PAPEN G., KRUSCHWITZ C., LOUGHMILLER P., GRIME B. & ENGELMAN J. 2000. First observation of long-lived meteor trains with resonance lidar and other optical instruments, *Geophys. Res. Lett.*, **27**, 1811–1814.

KERRIDGE J. F. & MATTHEWS M. S. 1988. *Meteorites and the Early Solar System*, University of Arizona Press, Tucson, AZ , 1197.

KOPP E. 1997. On the abundance of metal ions in the lower ionosphere, *J. Geophys. Res.*, **102**, 9667–9674.

LOVE S. G. & BROWNLEE D. E. 1991. Heating and thermal transformation of micrometeoroids entering the Earth's atmosphere, *Icarus*, **89**, 26–43.

LOVE S. G. & BROWNLEE D. E. 1993. A direct measurement of the terrestrial mass accretion rate of cosmic dust, *Science*, **262**, 550–553.

MASON B. 1971. *Handbook of Elemental Abundances in Meteorites, 1971.* Gordon and Breach, Newark, NJ.

MCNAUGHT R. H. & ASHER D. J. 1999. Leonid dust trails and meteor storms, *WGN, the Journal of the IMO*, **27**, 85–102.

MCNEIL W. J., LAI S. T. & MURAD E. 1998. Differential ablation of cosmic dust and implications for the relative abundances of atmospheric metals, *J. Geophys. Res.*, **103**, 10899–10911.

MCNEIL W. J., DRESSLER R. A. & MURAD E. 2001. The impact of a major meteor storm on the Earth's ionosphere: A modeling study, *J. Geophys. Res.*, **106**, 10447–10465.

MÉGIE G. 1988. Laser measurements of atmospheric trace constituents, in *Laser Remote Chemical Analysis*, edited by R. M. Measures, John Wiley & Sons, Inc., New York, NY, 333–408.

MÉGIE G. & BLAMONT J. E. 1977. Laser sounding of atmospheric sodium interpretation in terms of global atmospheric parameters, *Planet. Space Sci.*, **25**, 1093–1109.

MÉGIE G., BOS F., BLAMONT J. E. & CHANIN M.-L. 1978. Simultaneous nighttime lidar measurements of atmospheric sodium and potassium, *Planet. Space Sci.*, **26**, 27–35.

MOLAU S. 2001. Automated meteor observing, *Sky & Telescope*, **101, No. 5**, 132–136.

MORTON D. C. 1991. Atomic data for resonance absorption lines, I, Wavelengths longward of the Lyman limit, *Astrophys. J. Suppl. Ser.*, **77**, 119–202.

NOTSU K., ONUMA N., NISHIDA N. & NAGASAWA H. 1978. High temperature heating of the Allende meteorite, *Geochim. Cosmochim. Acta*, **42**, 903–907.

PLANE J. M. C., GARDNER C. S., YU J., SHE C. Y., GARCIA R. R. & PUMPHREY H. 1999. The mesospheric Na layer at 40° N: Modeling and observations, *J. Geophys. Res.*, **104**, 3773–3788.

SEARS D. W. G. & DODD R. T. 1988. Overview and classification of meteorites, in *Meteorites and the Early Solar System*, edited by J. F. Kerridge and M. S. Matthews, The University of Arizona Press, Tucson, AZ , 3–31.

SHE C.-Y. 1990. Remote measurement of atmospheric parameters: New applications of physics with lasers, *Contemp. Phys.*, **31**, 247–260.

SHE C.-Y & KRUEGER D. A. 2001. A new lidar is born at ALOMAR, *The CEDAR Post*, **41**, 8–10.

SHE C.-Y., YU J. R., HUANG J. W., NAGASAWA C. & GARDNER C. S. 1991. Na lidar measurements of gravity wave perturbations of wind, density, and temperature in the mesopause region, *Geophys. Res. Lett.*, **18**, 1329–1331.

SHE C.-Y., CHEN S., HU Z., SHERMAN J., VANCE J. D., VASOLI V., WHITE M. A., YU J. & KRUEGER D. A. 2000. Eight-year climatology of nocturnal temperature and sodium density in the mesopause region (80 to 105 km) over Fort Collins, CO (41° N, 105° W), *Geophys. Res. Lett.*, **27**, 3289–3292.

SPURNÝ P., BETLEM H., VAN'T LEVEN J. & JENNISKENS P. 2000. Atmospheric behaviour and extreme beginning heights of the thirteen brightest photographic Leonid meteors from the ground-based expedition to China, *Meteorit. Planet. Sci.*, **35**, 243–249.

STATES R. J. & GARDNER C. S. 1999. Structure of the mesospheric Na layer at 40° N latitude: Seasonal and diurnal variations, *J. Geophys. Res.*, **104**, 11 783–11 798.

UCHIUMI M., HIRONO M. & FUJIWARA M. 1985. The seasonal variation of night-time sodium layer at 33° N, in *Special Issue No. 36, Proc. 7th Symp. Coordinated Observations of the Ionosphere and the Magnetosphere in the Polar Regions*, Memoirs Nat'l Inst. Polar Res., National Institute of Polar Research, Tokyo.

UCHIUMI M., NAGASAWA C., HIRONO M., FUJIWARA M. & MAEDA M. 1993. Sporadic enhancement of the mesospheric sodium during the Perseids meteor shower, *J. Geomag. Geoelectr.*, **45**, 393–402.

VON ZAHN U. & HANSEN T. L. 1989. Reply to Comments by B. R. Clemesha and D. M. Simonich on a paper entitled 'Sudden neutral sodium layers: a strong link to sporadic E layers' by U. von Zahn and T. L. Hansen, *J. Atmos. Terr. Phys.*, **51**, 147–150.

VON ZAHN U. & HÖFFNER J. 1996. Mesopause temperature profiling by potassium lidar, *Geophys. Res. Lett.*, **23**, 141–144.

VON ZAHN U., GERDING M., HÖFFNER J., MCNEIL W. J. & MURAD E. 1999. Fe, Ca, and K atom densities in the trails of Leonids and other meteors: Strong evidence for differential ablation, *Meteorit. Planet. Sci.*, **34**, 1017–1027.

8
IN SITU MEASUREMENTS OF METEORIC IONS

By J O S E P H M. G R E B O W S K Y[†]
AND A R T H U R C. A I K I N[‡]

NASA, Goddard Space Flight Center, Greenbelt, MD 20771, USA

Metal ions found in the atmosphere above 60 km are the result of incoming meteoroid atmospheric ablation. Layers of metal ions are detected by sounding rocket *in situ* mass spectrometric sampling in the 80–130 km region, which coincides with the altitude region where meteors are observed. Enhancements of metal ion concentrations occur during meteor showers. Even outside of shower periods, the metal ion altitude profiles vary from measurement to measurement. Double layers are frequent at middle latitudes. More than 40 different meteoric atomic and molecular ions, including isotopes, have been detected. Atmospheric metal ions on average have an abundance that matches chrondritic material, the same composition as the early solar system. However, there are frequently local departures from this composition due to differential ablation, species dependent chemistry and mass dependent ion transport. Metal ions react with atmospheric O_2, O, O_3, H_2O and H_2O_2 to form oxygenated and hydrogenated ionic compounds. Metal atomic ions at high altitudes have long lifetimes. As a result, these ions, in the presence of Earth's magnetic field, are transported over long distances by upper atmospheric winds and ionospheric electric fields. Satellite measurements have detected metal ions as high as ~ 1000 km and have revealed circulation of the ions on a global scale.

8.1. Introduction

Extraterrestrial material is the source of metal ions in the Earth's atmosphere. Each year $\sim 10^8$ kg [Ceplecha (1992)] of material is intercepted by the Earth. The origin of this material is predominantly solar orbiting interplanetary debris from comets or asteroids that crosses the Earth's orbit. It contains a very small amount of interstellar material. On occasion the Earth passes through enhanced amounts of debris associated with the orbit of a decaying comet. This leads to enhanced meteor shower displays for up to several days. The number flux of shower material is typically several times the average sporadic background influx of material. Meteoric material is some of the earliest material formed in the solar system. By studying the relative elemental abundances of atmospheric metal ions, information can be gained on the chemical composition of cometary debris and the chemical makeup of the early solar system.

Using *in situ* sampling with rocket-borne ion mass spectrometers, there have been approximately 50 flights that made measurements of the metal ion abundances at altitudes between 80 and 130 km. It is this altitude range where incoming meteoric particles are ablated, the larger ones giving rise to visible meteor displays. In several rocket measurements isotopic ratios of different atomic ion mass components and metal molecular ion concentrations have been determined and used to identify unambiguously the measured species and to investigate the processes controlling the metal ion distributions.

The positive ion composition of the Earth's ionosphere was first sampled by an ion mass spectrometer flown on a rocket in 1956 [Johnson *et al.* (1958)]. In 1958 a rocket-borne ion

---

[†] Email: U5jmg@lepvax.gsfc.nasa.gov
[‡] Email: aikin@chapman.gsfc.nasa.gov

spectrometer identified, for the first time, a layer of metal ions [Istomin (1963)] near 95 km. These data were interpreted as evidence of an extraterrestrial rather than a terrestrial source. Istomin (1963) predicted: "It seems probable that with some improvement in the method that analysis of the ion composition in the E-region may be used for determining the chemical composition of those meteors which do not reach the ground. Particularly, we hope to get information about the composition difference between particles of different meteor showers and also sporadic and shower meteoroids." These visions categorized the aims of many subsequent rocket-borne ion mass spectrometer experiments in the lower ionosphere. Although the use of such measurements to deduce the composition of different classes of meteoroids has not been successful, the past four decades of rocket observations have provided powerful sets of data for advancing our understanding of meteor ablation, meteoric composition, metal neutral and ion chemistry as well as ionospheric dynamics.

In addition to meteor ion populations measured in the lower ionosphere over the past four decades, metallic ions have also been detected from satellites hundreds of kilometers above the peak of visible meteor activity between 90 and 100 km. The first detection was by Hanson & Sanatani (1970). This wasn't totally unanticipated because earlier ground-based observations of Ca^+ resonantly scattered radiation in the twilight F-region revealed traces of Ca^+ up to 280 km [Broadfoot (1967)]. Subsequent satellite ion composition measurements and resonant scattering observations explored these high-altitude metal ion distributions to reveal a persistent global circulation of long-lived meteoric ions that move up from the meteoroid ablation region. The effect of this circulation on the lower-altitude meteor layers has yet to be completely resolved, but it is evident that ion dynamics does play a significant role in the overall process by which infalling solar system particles leave their imprint on the ionosphere.

This chapter presents an overview of the *in situ* ion composition measurements of the metal ion distributions. It demonstrates how these observations have led to furthering our understanding of the interactions of meteoroids with the atmosphere and how the ablated material is processed to form persistent and complex metal ion layers.

8.2. Rocket Measurements of Meteoric Ions

8.2.1. *Early History and Instruments*

The first rocket-borne ion mass spectrometer measurements of metal ions [Istomin (1963)] were made at mid-latitudes and employed Bennett R-F ion mass spectrometers [Bennett (1950)]. This instrument is a time-of-flight device that uses a fixed accelerating electric potential at the entrance aperture. Each ion species entering is accelerated to a unique speed, depending upon its mass. Radio frequency (rf) electric potentials are applied to a series of internal grids and the frequency is swept. For a given frequency the ion species, whose entrance-accelerated velocity lets it move through the sequence of grids in phase-resonance with the rf cycle, is maximally accelerated and will have larger kinetic energy exiting the rf grid section than other ion species present. A positive retarding potential is applied to a grid in front of the ion collector to reject the non-resonant ions. For this instrument, cycling the rf through an entire mass scan cycle takes the order of a few seconds so that the spatial resolution obtainable on a fast moving rocket is somewhat coarse. As a result, the instrument is not well suited for measuring narrow layers, but it is a reliable instrument that has yielded important measurements, particularly from satellites.

For the rocket flight detailed by Istomin (1963) the temporal sampling allowed only 100 complete mass spectra to be obtained between 92 and 206 km, but five spectra detected the presence of mass 24 and 26 amu ions. These ions were identified as isotopes of Mg^+, since the ratio of the two ion currents collected was consistent with solar system Mg isotopic abundances. The ambient ionospheric species, NO^+ and O_2^+, were detected on all mass scans and these were the dominant ions. Because of the complex dynamics of ions in the vicinity of fast moving rockets in dense atmospheric regions, it is not a simple matter to convert the collected ion currents into absolute ion concentrations without an independent measurement. Ion spectrometers provide a measure of relative ion composition. A second instrument is flown, or ground-based measurements are taken, to measure the electron/ion density along the flight path. The sum of the collected ion currents of all species is normalized to this ion/electron density. The concentrations of individual ion species are then determined (using laboratory calibration, and/or theoretical modeling, of the spectrometer's response to ion mass) from its proportion of the sum of all currents. Istomin (1963) used a UHF dispersion interferometer experiment to measure the electron density and found that the largest Mg^+ ion density was $\sim 10^4$ cm^{-3}. Similar peak densities were observed on both the upleg and downleg trajectories near 103 and 105 km respectively, implying the presence of a persistent metal ion layer. In addition, Mg^+ was detected in one mass spectral scan near 120 km indicating the presence of multiple layers. The species Fe^+ was also seen detected near the Mg^+ maximum and one mass scan contained the signature of positive ions with mass 40, presumably Ca^+ (or MgO^+). The ratio of the latter ion concentration relative to the measured Mg^+ was $\sim \frac{1}{25}$, which is close to meteor composition for Ca^+/Mg^+. Describing two earlier rocket experiments, Istomin (1963) reported Fe^+ also at estimated concentrations of 10^4–10^5 cm^{-3} between 101 and 105 km. Mass 28 ions were detected near a Fe^+ occurrence that suggested it was Si^+ and not N_2^+ ions, which are often present in the ambient ionosphere.

These early Bennett ion spectrometer measurements provided the first measure of permanent ionization layers resulting from meteoritic deposition. Understanding of the processes underlying these layers has been refined over subsequent years by rocket measurements made with much better spatial resolution and sensitivity. Still Istomin (1963)discovered many features of the metal ion distributions and their sources that were observed in all subsequent flights. In addition the large measured concentrations of Mg^+ and Fe^+, exceeding 10^4 cm^{-3}, indicated their importance as potentially dominant ions in the ionosphere. Istomin's measured high dayside metal ion concentrations provided the first direct evidence that at night, where ambient ionosphere molecular ion concentrations decay, the lowest-altitude ionosphere layers could be dominated by the longer-lived metal ions.

Before beginning an in-depth review of current understanding of the meteor ion layers derived from *in situ* ion composition measurements, we will describe the next published sounding rocket experiment. This is done to provide background for the work–horse ion spectrometer instrument, the quadrupole, which has dominated sounding rocket measurements of metal ions up to the present day. Narcisi & Bailey (1965) reported on the first launch of an ion mass spectrometer with the quadrupole mass filter system. It measured positive and negative ions from 64 km to $\sim$115 km, overlapping some altitudes sampled by Istomin (1963). Ion mass selection in this instrument is created by four parallel (quadrupole) rods in a rectangular configuration, with its central axis perpendicular to the plane of the entrance aperture. Each opposite pair of rods has an electric potential difference (the bias voltage) applied to produce hyperbolic potential surfaces. By applying a sinusoidal voltage between the rods only a narrow range of mass/charge ions will traverse the length of the analyzer, and encounter the collector for a fixed bias voltage. Other ion species are unstable in the time-varying field configuration and are accelerated

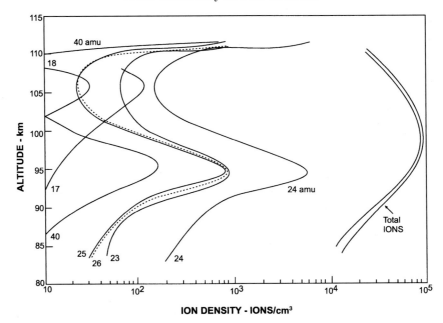

FIGURE 8.1. First complete profile of main metal ion layer [from Narcisi & Bailey (1965)]. The curves are labeled by the atomic masses of the measured positive ions. Mg^+ corresponds to 24, 25 and 26 amu. Total ion densities were measured by electrostatic analyzers. A second metal ion layer was skirted at apogee.

horizontally out of the central region. The mass spectrum is scanned by varying the rf and dc voltages, while keeping the voltage ratio between the rf amplitude and dc field constant. In order to make measurements in the altitude regime below 90 km, where ambient gas pressures are greater than 10^{-3} mm Hg, leading to performance-degrading collisions within the instrument, a high speed, high capacity vacuum pump is employed. This quadrupole experiment provided the first measure of the entire lower-altitude iono-spheric region where meteors are observed.

The Narcisi & Bailey (1965) measurements are shown in Figure 8.1. The high ambient molecular ion concentrations at the higher altitudes saturated the instrument, but the total ion density was available from onboard electrostatic probes. This plot was the first complete picture offered of a metal ion layer – one in which the Mg^+ (24 amu) ion concentration peaked at ~94 km with densities higher than 10^3 cm^{-3}. Similar layers were also present in the altitude profiles of other metal ion species. Ions with masses 25, 26, 23, and 40 amu were identified as two minor isotopes of Mg^+, Na^+ and either Ca^+ or MgO^+ respectively. The instrument only sampled from 1 to 46 amu, so that the presence of Fe^+ could not be ascertained. The Mg^+ isotope relative abundances compared favorably with solar system Mg abundances, as in the Istomin observations. Above 105 km all metal ion concentrations increased with altitude until apogee at 112 km. The 18 and 17 amu ions were due to water contamination. The peak of the main metal layer was below the altitude inferred by Istomin, whose experiment did not sample this low-altitude region.

These initial experiments at mid-latitudes, near midday, confirmed the presence of permanent metal ion layers of extraterrestrial origin in the lower ionosphere. They also showed that multiple layers are present. Subsequent flights provided details of the distri-butions and controlling factors, which will now be discussed.

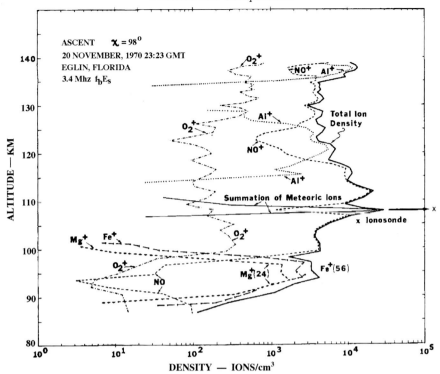

FIGURE 8.2. Mid-latitude nightside measurements of positive ions [from Philbrick *et al.* (1973). Presented with permission of Elsevier Science]. Two layers of metal ions were traversed. The major molecular ion concentrations were reduced when the metal ions peaked. The upper narrow layer was associated with blanketing sporadic E conditions measured by ground-based radio soundings. The species Al^+ was the product of a chemical release experiment.

8.2.2. *Metal Ion Layer Average Morphology*

The continuous global influx of the sporadic background of meteoroids might be expected to lead, through ablation, to a single contiguous global layer of neutral metal atoms. Metal atoms below 100 km become ionized predominantly by charge exchange [e.g. Swider (1969)] with ambient ionosphere species – most notably NO^+ and O_2^+ – as well as by photoionization. A simple picture, ignoring structured ion dynamics and atmospheric waves, is that there would be a single global-wide layer of meteoric ions in the ablation region. Sounding rocket experiments, such as the example in Figure 8.1 and another example shown in Figure 8.2, consistently find a layer of metal ions between 90 and 100 km, where most meteors are seen. The major metal ions in this main layer are Mg^+ and Fe^+. However, secondary layers of metal ions appear repeatably above 100 km. This is seen in Figure 8.2, which shows a narrow metal layer near 108 km with a peak density larger than that of the lower layer. Other examples are shown in Figure 8.3, which depicts two encounters of layers above 100 km. The extreme narrowness of these upper layers, at times ~1 km, a thickness much shorter than diffusion scales, and the presence of a layer above the major meteoroid ablation region, are indications that the upper layers are formed by processes different from those forming the lower layer.

Although the main meteor ionization layer below 100 km, composed predominantly of Fe^+ and Mg^+, is persistent, and its structure sometimes appears similar, as depicted in Figures 8.1, 8.2 and 8.3, its most distinctive property, as seen from sounding rocket

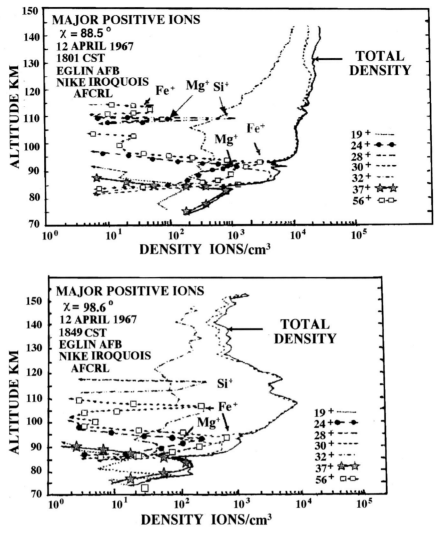

FIGURE 8.3. Measurements from two flights less than 1 hour apart, before (top) and after (bottom) sundown [Narcisi (1971)]. Within this short period the metal ion layers above 100 km completely changed. The metal ions below 95 km showed a reduction in density. The two plots are positioned so that their concentration scales match each other.

measurements, is its variability from flight to flight. Figure 8.4 shows two examples that contrast with the earlier figures. The left panel of the figure shows data from a flight that encountered no metal ion layer below 100 km. The right panel depicts data from a flight which detected an irregular layer for the species Mg^+ and none for Fe^+.

To investigate statistically the metal ion distributions Grebowsky *et al.* (1998) and Grebowsky & Pesnell (1999) computer-scanned all published sounding rocket metal ion density–altitude profiles. Interpolating each at 1 km altitude increments they established a data base that could be statistically analyzed. For this review we have added to the original database measurements that were not in the original studies [i.e., Earle *et al.* (2000); Goldberg & Blumle (1970)] and corrected one profile that had been erroneously scaled. There were 46 published ion spectrometer concentration profiles. A few publications, which presented only raw collected ion currents, were not used. Figure 8.5 contains

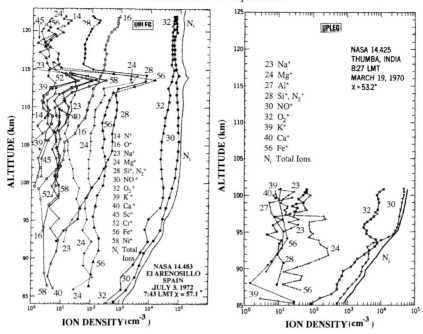

FIGURE 8.4. Examples where metal ions do not layer near 95 km. On left, a middle-latitude measurement [Goldberg (1975); Aikin *et al.* (1974), with permission from Elsevier Science], only the layer near 114 km is prominent. On right, a low-latitude measurement [Aikin & Goldberg (1973)], the metal ion peak is structured and Fe^+ is depleted relative to Mg^+. The measurement on the left was taken during the β-Taurids shower.

a plot of the total metal ion densities from the database. An order of magnitude spread is observed with an average (in 2 km bins) peak concentration between 90 and 95 km. This range of variability and the location of the average peak is also seen for individual metal species as seen on the right of Figure 8.5. The average altitude profiles of three species Fe^+, Mg^+ and Ca^+ were culled from all sounding rocket measurements for middle (30–60°) and high (>60°) latitudes. As seen, the heavy ion Fe^+ is typically the dominant metal species. All species peak in the same region.

A different perspective on the changes in the metal ion profiles from flight to flight is depicted in Figure 8.6. This is a plot of all the individual rocket flight profiles of Fe^+ interpolated to 1 km resolution. The plot is broken into a middle-latitude and high-latitude zone. This is done because metal ion production and loss processes depend upon the local ionosphere and atmosphere at a given time. Hence it is likely that the metal layers in the polar regions, characterized by long winter (summer) nights (days) and auroral energetic particle deposition, will differ from the middle-latitude behavior [e.g., Swider (1984)]. The middle- and high-latitude zones have different atmosphere and ionosphere composition vs. altitude profiles than the equatorial region. Thus low-latitude measurements might form another unique grouping. Unfortunately, there have been only four low-latitude rocket flights. One of these measurements is plotted in Figure 8.4 and two others (in Figure 8.10) will be discussed later. At all latitudes the main metal ion layer maximum moved about in the altitude interval between 90 and 100 km and the layer shape changed from flight to flight.

It is not possible with the limited rocket statistics to correlate the main layer peak structure with local time or latitude. The measurements took place over all seasons and

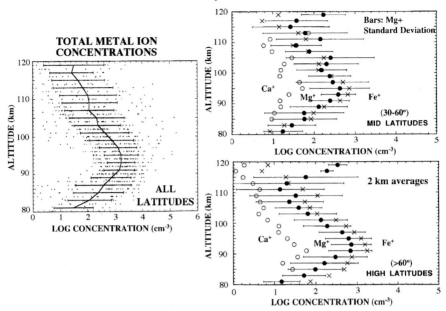

FIGURE 8.5. Left: total metal ion concentrations from all published positive ion measurements (1 km altitude interpolation). The points correspond to total metal ion densities when that was plotted in a publication otherwise it is the sum of Mg^+ and Fe^+, the two dominant metal ions. There is more than an order of magnitude scatter in the concentrations at all altitudes. Right: averages for three species for middle and high latitudes.

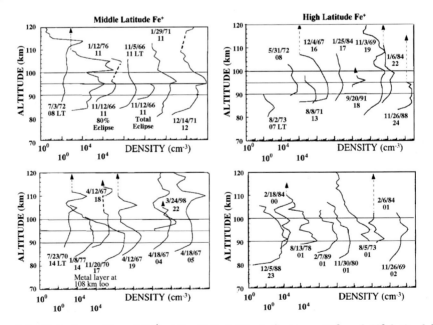

FIGURE 8.6. Altitude profiles of Fe^+. All middle-latitude (between $30°$ and $60°$ latitude) measurements are on the left. Measurements poleward of this zone are on the right. The density scales for individual profiles are offset by a factor of 100 from each other. Data are plotted sequentially in local time. Arrows indicate altitudes where data were taken but no Fe^+ reported. The horizontal lines have been inserted to provide references for altitude changes in the layer profiles.

many years, under widely ranging solar and magnetic activity conditions. Still, there are trends that appear in the data. Using the altitudes of the main layer maximum concentrations taken from published rocket ion composition plots, there is evidence for a decrease in the peak altitude from middle to high latitudes and a day-to-night trend in each zone. For all Fe^+ concentration maxima, the average height and standard deviation at middle latitudes (17 samples) is 95.0±2.5 km compared with 92.3±3.5 km at high latitudes (15 cases). This trend is also evident in the altitude averages depicted in Figure 8.4. Looking at middle-latitude measurements alone, the average dayside altitude (12 samples) is 93.9±1.3 km compared with the nighttime (5 samples) 97.5±2.8 km. For high latitudes (four cases) it is 89.9±4.2 km in the day and 93.1±3.0 km at night (11 cases). The same trends are seen for the total metal ion density when that parameter and not individual species concentrations were published. The geometrical average of the middle-latitude peak densities is 2×10^3 cm^{-3}. At high latitudes the average is 4.6×10^3 cm^{-3} and in both regions the average width of the peaks at their half point is ~5 km.

Another prominent feature seen in Figure 8.6 is secondary ionization layers at high altitudes in the middle-latitude region. At high latitudes the high-altitude layers are not evident, but some double-layer structures do appear below 100 km. Double-layer structures have not been observed on any of the equatorial experiments. The middle-latitude upper layers are more variable in time than the main meteor ionization peak.

In the lower ionosphere, atmospheric horizontal winds exert drag on the ions. The Earth's magnetic field (**B**) provides a barrier to cross **B** motions of the ions. The wind component along **B** can move the ions freely. For inclined magnetic field lines, a vertical ion motion can be produced. At very low altitudes the atmospheric pressure is so dense that the neutral winds can drag ions (velocity **v**) across magnetic field lines. In this case, the resulting **v** × **B** Lorentz force on the ions lead to vertical ion motions. Vertical shears in the winds in the correct sense lead to convergent ion flows and compression of the ion concentrations forming thin layers at the convergent nodes in the vertical ion drift. The physics behind this process was devised by Whitehead (1961). At high, auroral region latitudes, electric fields produced in the lower ionosphere from the magnetosphere are intensified. These fields transport and change the distribution of metal ions. These wind/electric field transport mechanisms operate down to ~90 km. Below this altitude collisions in the dense neutral atmosphere inhibit any magnetic/electric field associated effects on the ions. These mechanisms could play a prominent role in the latitudinal dependence of the metal ion profiles (indicated in Figure 8.6). A detailed description of the physics of these processes can be found in the review by Kelley (1989).

The contrasting variability of the upper-altitude layers compared with the main metal ion peak can be seen in Figure 8.3, which depicts a set of measurements separated by only 48 minutes, just before and after sunset. During the initial flight narrow layers were traversed near and above 110 km with the heavier ion Fe^+ layering several kilometers above Si^+ and Mg^+. Narcisi (1971) noted that a layer of Si^+ ions near 110 km is a prevalent feature of the dayside ionosphere, having been detected on several rocket flights. On the second flight, the Si^+ layer shifted to a higher altitude. The most prominent upper Fe^+ layer was now below 110 km. Several explanations for the observed changes are possible:

(a) the small metal ion layer, seen near this altitude on the previous flight grew;

(b) the upper layer moved downward; and

(c) the metal ions were distributed horizontally in patches.

The main meteor ion layer profile did not change dramatically, but its metal ion den-

TABLE 8.1. Meteoric ion composition relative to cosmic abundances

Avg. mass AMU	Element	Cosmic abundance[a]	b	c	d	e	f
23	Na	0.053	0.24	0.12	0.15	0.12	0.27
24	Mg	1.0	1.0	1.0	1.0	1.0	1.0
27	Al	0.079		0.03	0.063	0.036	0.019
28	Si	0.93	2.04	—	—	0.002	3.9
39	K	0.0035	0.016	0.0095	0.0063	0.011	0.006
40	Ca	0.057	0.07	0.04	0.034	0.045	0.024
45	Sc	0.00003	0.02	—	—	—	0.019
48	Ti	0.0022	—	—	—	0.0007	—
52	Cr	0.012	0.069	—	0.0077	0.016	0.026
56	Fe	0.84	3.04	1.64	0.53	1.37	0.93
59	Co	0.0021	—	—	0.0041	0.0016	—
60	Ni	0.046	0.086	0.05	0.029	0.037	—
66	Zn	0.0012	—	—	—	0.0011	—

[a] Abundances are total for a given element and normalized to total abundance of magnesium [Anders & Grevesse (1989)].
[b] Composition within a sporadic E-layer during the period of meteor shower associated with comet Enke [Goldberg & Aikin (1973)].
[c] Metal ion column densities based on five rocket flights [Kopp (1997)].
[d] Ten days after maximum of the Leonid meteor shower. Measurements within an ion layer at 95 km [Krankowsky et al. (1972)].
[e] Sampling for ion layer at 95 km [Zbinden et al. (1975)].
[f] Sampling for ion layer at 120 km [Zbinden et al. (1975)].

sities did drop between the two flights. At times there are spatial separations between separate ion species. This is more evident in the upper layers. In the main meteor layer and even in most of the prominent upper layers all metal species tend to peak in tandem. Hence the metal ion composition in very prominent layers may provide a valid basis for inferring the composition of the parent meteoroid bodies, as will be discussed next.

8.2.3. Composition of Deposited Material

Several ion composition studies have considered in detail the measured relative abundances of metal ions in the lower ionosphere. Figure 8.4 (left) is one such example, that shows the altitude distribution of many metal ion species (measured during a β-Taurids shower). The relative abundances in this case varied with altitude. Any abundance analysis must take this altitude variation into account, either by confining the sample to a specific altitude feature or by using total column densities. Also, ion mass spectrometers have a mass dependent response that must be considered. It is assumed that such corrections were taken into account in the published data.

Table 8.1 summarizes the relative abundances cited in the literature from different ion spectrometer flights. All measurements listed are normalized to the total concentration of magnesium ions. In addition to the elements listed in Table 8.1, Herrmann et al. (1978) identified the presence of vanadium and copper metal ions. For comparison purposes with the likely source, the relative abundances of chrondritic material is shown in the third column. Each of the other columns, labeled by author, is a list of results from publications that explicitly specified the ratio of measured ion abundances (either referring to concentrations in the peak of a layer or to the total vertical content) to chondritic abun dance ratios. There is approximate agreement in most cases with the chrondritic values. This indicates that the source particles on average have essentially the composition of the Sun and the early solar system. They are not of terrestrial or lunar origin, since the compositions of the Earth and Moon have been differentiated over time from formation values by a variety of processes. Meteoric material is undifferentiated.

A general conclusion can be drawn that the meteoric material from which the ions are derived, on average, has chrondritic composition. However, it is not possible to use precisely the ion data to deduce meteoroid composition without considering sources of error in the data, details of the physical processes associated with meteoric deposition, and the nature of the interaction of metallic neutral and ionic elements with atmospheric oxygen and water vapor. The metal ions are also subject to dynamical effects in the atmosphere including neutral wind drag in the presence of the terrestrial magnetic field and electric fields. Using data outside of prominent layers yields composition results which are less in agreement with chrondrites than when ion concentrations are employed from within a layer [Krankowsky *et al.* (1972)]. Either the metal ion composition within the layers is the result of immediate local seeding by ablation, or dynamics may be focusing ions into a layer from adjacent regions. In the latter case, the peak species concentrations are proxies of the net vertical content.

The relative composition has usually been determined from the maximum concentration region of one of the layers, either the main layer or a prominent one above 100 km. With the exception of Kopp (1997) all of the composition studies cited in Table 8.1 followed this approach. Kopp (1997) used the measured total column amount of each ion. In the latter study the composition was in extraordinary agreement with chrondritic abundances. Krankowsky *et al.* (1972) found measurements to be within 40% of chrondritic composition and within 30% of stony-meteorite values, with the exception of sodium, potassium and cobalt. These elements were also overabundant in other data sets, suggesting differential ablation rates for different metals. Some inferred elements such as calcium and scandium have shown relative abundances far in excess of cosmic or terrestrial values. This is possible evidence for the confusion of these species with other species with the same mass to charge ratio as the metal. For example, MgO^+ could be confused with Ca^+ and $SiOH^+$ with Sc^+.

Several metallic ion species have the same mass to charge ratio as oxides or hydroxides of other metal species or as ambient ionospheric species. Comparison of the measured concentrations of metal atom isotopic masses for a species can be used to resolve the ambiguity. Table 8.2 lists cosmic abundances for isotopes of metal species normalized to the total magnesium abundance. The table includes ion isotope measurements from Krankowsky *et al.* (1972) and from three separate rocket flights studied by Steinweg *et al.* (1992). With the exception of sodium in the Krankowsky *et al.* study, whose concentration was three times greater than the cosmic abundance, the relative concentrations are in approximate agreement with cosmic abundances. All the studies yielded reasonable agreement with solar system values for the metal ion isotope ratios, but the relative percentages of some of the observed metal species were a factor of ten below average solar system values.

Contrary to the dominance of Mg over Fe in chondrites, measurements reveal a typical overabundance of iron ions (as seen in Table 8.1) but there are exceptions. For example, in Figure 8.4 (right) Mg^+ is more dominant than Fe^+. Extracting the Fe^+/Mg^+ ratio at the main meteor layer peak from all published sounding rocket profiles the average turned out to be 2.4, with individual values ranging from 0.05 to 4.2. Mg^+ had the greatest

TABLE 8.2. Isotopic abundances of meteoric ions

AMU	Element	Cosmic abundance[a]	b	c	d	e
23	Na	0.05				
24	Mg	0.79	0.82	0.74		0.76
25	Mg	0.10	0.090	0.12		0.11
26	Mg	0.11	0.091	0.14		0.13
40	Ca	0.055	0.034			
42	Ca	0.00037	0.0011			
44	Ca	0.0012	0.0011			
52	Cr	0.012	0.0077			
54	Fe	0.049	0.049	0.063	0.066	0.059
56	Fe	0.77	0.53	0.91	0.93	0.91
57	Fe	0.023	0.028	0.031		0.031
58	Ni	0.68		0.74		0.75
60	Ni	0.26		0.26		0.25
61	Ni	0.056				
62	Ni	0.056				
64	Ni	0.056				

[a]Abundances are fraction of the total [Anders & Grevesse (1989)]. Total abundances are normalized to total magnesium abundance.
[b]Krankowsky et al. (1972).
[c]Steinweg et al. (1992) – Flight F2.
[d]Steinweg et al. (1992) – Flight F3.
[e]Steinweg et al. (1992) – Flight F4.

density for 21% of the cases. For shower measurements the average ratio was nearly the same, 2.2, with Fe^+ always dominant. There is an unexplained variability in Mg^+/Fe^+. Possible explanations suggested by McNeil et al. (2001) include variations in meteoroid composition or differential ablation. LIDAR observations [von Zahn et al. (1999)] and modeling [McNeil et al. (1998)] have demonstrated the importance of differential ablation for species with different volatilities. For example the peak Na deposition due to ablation is modeled to be more than 10 km above that of Mg. Hence the metal atom and associated ionospheric metal ion composition could change with altitude. However, the detailed rocket metal ion composition relations (Tables 8.1 and 8.2) are not altitude-dependent. Neutral metal atoms are converted to ions by charge exchange with ambient O_2^+ and NO^+ and the chemistry is affected by species-dependent chemical interactions with the atmosphere. These processes along with ionospheric dynamics will blur the altitude imprint of differential ablation on the metal ions but still leave differences between the relative abundances of the measured metal ions compared to meteoroid atomic atom composition.

8.2.4. Chemistry of Metal Ions

The multiplicity of ion species, combined with limited ion spectrometer mass resolution, can contribute to an uncertainty in the assignment of mass to a particular metal ion isotope. The chemical processing of ablation products to ions is another error source in determining the neutral composition of the meteoroids. Table 8.3 provides a list of all observed ion masses that have been related to metal ion molecular species. Possible duplicate ion identifications are indicated. With the exception of the possible confusion

TABLE 8.3. Metal compounds detected in the upper atmosphere

Compound	Mass (AMU)	Reference
K^+, NaO^+	39	Zbinden *et al.* (1975); Kopp *et al.* (1985)
Ca^+, $NaOH^+$, MgO^+	40	Zbinden *et al.* (1975); Kopp *et al.* (1985)
$MgOH^+$, $Na\cdot H_2O^+$	41	Zbinden *et al.* (1975); Kopp *et al.* (1985)
AlO^+	43	Zbinden *et al.* (1975)
SiO^+	44	Zbinden *et al.* (1975)
Sc^+, $SiOH^+$, $Al^+\cdot H_2O$	45	Zbinden *et al.* (1975); Kopp *et al.* (1985)
SO^+	48	Kopp *et al.* (1985)
$CaOH^+$	57	Kopp *et al.* (1985)
$MgOH^+\cdot (H_2O)$	59	Kopp *et al.* (1985)
$Na\cdot (CO_2)^+$	67	Kopp *et al.* (1985)
FeO^+	72	Kopp *et al.* (1985); Goldberg & Witt (1977)
$FeOH^+$	73	Goldberg & Witt (1977)
$Fe^+\cdot (H_2O)$	74	Kopp *et al.* (1985)
Co, $MgOH^+\cdot (H_2O)_2$	77	Kopp *et al.* (1985)
$FeO_2{}^+$	88	Goldberg & Witt (1977); Kopp *et al.* (1985)
$FeO^+\cdot (H_2O)$	90	Goldberg & Witt (1977)
$Fe^+\cdot (H_2O)_2$	92	Goldberg & Witt (1977)
$FeO_2{}^+\cdot (H_2O)$	106	Kopp *et al.* (1985)
$FeO^+\cdot (H_2O)_2$	108	Goldberg & Witt (1977)
$Fe^+\cdot (H_2O)_3$	110	Goldberg & Witt (1977)
$MgOH^+\cdot (H_2O)_4$	113	Kopp *et al.* (1985)
$FeO_2{}^+\cdot (H_2O)_2$	124	Goldberg & Witt (1977)
$FeO^+\cdot (H_2O)_3$	126	Goldberg & Witt (1977)
$Fe^+\cdot (H_2O)_4$, Fe_2O^+	128	Goldberg & Witt (1977)
$MgOH^+\cdot (H_2O)_6$	149	Kopp *et al.* (1985)
$MgOH^+\cdot (H_2O)_7$	167	Kopp *et al.* (1985)

of Si^+ and $N_2{}^+$, atomic metal ion species do not have the same mass as the major ionospheric ions $O_2{}^+$, NO^+, O^+, and $N_2{}^+$ or their isotopes. The appearance of narrow 28 amu ion layers, particularly within or near layers of other metal ions, can be taken as proof that Si^+ is the proper identification. An interesting case of ambiguity is the positive ion with mass 45 amu, which was first reported to be scandium by Goldberg & Aikin (1973) (see Table 8.3). The difficulty with this assignment is that the observed abundance is 600 times the cosmic abundance of Sc^+. The same mass, with similar abundance, has been observed in other investigations [Zbinden *et al.* (1975)]. One explanation is that the species is $SiOH^+$, which forms in the reaction [Fahey *et al.* (1981)]

$$Si^+ + H_2O \rightarrow SiOH^+ + H. \tag{8.2.1}$$

$SiOH^+$ disappears rapidly by dissociative recombination with electrons so that a significant amount of water must be present at the altitude of the metal ion layer to insure a continued presence of $SiOH^+$. Zbinden *et al.* (1975) attributed the water source to rocket outgassing.

Metal ions are converted to neutral metals either by radiative recombination with ambient electrons

$$X^+ + e \rightarrow X + h\nu \tag{8.2.2}$$

Or, by conversion into a molecular ion (such as XO^+) followed by dissociative recombination with electrons

$$XO^+ + e \rightarrow X + O, \tag{8.2.3}$$

$$XO_2^+ + e \rightarrow X + O_2. \tag{8.2.4}$$

Radiative recombination proceeds at a rate of about 1×10^{-12} cm^3 s^{-1} while dissociative recombination at 300 K is about 5×10^{-7} cm^3 s^{-1} for most molecular ions. Thus atomic ions have a lifetime five orders of magnitude longer than molecular ions. For an electron density of 1×10^5 cm^{-3}, a nominal dayside ionospheric density, the respective lifetimes of atomic and molecular ions are 10^7 and 20 seconds. At high altitudes, above the ablation region, where transformation of atomic ions into molecular ions by interaction with atmospheric molecules is not significant, metal ions prevail only in atomic form. Molecular ion concentrations decay more rapidly as the electron density increases. This effect is the reason that the narrow (Sporadic E) ionosphere layers are often comprised of a high percentage of metal ions relative to the ambient ions NO^+ and $O_2{}^+$.

At lower altitudes in the ablation region, and below, the atomic metal ions are rapidly converted by chemical reactions into molecular ions and rapidly disappear with decreasing altitude. Table 8.3 lists the variety of metal ion compounds that have been attributed to measured ion masses. Species range from simple oxides such as AlO^+, SiO^+, and FeO^+ to more complex oxygenated species (e.g. $FeO_2{}^+$). There are also hydrated constituents like $CaOH^+$ and $FeOH^+$. Further water can attach directly to form species such as $Fe^+(H_2O)$ and $Fe^+(H_2O)_2$. These molecules form typically in the cold summer mesopause region at high latitudes [Goldberg & Witt (1977)].

Laboratory studies [Plane (1991)] have shown that metal ions react with molecular oxygen to form oxides in the reaction

$$X^+ + O_2 + M \rightarrow XO_2^+ + M, \tag{8.2.5}$$

where X represents any metal such as Mg or Fe and M is a third body, which in the atmosphere is either O_2 or N_2. N_2 can be substituted for O_2 giving rise to XN_2 [Plane et al. (1999)]. Metal oxides can also be formed by

$$X^+ + O + M \rightarrow XO^+ + M, \tag{8.2.6}$$

and

$$X^+ + O_3 \rightarrow XO^+ + O_2. \tag{8.2.7}$$

Metal hydroxides can be formed by

$$X^+ + H_2O_2 \rightarrow XOH^+ + OH. \tag{8.2.8}$$

The ion XO_2^+ is produced by

$$XO^+ + O + M \rightarrow XO_2^+ + M \tag{8.2.9}$$

while

$$XO_2^+ + O \rightarrow XO^+ + O_2 \tag{8.2.10}$$

converts $XO_2{}^+$ to XO^+. In the case of elements such as sodium and silicon, formation of metal hydroxyls occur by

$$XO^+ + H_2O \rightarrow XOH^+ + OH \tag{8.2.11}$$

These processes account for the metal molecular ions measured in the ablation zone below 100 km.

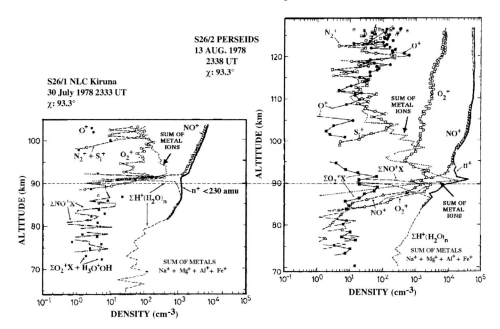

FIGURE 8.7. Two high-latitude measurements through noctilucent cloud regions [Kopp *et al.* (1985)]. On the bottomsides of the metallic ion layers (near 90 km, indicated by the dotted line), the metal ion concentrations drop more than an order of magnitude in 1–2 km, where hydrates and negatively charged aerosols (not shown here) were detected. Interactions of the metal ions with negative aerosols is responsible for the metal ion loss. The measurements were taken from the same location before and during the Perseid shower.

Metal compounds are absent above ~100 km and become dominant over atomic metal ions below 85 km. Above 100 km atomic oxygen increases dramatically. In this environment, reaction with O, i.e.

$$XO^+ + O \rightarrow X^+ + O_2 \qquad (8.2.12)$$

in addition to dissociative electron–ion recombination reactions destroys XO^+ and XO_2^+. Below 85 km the atmosphere electron density is reduced, decreasing the effectiveness of dissociative recombination in destroying XO^+ and XO_2^+. Also below 85 km, O_3, O_2 and H_2O increase in concentration, leading to enhanced production of XO^+, XO_2^+ and XOH^+. The combination of increased production and decreased loss leads to increased metal oxide ion concentrations below 85 km. However, the atomic oxygen contribution could be sufficient to prevent significant concentrations of ions such as XO^+ from appearing.

At low altitudes atmospheric aerosols lead to the disappearance of metal ions through the production of clustered metallic ion species [e.g., Kopp & Herrmann (1984)] which rapidly become neutralized. Figure 8.7 shows a case where the bottom of the main meteor ion layer overlaps the upper boundary of the hydrated ion layer (the molecular ion with mass 37 amu corresponds to the cluster $H^+ \cdot (H_2O)_2$). Some particularly sharp cutoffs in the metallic ion layers were measured at high latitudes above noctilucent clouds [Kopp *et al.* (1985)] as seen in Figure 8.7. The origin of the decrease of ionization in the vicinity of noctilucent clouds is the attachment of free electrons and ions to cloud particles, increasing the loss rate of the metal ions.

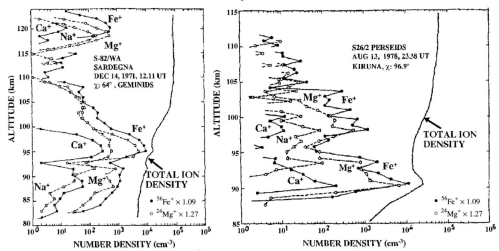

FIGURE 8.8. Two sets of ion measurements associated with meteor showers [from Kopp (1997)]. Measurements on the left (at a middle latitude) were obtained during the maximum of the Geminid shower and that on the right (at a high latitude), one day following the maximum activity of the Perseid shower. The main peak metallic ion densities observed were higher than the average non-shower concentrations.

8.2.5. *Meteor Showers*

One goal of sounding rocket ion composition experiments has been to infer the composition of the parent bodies of meteor showers. Model calculations by McNeil *et al.* (2001), McNeil (1999) and Grebowsky & Pesnell (1999) detail how individual streams might affect the ionosphere. However, measurements are needed to establish the extent of meteor shower effects. An increase in the number of individual ionization trails is associated with a shower, but this ionization quickly disperses and may not lead to a lasting enhancement of ion concentration in the ionosphere. However, during the event, additional ablated neutral atoms are added to the atmosphere. The subsequent ionization of this new material, either by photoionization or charge transfer processes with ambient ionosphere species, has the potential to produce measurable lasting ionospheric effects separable from the persistent layers produced by the sporadic background of meteoroids. It has been commonly assumed in previous publications that the metal ion distributions observed during a shower are due to the shower itself. As will be shown, the rocket evidence points in this direction, but statistics are still not sufficient, in terms of the number of experiments and the control of the conditions under which the measurements were made, to remove all ambiguity.

Figure 8.8 shows measurements taken during, or closely after two major showers. One of the samples is at middle latitudes and the other in a high latitude, auroral, energetic particle precipitation zone. The maximum Fe^+ and Mg^+ metal concentrations for both events were near 10^4 cm^{-3}. Compared with the ensemble of all Fe^+ densities (Figure 8.5), they fall within the upper quartile of all observed concentrations below 100 km. More evidence for a shower effect is seen in Figure 8.7. The data on the right were obtained during a Perseids shower and the measurements on the left were obtained from the same location a few weeks earlier. The shower-period peak metal ion density was almost two orders of magnitude above that of the earlier period. As appealing as it is to conclude a shower effect is present from these individual measurements, there are still uncertainties. For example, the high-latitude shower event in Figure 8.7 was characterized by significant

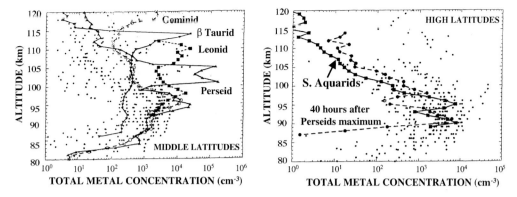

FIGURE 8.9. A comparison of individual profiles of the total metal ion density for showers superimposed on 1 km interpolated values for non-shower periods [from Grebowsky *et al.* (1998), with permission of Elsevier Science]. Mid-latitude observations (left) show that shower measurements have higher peak densities – particularly in the upper-altitude layers. The high-latitude observations do not show a clear trend.

enhancement of the NO^+ concentration compared with the non-shower period and the bottom of the pre-shower metal ion layer was apparently eroded by losses associated with aerosols. Hence, background ambient conditions were very different between the two flights making a comparison for meteoric effects uncertain.

In an attempt to resolve this issue Grebowsky *et al.* (1998) compared published shower measurements of the total concentrations of metal ions with the ensemble of all published metal concentrations. The results are shown in Figure 8.9. For middle latitudes there were only four measurements near or immediately following showers [Narcisi (1968); Goldberg & Aikin (1973); Zbinden *et al.* (1975); Herrmann *et al.* (1978)]. These shower profiles all had prominent high-altitude layers with metal ion concentrations exceeding quiet time layer concentrations at the same altitudes. In the main metal ion layer region, three of the shower events also showed enhanced metal ion densities compared with the non-shower data. However, the flight associated with the β-Taurids (Figure 8.4, left) didn't find a metal ion layer below 100 km. At high latitudes (Figure 8.9) the two shower-period profiles [Kopp *et al.* (1985); von Zahn *et al.* (1989)] did not have the highest densities. Interestingly, all six shower flights showed double layer structures. The evidence points toward an enhancement of the metal ion layer concentrations during showers at middle latitudes. However, the number of data samples is limited and the effects of the different ionospheric and atmospheric conditions under which the rocket measurements were made have not been investigated. The detailed ion composition studies, discussed earlier, did not show any significant differences between shower and non-shower periods. This point was made by Kopp (1997) as shown by the data presented in Tables 8.1 and 8.2; the Zbinden *et al.* (1975) and Goldberg & Aikin (1973) measurements were taken during shower periods. Further studies are needed to establish definitively the impact of the showers on the ionospheric metal ion distributions.

8.2.6. *Sources of Variability in Metal Ion Layers*

It is a common assumption in the analyses of the physics of the meteor ion layer that the sporadic background of incoming particles is in an equilibrium state in which day to day variations in their properties are not significant. Meteor shower events, on the other hand, introduce discrete time dependent bursts of different particle populations into the atmosphere. This can have an impact on the metal ion distributions. On the other hand,

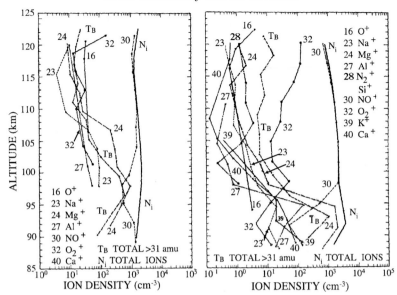

FIGURE 8.10. Upleg positive ion measurements [Aikin & Goldberg (1973)] over Thumba, India on the night of March 9–10, 1970 at 1938 MLT (left) and 0108 LMT (right). Metal layer moved downward ~7 kilometers – its displacement was consistent with magnitude of coincidentally observed downward drift of higher-altitude F-layer.

lidar, ground-based radio/radar soundings, and *in situ* ion composition studies have shown that secular and dynamic plasma and atmospheric processes have direct control of the lower ionosphere layers. These are sources for much of the observed structure in the metal ion distributions and they must be considered when looking for meteor shower effects in rocket-borne ion composition measurements.

Atmospheric/ionospheric dynamics are responsible for the complex narrow, time varying layers in the lower ionosphere, leading to complex "sporadic-E" structures in which metal ions are often the dominant ion species. One clear example of the temporal impact of ionospheric dynamics on the main meteor ion layer is shown in Figure 8.10. The metal layer, with Mg^+ dominant, dropped 7 km in the $4\frac{1}{2}$ hours between the two launches. The metal ion vertical content did not change, while the peak metal ion density increased and the concentration of ambient NO^+ decreased. The altitude decrease was in phase with the downward F region drift measured at the same time by an ionosonde, and the NO^+ density decrease was consistent with enhanced dissociative recombination due to the increasing electron density resulting from enhancements metal ion concentrations. Hence, the metal ion layer was compressed as it moved down.

Ionospheric electric fields and atmospheric waves not only move the meteor layers around, but they can have an ion mass spectrometer effect on the metal species as well. That is, ion species can be separated in space. Examples of this are shown in Figure 8.11. In both cases there is ~1 km separation between the Fe^+ and the lighter Mg^+ layers. This behavior was predicted by Chimonas (1969) as a characteristic for metal layers formed and transported downward in phase with the nodes of the horizontal wind. The neutral wind is more effective in dragging the lower mass ions. The presence of this process could at times compromise the use of local measurements of relative metal ion composition as indicators of meteoroid composition.

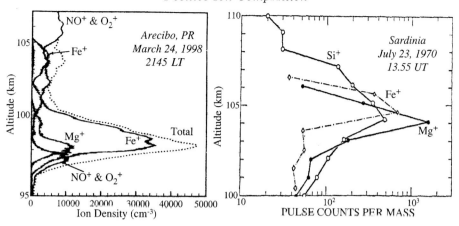

FIGURE 8.11. Two middle-latitude examples of metal ion species whose layers are separated in altitude. In both cases, the lighter ion, Mg^+, peaks at lower altitudes than the heavy ion Fe^+. This is what is expected for the transport of plasma to a node in the vertical shear of the horizontal wind from above and below. Data on left are from Earle *et al.* (2000). The data on the right [from Johannessen & Krankowsky (1974), with permission from Elsevier Science] are in the form of uncorrected count rates.

The effects of dynamics are predominant for the upper metal ion layers but also influence the main meteor ionization layer [e.g. see models of McNeil *et al.* (1996); Carter & Forbes (1999)]. The topside slope of the latter layer can vary on short time scales compared with the ablation/chemistry/diffusion processes that would produce the layer under quiescent conditions. On the other hand, the bottomside of the main metal ion layer near and below 90 km is not sensitive to externally driven ion dynamics. This side of the layer is dominated by local ion-neutral chemistry as detailed in section 8.2.4. The location of the lower boundary of the metal ion layer varies from flight to flight (see Figure 8.6), but the bottomside altitude contours contain no prominent irregular structure in contrast to the topside region. The chemistry is controlled by hydrogen and oxygen components which change with season, time of day, and atmospheric/auroral activity [e.g., in Narcisi (1973); Arnold & Krankowsky (1977) and Kopp & Herrmann (1984)]. Hence, the bottom of the main layer has distinctly different sources of variability than the upper side. The combination of dynamics and variations in chemistry lead to complex metal ion layer behaviors.

8.3. Satellite *In Situ* Measurements of Meteoric Ions

Sounding rockets launched at middle latitudes frequently traverse metal ion layers above 100 km, the result of long-lived atomic metal ions that are entrapped in convergent ion velocity nodes produced by atmospheric waves. Since the main metal ion production is at lower altitudes, upward transport is required, unless the descending layer nodes are directly seeded by the immediate ionization of meteoric debris at high altitudes. The latter effect could take place by impact ionization of very fast particles that, in addition to ablating at higher altitudes than slower particles, also have more energy available to produce ionization by impact ionization of the ablated gases. Ambient ionospheric ions can also charge exchange with ablated neutral metal atoms. This high-altitude seeding could be important for high-speed meteor showers, such as the Leonids and Perseids with average speeds of the order of 60–70 km/s (compared with the ∼20 km/s average speeds of the sporadic meteoroids). If local production of metal ions in the convergent drift nodes is not significant, the metal ion layers above 100 km must consist of long-lived metal ionization that had been previously transported upwards from the ablation region, before being entrapped in narrow layers by atmospheric waves. Evidence for the importance of this upward transport has been found in satellite observations of metal ions at much higher altitudes than the sounding rocket observed layers.

The near-Earth detection of metal ion layers in the vicinity of ~100 km was not really surprising since this is the region where ionization is evident in meteors. Early ground-based observations provided evidence for the presence of meteoric ions at much higher altitudes. Resonantly backscattered solar radiation emissions from Ca^+ were seen originating from altitudes as high as 280 km [Broadfoot (1967)]. The first *in situ* measurements of metal ions from a satellite, (OGO-6), found Fe^+ concentrations exceeding 100 cm^{-3} near 500 km [Hanson & Sanatani (1970)]. This discovery was made, not with an ion mass spectrometer, but by interpreting changes in the current collected on an ion trap experiment as a retarding potential was enhanced to eliminate the collection of low mass ions. A follow-up study substantiated the presence of Fe^+ and found traces at altitudes as high as 1000 km [Hanson *et al.* (1972)]. These initial Fe^+ high-altitude detections were at low latitudes and were attributed to electrodynamical lifting of the ions from the meteor ablation source region. Satellite airglow measurements of Mg^+ [e.g. Gérard & Monfils (1978), Fesen & Hays (1982)] subsequently revealed the more prominent morphological features of metal ions at low latitudes. A later study [Grebowsky & Brinton (1978)] discovered similar large concentrations of Fe^+ at altitudes of several hundred kilometers at middle and high latitudes.

Figure 8.12 shows a satellite measurement in which Fe^+ was found to be the dominant ion near 234 km. These data were taken from a satellite with a nearly circular orbit. A more descriptive example is shown in Figure 8.13, which uses data from two consecutive eccentric orbits of Atmosphere Explorer C. Fe^+ data were obtained with BIMS (Bennett Ion Mass Spectrometer) and Mg^+ data with MIMS (Magnetic Ion Mass Spectrometer). The latter instrument [Hoffman *et al.* (1973)] used a magnetic field to separate ion masses by their differing gyroradii. BIMS [Brinton *et al.* (1973)] had reduced sensitivity and different temporal resolution of the metal ions than did MIMS. The absence of Fe^+ when Mg^+ was present could have been due to an instrumental effect. Metal ions were encountered from the spacecraft's perigee altitude near 140 km to above 200 km with concentrations comparable to those of rocket measurements in the main meteor ion layer. The Mg^+ concentration tapers off as the ionospheric density (mostly O^+) approaches its peak value in the F-Layer. There was a large change in the Mg^+ concentration at low altitudes in the 2 hours between the orbits but the metals persisted continuously below 240 km along the satellite track.

Satellite ion composition measurements are much more comprehensive than sounding rocket measurements and provide statistically meaningful studies of the global, high-altitude metal ion distributions. Kumar & Hanson (1980) provided the first global overview of their altitude distribution by studying several metal ion species from Atmosphere Explorer, Dynamics Explorer and OGO 6 spacecraft. The metal ions were encountered in all latitude zones, with the highest percentage of detection from 200 to above 300 km, corresponding to the typical bottom side of the F-layer. The metal ions were found at altitudes exceeding 300 km in the equatorial region more often than at other latitudes.

The global distribution of the high-altitude metal ions is depicted in Figure 8.14. This is a plot of the locations where Fe^+ ions were detected with concentrations exceeding 30 ions/cm^3, between 150 and 400 km, from Atmosphere Explorer C (with a high orbital inclination) and Atmosphere Explorer E (a near equatorial orbit). A strong dawn–dusk local time asymmetry in the low-latitude occurrence frequencies is apparent with more

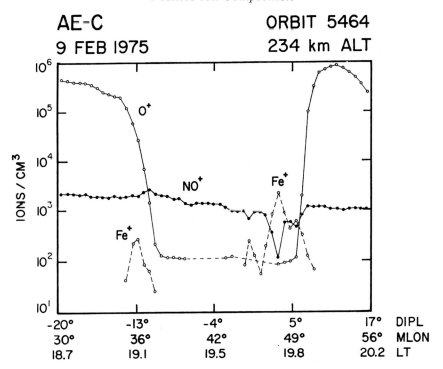

FIGURE 8.12. Low-latitude measurements of major ions at 234 km by the Bennett Ion Mass Spectrometer on Atmosphere Explorer C. Even at this high altitude, the metal ion Fe^+ is detected and in one region is the dominant ion. Data were obtained from National Space Sciences Data Center A.

patches traversed in the afternoon–dusk regions. This was first noted in the study of Kumar & Hanson (1980). At higher latitudes Fe^+ tends to be concentrated in a mid-latitude band at night between 50 and 60 degrees magnetic latitude, which extends through dawn. There is another separate intensified band on the dayside of the polar cap.

Mechanisms for transporting the long-lived metal ions from their source region below 100 km upward have been identified. The processes are depicted schematically in Figure 8.15. Low-latitude metallic ions can be pulled upwards out of the main meteor layer region by either: 1. a vertical polarization electric field associated with the equatorial dynamo; or 2. the Lorenz force $U_{east} \times B$ (U_{east} is the eastward component of the neutral wind and B is the magnetic field vector). Once the ions move to a region where the ion collision frequency is less than the ion gyrofrequency, the neutral wind and electric field can no longer drive the ions across the magnetic field. However, the ions can still be driven upwards across the field lines by the $E \times B$ convection due to the dynamo electric field. The metal ions eventually will fall to lower altitudes. The prevailing horizontal neutral wind can skew the high-altitude ion distribution in the windward direction by dragging the ions along the magnetic field lines. These mechanisms were originally suggested by Hanson & Sanatani (1970) and have since been treated in detailed modeling studies of the low-latitude Mg^+ distributions [e.g., Hanson *et al.* (1972); Fesen *et al.* (1983); Carter & Forbes (1999)].

At middle and high latitudes (right side of Figure 8.15), the mechanisms for uplifting the metal ions from their source region below 100 km are basically the same as at low latitudes (i.e., E field and $U_{east} \times B$ transport in regions where the magnetic field lines

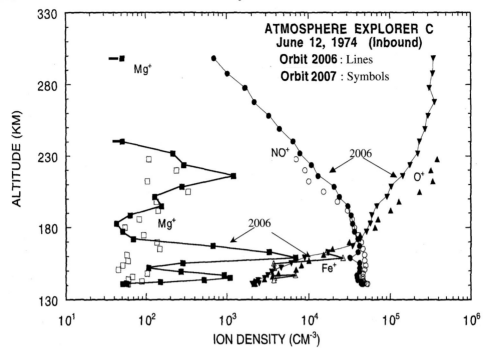

FIGURE 8.13. Measurements on two consecutive eccentric orbits of AE-C just on the dayside of dawn at middle latitudes. Mg^+ was measured by the Magnetic Ion Mass Spectrometer (MIMS). Fe^+ was detected by the Bennett Ion Mass Spectrometer (BIMS) only on orbit 2006. Metal ions were detected continuously through the bottomside of the F-layer (composed predominantly of O^+) down to perigee. Data were obtained from National Space Science Data Center-A

are inclined). When ions enter the region where their collision frequency is less than their gyrofrequency, equatorially directed atmospheric winds can further drag the ion, not across the magnetic field lines, but along the inclined magnetic field lines upward into the F-region. The upward motion is opposed by the drag of the downward diffusive flow of the major ion comprising the F-layer, O^+, so that the metal ions tend to slow and stop on the bottomside of the layer. At high polar latitudes, where large magnetospherically induced electric fields prevail, the convective flow of the ions perpendicular to **B** will raise the ions to even higher altitudes. The metal ions will collapse downward when they move into a region where the neutral wind and/or electric field are no longer capable of holding them up by providing an upward force which is directed against gravity. These mechanisms were developed to explain the trends in the high-latitude metal ion distributions shown in Figure 8.14 [e.g., Grebowsky & Brinton (1978); Kumar & Hanson (1980); Grebowsky & Pharo (1985); Bedey & Watkins (1997)]. Due to the sparseness of observations, three dimensional ion distributions are not yet sufficiently known to develop high-latitude models as detailed as those for the low latitudes, where metal ions are almost always present at high altitudes in the ionosphere.

The dynamical models for the high-altitude metal ion distributions have concentrated primarily on only one of the two major metal species, Mg^+ or Fe^+. However, all atomic metal species that are detected in the ablation region are anticipated to be present at high altitudes. The satellite study by Kumar & Hanson (1980) identified the presence of Al^+, Na^+, Si^+ and Ca^+ and calculated their measured relative abundances from those individual mass spectra (8 second scan period) from MIMS on AE-C and AE-D when Fe^+

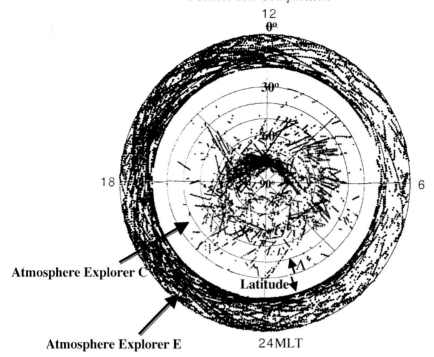

FIGURE 8.14. Measurement locations of Fe^+ ions with concentrations greater than 30 cm^{-3} from AE-C and AE-E between 150 km and 400 km. It is a composite of AE-E low-latitude data, obtained from the National Space Science Data Center-A, superimposed on published AE-C data [Grebowsky & Pharo (1985)] poleward of 30° magnetic latitude. The coordinate of magnetic latitude (Λ) is used in plot for >30°, while the absolute value of the geodetic latitude is used for lower latitudes.

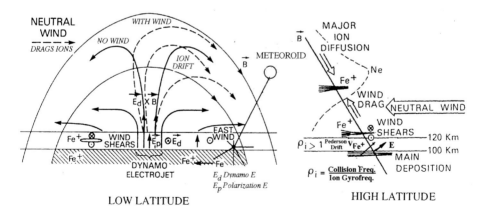

FIGURE 8.15. Mechanisms responsible for transporting metal ions upward from the meteor ablation region. At low latitudes ion vertical motions are associated with ionospheric dynamo electric fields produced by atmospheric winds dragging ions across the magnetic field. At high latitudes magnetospherically induced electric fields are important. At low altitudes, where the ratio, ρ, of the ion-neutral collision frequency to ion gyrofrequency exceeds 1, **E** fields can transport positive ions vertically across **B**. Above ~130 km the neutral wind drags the ions along the magnetic field lines. Downward diffusion flows of O^+ near the F-region peak impedes the upward wind drag of the ions. These figures are modifications of those appearing in Grebowsky & Reese (1989) and Grebowsky & Pharo (1985), with permission from Elsevier Science.

exceeded 500 ions/cm^3. Observed variability in the relative composition of the metal ions was very large, making the significance of mean values somewhat suspect and indicating that transport may have spatially separated the different metal ion species. For example, the Fe$^+$/Mg$^+$ ratio on AE-C was found to vary from 0.02 to 24 with an average of 1.5, and on AE-D the extremes were 0.17 to 34.2 with a mean value of 3.2. The values for Al$^+$/Mg$^+$ and Si$^+$/Mg$^+$ were approximately an order of magnitude lower.

An unresolved issue is the importance of this high-altitude reservoir of long-lived atomic metal ions to the concentration and composition of the main meteor ionization layer. If the upward transport processes separate species in space, then they might have an impact on the main meteor layer composition. The high-altitude metal ion content is comparable to that in the low-altitude layer. Hence, if there is species separation at high altitudes, the question arises as to whether this species separation could introduce compositional changes in the low-altitude layer when they move downward. Could this play a role in the unexplained wide range of Fe$^+$/Mg$^+$ values in the main layer discussed earlier?

8.4. Summary

Rocket-borne ion mass spectrometric measurements discovered permanent metal ion layers in the 85–120 km altitude range. This is the altitude region where meteors are produced and layers of neutral metals are observed. Analysis of the relative abundance of the metals and their isotopes shows general agreement with solar system chrondritic material. The discrepancies are attributed to ion redistribution, ambiguous ion identification, and metal–atmosphere chemistry. One of the most prominent characteristics of the metal ion profiles measured in the rocket experiments is the variability in the distributions from flight to flight. This is due to:

(a) ionosphere/atmosphere dynamics;

(b) atmospheric composition changes; and/or

(c) changes in the incoming distributions of the meteoroids.

The first redistributes the metal ionization. (This is most evident in satellite observations of high-altitude metal ions with vertical contents often comparable to those in the meteor ablation region.) The second leads to changes in chemistry of the metal ions, and the third could introduce variations in the metallic atom deposition rates or, in the case of changing mass or velocity distributions of the infalling material, lead to changes in the altitude where the ablation occurs. Observations taken during meteor shower periods tend to show enhancements in metal ion concentrations. However, the sounding rocket studies have, thus far, not unambiguously separated metal ion distribution perturbations resulting from atmospheric and ionospheric changes from those introduced by changes in the incoming meteoroid properties.

Acknowledgements: The authors would like to acknowledge the helpful review of the manuscript by W. Dean Pesnell of Nomad Research, Inc.

REFERENCES

AIKIN A. C. & GOLDBERG R. A. 1973. Metallic ions in the equatorial ionosphere, *J. Geophys. Res.*, **78**, 734–745.

AIKIN A. C., GOLDBERG R. A. & AZCARRAGA A. 1974. Ion composition during the formation of a mid-latitude E$_s$ layer, *Space Res.*, **XIV**, 283–288.

ANDERS E. & GREVESSE N. 1989. Abundance of the elements: meteoritic and solar, *Geochim. Cosmochim. Acta*, **53**, 194–214.

ARNOLD F. & KRANKOWSKY D. 1977. Ion composition and electron- and ion-loss processes in tho Earth's atmosphere, in *Dynamical and Chemical Coupling*, edited by B. Grandal and J. A. Holtet, Reidel Publishing, Dordrecht, The Netherlands, 93–127.

BEDEY D. F. & WATKINS B. J. 1997. Large scale transport of metallic ions and the occurrence of thin ion layers in the polar ionosphere, *J. Geophys. Res.*, **102**, 9675–9681.

BENNET W. H. 1950. Radiofrequency mass spectrometer, *J. Appl. Phys.*, **21**, 143–147.

BRINTON H. C., SCOTT L. R., PHARO III M. W. & COULSON J. T. C. 1973. The Bennett ion-mass spectrometer on Atmosphere Explorer-C and -E, *Radio Sci.*, **8**, 323.

BROADFOOT A. L. 1967. Twilight Ca^+ emissions from meteor trails up to 280 km, *Planet Space Sci.*, **15**, 503–513.

CARTER L. N. & FORBES J. M. 1999. Global transport and localized layering of metallic ions in the upper atmosphere, *Ann. Geophysicae*, **17**, 190–209.

CEPLECHA Z. 1992. Influence of interplanetary bodies on Earth, *Bull. Amer. Astron. Society*, **24**, 952.

CHIMONAS G. 1969. Ion separation in temperate zone sporadic E and the layer shape, *J. Geophys. Res.*, **74**, 4189–4195.

EARLE G. D., KANE T. J., PFAFF R. F. & BOUNDS S. R. 2000. Ion layer separation and equilibrium zonal winds in midlatitude sporadic E, *Geophys. Res. Letts.*, **27**, 461–464.

FAHEY F. E., FEHSENFELD F. C., FERGUSON E. E. & VIEHLAND L. H. 1981. Reactions of Si^+ with H_2O and O_2 and SiO^+ with H_2 and D_2, *J. Chem. Phys.*, **75**, 669–674.

FESEN C. G. & HAYS P. B. 1982. Mg^+ Morphology from visual airglow experimental observations, *J. Geophys. Res.*, **87**, 9217–9223.

FESEN C. G., HAYS P. B. & ANDERSON D. N. 1983. Theoretical modeling of low-latitude Mg^+, *J. Geophys. Res.*, **88**, 3211–3223.

GÉRARD J.-C. & MONFILS A. 1978. Satellite observations of the equatorial MgII dayglow intensity distribution, *J. Geophys. Res.*, **79**, 2544–2550.

GOLDBERG R. A. 1970. Silicon ions below 100 km – a case for SiO_2^+, *Radio Sci.*, **10**, 329–334.

GOLDBERG, R. A. & AIKIN A. C. 1973. Comet Encke: Meteor metallic ion identification by mass spectrometer, *Science*, **180**, 294–296.

GOLDBERG R. A. & BLUMLE L. J. 1973. Positive ion composition from a rocket-borne mass spectrometer, *J. Geophys. Res.*, **75**, 133–142.

GOLDBERG R. A. & WITT G. 1977. Ion composition in a noctilucent cloud region, *J. Geophys. Res.*, **82**, 2619–2627.

GREBOWSKY J. M. & BRINTON H. C. 1978. Fe^+ ions in the high latitude F-region, *Geophys. Res. Lett.*, **5**, 791–794.

GREBOWSKY J. M., GOLDBERG R. A. & PESNELL W. D. 1998. Do meteor showers significantly perturb the ionosphere? *J. Atm. Sol. Terr. Phys.*, **60**, 607–615.

GREBOWSKY J. M. & PESNELL W. D. 1999. Meteor showers: modeled and measured effects in the atmosphere, *AIAA Paper No. 99-0503*, American Institute of Aeronautics and Astronautics.

GREBOWSKY J. M. & PHARO, III M. W. 1985. The source of midlatitude metallic ions at F-region altitudes, *Planet. Space Sci.*, **33**, 807–815.

GREBOWSKY J. M. & REESE N. 1989. Another look at metallic ions in the F region, *J. Geophys. Res.*, **94**, 5427–5440.

HANSON W. B. & SANATANI S. 1970. Meteoric ions above the F2 peak, *J. Geophys. Res.*, **75**, 5503–5509.

HANSON W. B., STERLING D. L. & WOODMAN R. F. 1972. Source and identification of heavy ions in the equatorial F layer, *J. Geophys. Res.*, **77**, 5530–5541.

HERRMANN U., EBERHARDT P., HIDALGO M. A., KOPP E. & SMITH L. G. 1978. Metal ions and isotopes in sporadic E-layers during the Perseid meteor shower, *Space Res.*, **XVIII**, 249–252.

HOFFMAN J. H., HANSON W. B., LIPPINCOTT C. R. & FERGUSON E. F. 1973. The magnetic ion mass spectrometer on Atmosphere Explorer, *Radio Sci.*, **8**, 312–315.

ISTOMIN V. G. 1963. Ions of extra-terrestrial origin in the Earth atmosphere, *Space Research*, **3**, 209–220.

JOHANNESSEN A. & KRANKOWSKY D. 1974. Daytime positive ion composition measurement in the altitude range 73–137 km above Sardinia, *J. Atmos. Terr. Phys.*, **36**, 1233–1247.

JOHNSON C. Y., HEPPNER J. P., HOLMES J. C. & MEADOWS E. B. 1955. First investigation of ambient positive-ion composition to 219 km by rocket-borne spectrometer, *Ann. Geophys.*, **14**, 475–482.

KELLEY M. C. 1989. *The Earth's Ionosphere*, Academic Press, London. 487 pp.

KOPP E. 1997. On the abundance of metal ions in the lower ionosphere, *J. Geophys. Res.*, **102**, 9667–9674.

KOPP E. & HERRMANN U. 1984. Ion composition in the lower ionosphere, *Ann. Geophys.*, **2**, 83–94.

KOPP E., EBERHARDT P., HERRMANN U. & BJORN L. G. 1985. Positive ion composition of the high-latitude summer D region with noctilucent clouds, *J. Geophys. Res.*, **90**, 13,041–13,053.

KRANKOWSKY D., ARNOLD F., WIEDER H. & KISSEL J. 1972. The elemental and isotopic abundance of metallic ions in the lower E-region as measured by a cryogenically pumped quadrupole mass spectrometer, *Int. J. Mass Spect. Ion Phys.*, **8**, 379–390.

KUMAR S. & HANSON W. B. 1980. The morphology of metal ions in the upper atmosphere, *J. Geophys. Res.*, **85**, 6783–6801.

MCNEIL W. J. 1999. Problems in the prediction of meteor shower effects on the atmosphere, *AIAA Paper No. 99-0506*, American Institute of Aeronautics and Astronautics.

MCNEIL W. J., DRESSLER R. A. & MURAD E. 2001. The impact of a major meteor shower on the Earth's ionosphere: A modeling study, *J. Geophys. Res.*, **106**, 10,447–10,465.

MCNEIL W. J., LAI S. T. & MURAD E. 1996. A model for meteoric magnesium in the iono-sphere, *J. Geophys. Res.*, **101**, 5251–5259.

MCNEIL W. J., LAI S. T. & MURAD E. 1998. Differential ablation of cosmic dust and impli-cations for the relative abundances of atmospheric metals, *J. Geophys. Res.*, **103**, 10,899–10,911.

NARCISI R. S. 1968. Processes associated with metal-ion layers in the E-region of the ionosphere, *Space Res.*, **8**, 647–658.

NARCISI R. S. 1971. Composition studies of the lower ionosphere, in *Physics of the Upper Atmosphere*, edited by F. Verniani, Editrice Compositore, Bologna, Italy. 11–59.

NARCISI R. S. 1973. Mass spectrometer measurements in the ionosphere, in *Physics and Chem-istry of the Upper Atmosphere*, edited by B. M. McCormac, D. Reidel Publishing Co., Dordrecht, Holland, 171–183.

NARCISI R. S. & BAILEY A. D. 1965. Mass spectrometric measurements of positive ions at altitudes from 64 to 112 kilometers, *J. Geophys. Res.*, **70**, 3687–3700.

PHILBRICK C. R., NARCISI R. S., GOOD R. E., HOFFMAN H. S., KENESHEA T. J., MCLEOD M. A., ZIMMERMAN S. P. & REINISCH B. W. 1973. The Aladdin Experiment – Part II, composition, *Space Res.*, **VII**, 441–449.

PLANE J. M. C. 1991. The chemistry of meteoric metals in the Earth's upper atmosphere, *Int. Rev. Phys. Chem.*, **10**, 55–106.

PLANE J. M. C., COX R. M. & ROLLASON R. J. 1999. Metallic layers in the mesopause and lower thermosphere region, *Adv. Space Res.*, **24(11)**, 1559–1570.

STEINWEG A., KRANKOWSKY D., LÄMMERZAHL P. & ANWEILER B. 1992. Metal ions in the auroral lower E-region measured by mass spectrometers, *J. Atmos. Terr. Phys.*, **54**, 703–714.

SWIDER W. 1969. Processes for meteoric elements in the E-region, *Planet. Space Sci.*, **17**, 1233–1246.

SWIDER W. 1984. Ionic and neutral concentrations of Mg and Fe near 92 km, *Planet. Space Sci.*, **32**, 307–312.

VON ZAHN U., GOLDBERG R. A., STEGMAN J. & WITT G. 1989. Double peaked sodium layers at high latitudes, *Planet. Space Sci.*, **37**, 657–667.

VON ZAHN U., GERDING M., HÖFFNER J., MCNEIL W. J. & MURAD E. 1999. Iron, calcium, and potassium atom densities in the trails of Leonids and aother meteors: Strong evidence for differential ablation, *Meteor. Planet. Sci.*, **34**, 1017–1027.

WHITEHEAD J. D. 1961. The formation of the sporadic E-layer in the temperate zones, *J. Atmos. Terr. Phys.*, **20**, 49–58.

ZBINDEN P. A., HIDALGO M. A., EBERHARDT P. & GEISS J. 1975. Mass spectrometer mea-surements of the positive ion composition in the D- and E-regions of the ionosphere, *Planet. Space. Sci.*, **23**, 1621–1642.

9

COLLECTED EXTRATERRESTRIAL MATERIALS: INTERPLANETARY DUST PARTICLES, MICROMETEORITES, METEORITES, AND METEORIC DUST

By FRANS J. M. RIETMEIJER[†]

Institute of Meteoritics, Department of Earth and Planetary Sciences, University of New Mexico, Albuquerque, NM 87131, USA

An average meteoric vapor composition in the Earth's atmosphere is maintained by ablation of (1) ordinary chondrite and lesser amounts of iron meteoroids, (2) unique CM-type micrometeoroids from the asteroid belt, and (3) carbon-rich chondritic interplanetary dust particles from comet nuclei and carbonaceous and 'ultra-carbonaceous' bodies in the outer asteroid belt. The OC meteorite-delivering meteoroids are mostly near-Earth asteroids. Calculated average mesospheric Fe, Ca and Na abundances based on the relative masses of these meteoroids agree well with observed metal abundances and predict high middle and upper atmospheric silicon and carbon abundances. Quantitative determination of latter will constrain the atmospheric interactions and compositions of meteors with very high beginning heights of ablation. Fireball and meteor compositions, differential meteor ablation, their physical properties and the sizes of their constituent grains can be constrained using the physical, chemical and petrographic properties of collected extraterrestrial materials.

9.1. Introduction

9.1.1. Definitions

I will highlight petrologic, physical and chemical properties of collected extraterrestrial materials that survived entry into the Earth's atmosphere to discuss (1) types of incoming meteoroids, (2) atmospheric interactions during deceleration, (3) thermal modification of pre-entry properties, and (4) the contributions of their ablation products to the mesospheric metal abundances.

Meteoroids are objects traveling through interplanetary space after being liberated from parent bodies that include comets, asteroids, Mars and the Moon. When decelerating in the upper atmosphere by collisions with air molecules, kinetic energy is dissipated as thermal radiation and as heat for meteoroid melting as well as evaporation. During this brief (5–15 s) thermal event meteoroids smaller than ~ 400 μm behave as isothermal bodies [Elford *et al.* (1997)]. The interior temperature of larger (>cm-sized) meteoroids is unaffected underneath the black fusion crust (~ 3 mm thick) that develops at the body's surface [Ramdohr (1967); Blanchard & Cunningham (1974)] following significant mass loss by melting and evaporation. A thin (a few nm) Fe-oxide rim of probable similar origin to this fusion crust can be found on unmelted or partially melted micrometeorites [Engrand & Maurette (1998); Taylor *et al.* (2000)] and interplanetary dust particles [Rietmeijer (1998a)].

Large meteoroids produce *meteors* or *fireballs* with a luminous persistent train. Smaller meteors experience flash-heating that causes evaporation ('shooting stars') or melting,

[†] Email: `fransjmr@unm.edu`

either completely or partially. Yet another fraction survives more-or-less intact but with thermal, so-called dynamic pyrometamorphic, modification of the original meteoroid [Rietmeijer (1998a); Rietmeijer (1998b)]. Cometary dust traveling at more than ∼20 km/s decelerates at 120–100 km altitudes and at ∼80 km altitude reaches its 'rest velocity', also described as 'terminal' or 'settling' velocity. Radar observations show ablation may be initiated at ∼140 km altitude [Elford et al. (1997)]. Ablation of several Leonid meteors (71 km/s) began at ∼ 155 km as a distinct 'diffuse phase' before developing a typical meteor train [Spurný et al. (2000)]. These high altitudes of beginning meteor ablation suggest the presence of low-boiling organic and inorganic compounds. Ablation of slower-moving asteroidal debris (∼12 km/s) typically begins in the mesosphere at >60 km altitude to end, often explosively, in the stratosphere below ∼30 km altitude [Halliday et al. (1996); Rietmeijer (2000)].

Meteors that survive deceleration fall towards the Earth's surface where they can be collected as *micrometeorites* (MMs) and *meteorites*. A meteorite is called a 'fall' when, after a fireball was witnessed, there was immediate collection of meteorite fragments. Most meteorites are incidental 'finds' or part of systematic recovery campaigns in environments where geological processes acted to concentrate individual 'falls'. For example, Antarctic meteorites were concentrated as a result of land-ice flow and ablation and Saharan meteorites were concentrated by aeolian ablation. Fossil meteorites occur in sedimentary rocks of geological age [Thorslund & Wickman (1981); Nystrom et al. (1988)] and are also found on other solar system bodies, such as the Bench Crater meteorite on the Moon [Zolensky (1997)].

Micrometeorites (<500 µm) are mostly recovered from land-ice in Antarctica and Greenland because the concentration of MMs is readily achieved by melting of the ice [Maurette et al. (1994)], and deep-sea sediments [Brownlee (1985)]. Recovery from other deposits such as lithified sediments typically requires robust concentration procedures [Maurette et al. (1987); Robin et al. (1990); Kettrup et al. (2000)]. They are mostly spheres known as cosmic spherules but Antarctic MMs in particular also include partially melted, highly vesicular and unmelted MMs often with fragile structures [Engrand & Maurette (1998); Taylor et al. (2000)].

Meteoroids typically smaller than MMs are collected in the stratosphere between 17–19 km altitudes when still settling towards the Earth's surface [Rietmeijer (1998a)]. This stratospheric dust is correctly referred to as *interplanetary dust particles* (IDPs). It includes primitive unmelted aggregates, partially melted IDPs and massive spherules as well as spall fragments and often hollow melt droplets, so-called secondary spheres, from ablating meteors [Brownlee (1985); Flynn (1994a)]. Condensation of meteoric vapors will result in the formation of *meteoric dust*, while metals in these vapors are the sources for the metals in the mesosphere.

Meteoriticists accept the Orgueil meteorite composition, a CI meteorite (see below), as the bulk composition of the solar system [Anders & Grevesse (1989)]. The CI composition matches the solar photosphere abundances except for the light elements, C, H, O and N, which are lower in the meteorites. This composition is a standard for comparison of other solar system materials. The formulae of minerals used in this chapter are shown in Table 9.1.

9.1.2. Influx

The near-Earth meteoroid flux curves are smooth continuous functions of objects varying in size from kilometer-sized bodies down to micrometer-sized dust [Hughes (1994)]. A gap between dust and meteoroids and large meteoroids, asteroids and comets in the accretion or global terrestrial influx curves (Table 9.2) reflects different atmospheric interactions

TABLE 9.1. Minerals in meteorites, micrometeorites and interplanetary dust particles discussed here

Name	General formula	End-members	Formula
Silicates			
Silica	SiO_2		
Olivine solid solution series	$(Mg,Fe)_2SiO_4$	Forsterite (Fo)	Mg_2SiO_4
		Fayalite(Fa)	Fe_2SiO_4
Pyroxene solid solution series	$(Mg,Fe)_2Si_2O_6$	Enstatite (En)	$Mg_2Si_2O_6$
		Ferrosilite(Fs)	$Fe_2Si_2O_6$
Ca-rich pyroxene	$Ca(Mg,Fe)Si_2O_6$		
Feldspar solid solution series	$(Na,K)AlSi_3O_8-CaAl_2Si_2O_8$		
Wüstite	FeO		
Magnetite	Fe_3O_4		
Hematite	$\alpha\text{-}Fe_2O_3$		
Maghémite	$\gamma\text{-}Fe_2O_3$		
Magnesioferrite	$MgFe_2^{+3}O_4$		
Chromite	$FeCr_2O_4$		
Perovskite	$CaTiO_3$		
Goethite	$FeOOH$		
Ferrihydrite	$Fe_2O_3\cdot2FeOOH\cdot2.6H_2O$		
Fe,Ni-sulfides			
Troilite	FeS		
Pyrrhotite	Fe_7S_8		
Pentlandite	$(Fe,Ni)_9S_8$		
Fe,Ni-metal:kamacite	$\alpha\text{-}Fe,Ni(<6\ wt.\%\ Ni)$		

TABLE 9.2. The mass-based meteoroid classification [Hughes (1994); Steel (1997)]

Mass range (kg)	Classification
$>10^{20}-10^8$	Comets, comet fragments and asteroids
10^8-10^{-2}	Fireballs (meteorite-producing meteoroids)
$10^{-2}-10^{-6}$	Meteors(complete meteoroid evaporation)
$10^{-6}-<10^{-24}$	Micrometeors, comet dust: $\sim95\%$ of incoming meteoroids

for incoming size/mass groups, such as 'conventional' meteorites and MMs during deceleration rather than a genetic discontinuity. The gap occurs at 0.1–1 kg for objects ~4–10 cm in diameter (assuming $\rho = 2.5$ g/cm^3). The total global annual influx is estimated to be $(1.7$–$2.2) \times 10^8$ kg with $\sim80\%$ due to an occasional large object [Hughes (1994); Steel (1997)]. The heaviest meteoroids include almost equal fractions from comets and asteroids. Meteoroids 4×10^3 kg to 2×10^6 kg are structurally weak comet-like objects, 2–15 m in diameter ($\rho = 1$ g/cm^3) [Ceplecha *et al.* (1997)]. Most millimeter-size meteors are porous, crumbly objects made of loosely conglomerated sponge-like material [Verniani (1969)]. The mass accretion of meteoroids ~20–500 μm in diameter is $(40\pm20) \times 10^6$ kg/year [Love & Brownlee (1993)] from progenitors ~30 μm to ~1 millimeter in size.

Catastrophic meteor disruption in the lower stratosphere is well-documented [Halliday *et al.* (1996)]. The fragments will continue to ablate with variable, as yet unknown, effi-

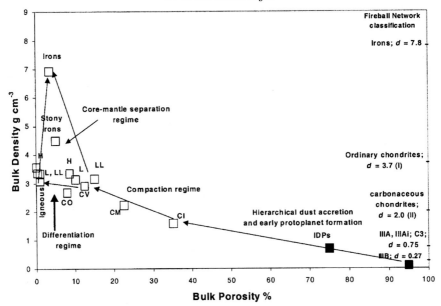

FIGURE 9.1. Measured bulk density (g/cm^3) and porosity of meteorite groups (open squares) and IDPs (solid squares) (modified after Rietmeijer & Nuth (2000a), which also provides a complete listing of the original data). For comparison the Fireball Network classification and its assumed meteoroid density (after [Ceplecha *et al.* (1998)]) is shown on the right side of the diagram, whereby the IIIA, IIIAi and C3 groups refer to 'regular cometary material' and the IIIB group refers to 'soft cometary material' [Ceplecha *et al.* (1998)].

ciency at lower velocities in a denser atmosphere. Fragments may include millimeter-sized chondrules and Ca, Al-rich inclusions from stony chondrite meteoroids of which the tensile strength will be a function of accretion history and parent body alteration including shock impact metamorphism. Weak structural bonds determining meteor fragmentation behavior can be the interfaces of aggregated entities and shock-induced fractures. The shower from the final break-up of the Mbale meteorite produced a strewnfield of large angular fragments with smaller fragments that broke up at higher altitudes as a result of as yet unappreciated physical atmospheric interactions [Jenniskens *et al.* (1994)]. An analysis of the mass distribution of fragments in sixteen different meteorite showers showed two different types of scaling exponents, viz. (1) a simple power law that was attributed to a single fragmentation event, and (2) a multiple scaling regime indicating multiple major fragmentation events [Odderschede *et al.* (1998)]. Showers of ordinary stony chondrites can show both single and multiple fragmentation while stony-iron and iron meteors display single-fragmentation. Multiple-fragmentation of the Murchison (CM) stony meteor indicates a 'structurally weak' bolide [Odderschede *et al.* (1998)]. There is a serious lack of information on tensile strength, density and other physical properties of asteroids and comets. This type of information could be obtained from laboratory analyses of the collected meteorites, MMs and IDPs. This bleak picture is rapidly improving with regard to meteorite density and porosity (Fig. 9.1) that broadly reflect a general trend in dust accretion and the evolution of structural properties of protoplanets [Rietmeijer & Nuth (2000a)]. This is also the case for elastic and shear moduli, compressive strength and seismic wave velocities in meteorites [Flynn *et al.* (1999)].

9.2. Constraints on Sources of Extraterrestrial Materials

9.2.1. *Primitive Protoplanets*

The major causes of releasing meteoroids from their parent bodies are (1) catastrophic disruption [Keil *et al.* (1994)], (2) crater-producing impacts such as seen on the asteroids Eros, Mathilda, Gaspra and Ida, and (3) ice sublimation on active comet nuclei during perihelion. Other causes include volcanism and electrostatic dust levitation [Flynn (1994a)]. Radar mapping of asteroids [Hudson & Ostro (1994)], fly-by missions to asteroids and P/comet Halley, ground-based telescopic and Hubble Space Telescope observations, e.g. fragmentation of the comet LINEAR nucleus and the 'string-of-pearls' comet Shoemaker–Levy, have dramatically altered our ideas about the physical properties of protoplanets. They are mostly structurally heterogeneous bodies or rubble piles held together by self-gravitation among the boulders, 'glued together' by dirty-ice (comet nuclei; icy asteroids), or some form of post-accretion activity such as regolith-formation [Wilson *et al.* (1999a)]. The latter will be a complex interplay between thermal and aqueous alteration and explosive protoplanet disruption and re-accretion [Muenow *et al.* (1995); Wilson *et al.* (1999b)]. Conventional wisdom that meteorites are asteroidal and dust is cometary is no longer strictly valid in the light of rubble pile models. The finding of collimated jets on P/comet Halley [Vaisberg *et al.* (1986)] showed that the released dust is from the dirty-ice 'glue' in between boulders. It is not clear if dust from the refractory carbonaceous mantle on a nucleus is also present among the ejected dust. Sources in the solar system that could produce meteoroids in Earth-crossing orbits include:

1. Active short- and long-period comets: (a) Oort cloud comets, and (b) Kuiper Belt Objects [Brownlee (1985); Flynn (1996); Shearer *et al.* (1998)],
2. Main belt Asteroids [Kerridge & Matthews (1988); Bell *et al.* (1989)],
3. Near-Earth Asteroids (NEAs) including the numerous small (<50 m) Earth-crossing objects [Chyba (1993)] or 'small Earth-approachers' [Rabinowitz *et al.* (1993)]. About half may be cometary in origin [Weisman *et al.* (1989)] and include active comets, 'extinct' comets, 'dark' asteroids such as Pholus [Hartmann *et al.* (1987)] and 'active asteroids' [McFadden *et al.* (1993)],
4. Mars [McSween & Treiman (1998)],
5. The Moon [Shearer *et al.* (1998)],
6. The Martian satellites Phobos and Deimos and the Jovian and Saturnian satellites, and
7. Dust rings around giant planets [Grün *et al.* (1993)] and the Sun [Dermott *et al.* (1994)].

Do we know the source of a collected meteorite, micrometeorite or IDP? The answer to this question has to be 'No, but'.

9.2.2. *Meteorites*

Meteoroids follow a chaotic path to 1 AU after release from a parent body [Wisdom (1985)], which makes it virtually impossible to retrace the sources that are believed to be located in the asteroid belt between ~2 and ~4.5 AU. Only the origin of Lunar meteorites is uniquely determined by comparison with the lunar rocks brought to Earth during the Apollo project. A Martian origin of the Shergottite–Nakhlite–Chassignite meteorite suite is still unproven but rests on a strong case of circumstantial evidence including a Viking measurement of the Martian atmosphere. Other meteorite-parent body links are tenuous. In the case of the asteroid 4-Vesta a similarity between its infrared (IR) reflectance spectrum and those for the howardite–eucrite–diogenite meteorite suite is the only indication of a parent body relationship. The lack of asteroid and comet nucleus samples is a serious problem in the search for parent bodies.

Establishing parent–meteorite relationships relies on IR spectral matches in combination with albedo data. The asteroid belt shows a stratification of spectral properties as a function of heliocentric distance [Bell *et al.* (1989)], which is intuitively appealing and simple. Meteoriticists readily accepted it. Matching of asteroid and meteorite reflectance spectra shows that the most evolved asteroids, such as 4-Vesta, are closest to the Sun and that primitive carbonaceous chondrite asteroids are more distal whereas the more primitive ultra-carbonaceous icy-asteroids are even farther from the Sun [Bell *et al.* (1989)]. This simple spatial distribution of asteroids has a major problem. Namely, the ordinary chondrites that make up ~80% of all meteorites have no spectral match among bodies in the asteroid belt. Space weathering at asteroid surfaces that obliterates the IR reflectance properties of the pristine rocks was proposed to get around this conundrum. Advances were made to support this hypothesis [Pieters *et al.* (2000)] but Meibom & Clark (1999) have argued that ordinary chondrites were never abundant in the asteroid belt. Where then do they come from? These meteorites still reach the Earth and there must be reservoirs to replenish them on a time scale less than the orbital lifetime. Alternative sources for ordinary chondrites might be sought among NEAs that originally resided in the asteroid belt. An important implication of meteoroids from NEAs is that they have higher atmospheric entry velocities than meteoroids from the asteroid belt as well as experience higher entry temperatures. Rietmeijer (2000) found that entry velocities of meteors from the Canadian camera network survey, known as the Meteorite Observation and Recovery Program (MORP) [Halliday *et al.* (1996)], range from 22–36.5 km/s compared with ~11 km/s expected for meteoroids from the asteroid belt. Without orbital and velocity data it is impossible to determine a meteor's origin. These data might be obtained from space-based measurements. Or, by monitoring the luminous fireball train of an incoming meteoroid such as the Lost City, Innesfree, Pribram, Peekskill and Mbale [reviewed in Rietmeijer (2000)] and Glanerbrug ordinary chondrite meteorites [Jenniskens *et al.* (1992)], and the Tagish Lake carbonaceous chondrite meteorite [Brown *et al.* (2000)].

9.2.3. *MMs*

Micrometeorites range from texturally unmodified fine-grained aggregates, to highly vesicular MMs due to degassing of volatile components to quenched-melt spheres [Brownlee *et al.* (1983)]. They lost all information on their pre-entry orbits and velocities, which means that a connection to parent bodies can only be inferred from petrographic and chemical properties. Kurat *et al.* (1994) and Brownlee *et al.* (1997) found that ~95% of MMs have an affinity with CM-type meteorites. The remainder has affinities to CR-type carbonaceous [Engrand & Maurette (1998)] and ordinary chondrites [Robin *et al.* (1990); Beckerling & Bischoff (1995); Brownlee *et al.* (1997)]. The overwhelming CM-type affinity might indicate that CM parent bodies are more common in the asteroid belt than can be inferred from the number of CM meteorites or that the physical properties of CM material favor survival of meteoroids that are <1 mm in diameter.

9.2.4. *IDPs*

Mackinnon & Rietmeijer (1987) showed that chondritic IDPs differ significantly in form and texture from the components of 'conventional' meteorites and contain some mineral assemblages that do not occur in any meteorite class. Thus, the IDPs and meteorites are from different parent bodies. Because of orbital dynamics, a meteorite-sized object from most distant icy-asteroids will have a very low probability of crossing the Earth's orbit but dust from these outermost fringes of the asteroid belt could reach the inner solar system as IDPs. The reflectance spectra (450–800 nm) including 'spectral reddening' of carbon-bearing aggregate IDPs resemble these properties of C, P and D IR-class outer-belt asteroids [Bradley *et al.* (1996)] and those of many NEAs [Hartmann *et al.* (1987)]. One needs additional information to link IDPs to asteroids, Oort cloud comets, or Kuiper Belt objects. Among other things, this information could be atmospheric entry velocity [Brownlee *et al.* (1995)], a meteor shower association [Rietmeijer & Jenniskens (2000)], or a unique density of solar flare tracks [Flynn (1996)].

9.2.5. *Origins of CI- and CM-type Materials*

The CI meteorites and CI-type IDPs are hydrated 'chunks of clay' that are mostly free of inclusions of any kind. An aqueous fluid necessary for chemical transport to form layer silicates by hydration of olivines and pyroxenes will aid the collapse of an originally porous structure. Layer silicate formation requires a parent body wherein temperatures were high enough and were maintained long enough for ice to melt. With our current understanding of small solar system bodies, this environmental constraint points to asteroids as sources for CI meteorites. Aqueous alteration could also have occurred in the past or still occur during perihelion passage in active comet nuclei but only in dirty-ice environments by a process known as hydrocryogenic alteration. Reactions occur at temperatures below the melting point of ice due to a nanometer-thin layer at the interface of dust and ice that is the conduit for dissolved chemical species [Rietmeijer (1998a)]. Hydrocryogenic alteration relaxes the thermal constraints on hydration and allows an early onset of aqueous alteration in icy protoplanets. Gombosi & Houpis (1986) suggested that boulders in a comet nucleus 'rubble pile' resemble CI-like material, which is consistent with CI-like fireballs entering the Earth's atmosphere with cometary orbits and velocities [Ceplecha *et al.* (1998)]. The structurally coherent CI materials might actually range from pebbles [Spurný & Borovička (1999)] to objects of many hundreds of meters in diameter [Gombosi & Houpis (1986)]. Rietmeijer & Nuth (2000a) suggested that some might be proto-CI material that is more Si-rich than (hydrated) CI material.

The IR reflectance spectra of compact chondritic aggregate IDPs, which are partially or completely hydrated, resemble C-class asteroids [Bradley *et al.* (1996)] in the outer asteroid belt that include the carbonaceous chondrite parent bodies. Scarcity of type-CI IDPs and MMs and CI meteorites could be an artifact of recovery because they are highly susceptible to alteration at the Earth's surface. Or it could support disintegration, evaporation, or both, of CI-meteoroids during atmospheric entry and deceleration. The IR spectral properties and low albedo of CM meteorites and C-type asteroids suggest that CM material might dominate the central region of the asteroid belt. While CM meteorites are rare, CM-type MMs are common consistent with CM meteors being structurally weak materials [Odderschede *et al.* (1998)]. The stratospheric entry times for CM-type IDPs are correlated with dates of CM meteorite falls [Rietmeijer (1996a)]. The size bins for CM materials, viz. (1) centimeter-sized meteorites, (2) 20–500 μm-sized MMs, and (3) unmelted IDPs <30 μm, might be correlated with a low mechanical strength, the nature of the removal mechanisms from the parent bodies, or both.

9.3. Meteorites

9.3.1. *Classification*

Reviews on all aspects of meteorites plus excellent photographs of meteorite features can be found in Dodd (1981) and Kerridge & Matthews (1988). Iron (I) and stony-iron (S-I) meteorites are subdivided into many groups based on variations of petrographic

CHONDRITES

Carbonaceous (C)						Ordinary (O)			Enstatite (E)	
CI	CM	CR	CO	CV	CK	H	L	LL	EH	EL
1	2	2	3	3	3-6	3-6	3-6	3-6	3-5	3-6

FIGURE 9.2. Chondrite meteorite classification in three classes subdivided into several groups each with their own petrographic type of 1 to 6 [modified after Brearley & Jones (1998)].

textures, mineralogy, minor, trace and rare-earth element distributions and stable isotope chemistry such as oxygen isotopes. The interrelationships among these groups reveal a complex history [Mittlefeldt *et al.* (1998)]. In a nutshell, I- and S-I meteorites are genetically related in a scenario of core–mantle formation on asteroids, similar to what occurred in the Earth, whereby I-meteorites represent the core and S-I meteorites the core–mantle boundary. Stony achondrites are evidence for igneous activity and volcanism on asteroids. They too include a variety of different classes.

Chondrites are divided on the basis of composition and petrographic properties into three classes: carbonaceous (C), ordinary (O) and enstatite (E) chondrites. Each class is subdivided into closely related groups with distinct bulk and oxygen isotopic compositions, such as CI and CM, which refer to the 'type-meteorites' Ivuna and Mighei (Fig. 9.2). The OC meteorites include three groups based on total iron content, viz. H (high-Fe), L (low-Fe), LL (very low-Fe). The enstatite chondrites similarly include two groups (Fig. 9.2). The primitive CC class includes meteorites that show the two distinct evolutionary paths, aqueous (hydrous) alteration or thermal metamorphism, taken by their parent bodies [Brearley & Jones (1998)].

The type 3 CO and CV meteorites are believed to be the original material that yielded types 2 (CM) and 1 (CI) meteorites as a result of increasing intensity of aqueous alteration and the types 4 to 6 due to increasing levels of thermal metamorphism (Fig. 9.2). The commonly accepted notion that CM and CI meteorites represent increasing levels of CO/CV meteorite hydration implies the efficient dissolution of chondrules by aqueous alteration. This efficiency probably doesn't exist to produce CI meteorites. The logical consequence is that CI meteorites are genetically disconnected from the other chondrites and represent a unique and different provenance (section 9.2.5).

As one might expect for natural materials, the distinction between individual groups and types can become diffuse. Increased intensity of thermal alteration in metamorphic O, E and CO meteorites can be traced by changes in the compositions of silicate minerals, re-crystallization of matrix minerals, devitrification of chondrule glass and increasingly diffuse chondrule–matrix boundaries [Brearley & Jones (1998)]. Densification is one effect of increasing levels of thermal metamorphism and/or aqueous alteration, probably along with a commensurate increase in tensile strength. Shock metamorphism [Brearley & Jones (1998)] tends to structurally weaken rocks near the surface of asteroids although impact melts might act to strengthen shocked rocks. These processes interact chaotically over the course of a lifetime of a meteoroid-to-be. It will be difficult to predict the physical properties of individual meteoroids. It is a cause for concern because meteorite survival may be biased to favor particular physical properties.

9.3.2. *Petrography and Chemistry*

The C, O and E chondrites are lithified aggregates of fine-grained matrix, chondrules, Ca/Al-rich inclusions (CAIs) and Fe/Ni metal grains, each with its own history prior

TABLE 9.3. Approximate vol. % of components in chondrites (after [Brearley & Jones (1998)]).
Chondrule diameters in millimeters are shown in square brackets

	Matrix	Chondrules	CAIs	Fe, Ni-metal
CI	>99	≪1	≪1	0
CM	70	20 [0.3]	5	0.1
CR	30–50	50–60 [0.7]	0.5	5–8
CO	34	48 [0.15]	13	1–5
CV	40	45 [1.0]	10	0–5
H	10–15	60–80 [0.3]	0.1–1?	10
L	10–15	60–80 [0.7]	0.1–1?	5
LL	10–15	60–80 [0.9]	0.1–1?	2
EH	<2–5?	60–80 [0.2]	0.1–1?	8
EL	<2–5?	60–80 [0.6]	0.1–1?	15

to aggregation. Chondrules are sub-millimeter high-temperature ($\sim 1500\,^{\circ}$C) quenched-melt spheres [Brearley & Jones (1998); Jones *et al.* (2000)]. The CAIs, the oldest (4.559 Ga) materials identified in the solar system, are evaporation residues of high-temperature refractory minerals [Jones *et al.* (2000)]. They can be spheres or irregular objects ranging from <1 mm to >1 cm in diameter. The largest CAIs, >1 cm, up to 2.5 cm, are found in CV meteorites. The proportions of these components vary from fine-grained matrix of CI meteorites to O- and E-chondrites with high chondrule contents (Table 9.3).

Average bulk compositions for OC, EC, CC, HED achondrites and stony-iron meteorites are listed in Table 9.4. Iron and nickel are the only major elements in I-meteorites. The use of bulk compositions is justified because the subtle compositional differences among groups will be lost in fireballs and meteors. The CI bulk and matrix as well as the CM bulk composition (Table 9.5) show the proposed effect of hydration of proto-CI material whereby according to Rietmeijer and Nuth (2000a) SiO_2 was removed in an aqueous fluid. Flynn *et al.* (1995) found that the matrix of the OC meteorite Semarkona had a slightly enriched Ni content (rel. to CI), was depleted in sulfur ($\sim 0.5 \times$ CI) and carbon ($0.4 \times$ CI). In a study of ten different UC meteorites, Makjanic (1990) found highly variable carbon contents in the matrix ranging from 0.03 to 1.27% C, corresponding to an average CI-normalized carbon content of 0.09 with a range of 0.01 to 0.4.

The lithophile (silicate-forming) element abundances in chondrules are generally similar to the host-chondrite composition but chondrules are on average depleted in siderophile (metal-forming) and chalcophile (sulfide-forming) elements [Grossman *et al.* (1988)]. Fireball and meteor chemistry is, within the error bars of data reduction, determined by the bulk compositions (Table 9.4). Even after fragmentation when chondrules and matrix, a lithified ultrafine-grained aggregate, interact independently with a denser atmosphere, the bulk composition remains a good proxy for the ablating materials. Iron is mostly present in the matrix, chondrules and metal inclusions. The CAIs are rich Al, Ca and Mg while Ti is common. Their refractory minerals (Table 9.6) have boiling points ranging from $\sim 1,000\,^{\circ}$C to $\sim 1,400\,^{\circ}$C (at 10^{-4} atm). After meteor fragmentation these CAIs might become millimeter- to centimeter-sized meteors.

TABLE 9.4. Si-normalized bulk compositions (atomic) for several meteorite types: S-I meteorite (average calculated by the author for a 50/50 mixture of pallasite olivine [Mittlefeldt *et al.* (1998); table 8] and I-meteorite), average HED achondrites [from data compiled by Mittlefeldt *et al.* (1998); Table 34], and EC, OC, and CC compositions [Dodd (1981); Table 2.1]

	Achondrites		Stony meteorites Chondrites		
	S-I	HEDs	Enstatite	Ordinary	Carbonaceous
C/Si	n.a.	n.a.	0.08	Trace	0.31
H/Si	n.a.	n.a.	Trace	Trace	2.04
Mg/Si	1.76	0.40	0.76	0.94	1.06
Fe/Si	2.88	0.29	0.75	0.70	0.83
S/Si	n.d.	n.a.	0.21	0.10	0.22
Al/Si	n.d.	0.22	0.05	0.07	0.09
Ca/Si	n.d.	0.16	0.04	0.05	0.07
Na/Si	n.d.	0.012	0.05	0.05	0.04
Ni/Si	0.14	n.a.	0.040	0.037	0.044
Cr/Si	0.001	0.01	0.069	0.009	0.012
Mn/Si	0.006	0.009	0.004	0.008	0.006
Ti/Si	n.d.	0.006	0.001	0.002	0.003
K/Si	n.d.	n.a.	0.003	0.003	0.003
P/Si	0.095	n.a.	0.011	0.008	0.008

TABLE 9.5. Si-normalized atomic ratios of CI matrix, bulk CI and CM chondrites [Brownlee (1978)], opaque matrix of OC meteorites [Huss *et al.* (1981)] and anhydrous proto-CI material [Rietmeijer & Nuth (2000a)]

	CI matrix	Bulk CI	Bulk CM	UOC matrix	Proto-CI
C/Si	0.6–0.8	0.7	No data	See text	0.52
Mg/Si	0.92	1.06	1.04	0.36	0.80
Fe/Si	0.54	0.90	0.84	1.04	0.67
S/Si	0.13	0.46	0.23	See text	0.35
Al/Si	0.09	0.085	0.08	0.12	0.07
Ca/Si	0.011	0.071	0.072	0.06	0.04
Na/Si	0.016	0.060	0.035	0.083	0.05
Ni/Si	0.047	0.051	0.046	See text	0.03
Cr/Si	0.012	0.013	0.012	0.011	0.010
Mn/Si	0.005	0.009	0.006	0.012	0.006
Ti/Si	0.001	0.002	0.002	0.002	0.002
K/Si	0.015	0.010	0.004	0.024	0.003

9.3.3. *Collection*

In 1997, >12,000 meteorites were recovered but with many ongoing recovery campaigns this number has climbed steadily. The petrographic, chemical and isotopic properties often make it possible to trace individual meteorites, and fragments from a meteorite shower, to a common parent body. A simple petrographic division recognizes iron (I),

TABLE 9.6. Refractory minerals in CAIs

Spinel	$MgAl_2O_4$
Diopside	$CaMgSi_2O_6$
Fassaite	$Ca(Mg, Ti, Al)(Si, Al)_2O_6$
Anorthite	$CaAl_2Si_2O_8$
Gehlenite	$Ca_2Al_2SiO_7$
Corundum	Al_2O_3
Perovskite	$CaTiO_3$
Hibonite	$CaAl_{11}Ti_{0.5}Mg_{0.5}O_{19}$

TABLE 9.7. Relative proportions [McSween (1999)] and mass [Bagnall (2000)] fractions (%) of meteorite classes for 776 non-Antarctic meteorite 'falls' compared with the proportion of 8425 Antarctic 'finds'.

Meteorite Class	'Falls'	'Finds'	Mass
Iron(I)meteorites	3.7	0.7	50.00
Stony-iron (S-I) meteorites	1.4	0.3	0.15
Stony (S) meteorites			
Moon	0.0	0.1	No data
Mars	0.5	0.1	0.03
Howardites–Eucrites–Diogenites (HED)	6.0	3.2	0.87
Other achondrites	1.6	0.5	2.40
Enstatite chondrites (EC)	1.7	0.6	0.45
Ordinary chondrites (OC)	81.4	92.7	36.90
Carbonaceous chondrites (CC)	3.7	1.8	9.10
Total	100.0	100.0	99.90

stony-iron (S-I) and stony (S) meteorites. Stony ordinary chondrites are the most abundant meteorites reaching Terra firma followed by HED, Iron and CC meteorites (Table 9.7). They account for >90% of the influx of ~1 cm to meter-sized objects. The EC, OC and CC meteorites are compact lithified aggregates that we were not melted on their parent body. During deceleration in the atmosphere they might be more susceptible to fragmentation than for example compact igneous and basaltic S-meteorites from completely and partially melted asteroids.

Antarctic meteorite 'finds' can be treated as 'falls' because they have not suffered the differential weathering experienced by 'finds' on land that favors survival of iron meteorites over, for example, chondrite meteorites. The 'finds' confirm the relative abundances of non-Antarctic 'falls' (Table 9.7). The total mass of non-Antarctic 'finds' is at least 5×10^4 kg. Chondrites represent ~85% of all meteorites but only ~45% of the incoming meteorite mass (Table 9.7). The average masses for individual classes clearly highlight the overall massive nature of I-meteorites compared to the others when excluding the Allende meteorite (Table 9.8). This exercise should be taken with great reservation. It is based on recovered fragments of meteorite 'falls' on land (~75% of the Earth's surface is covered by water) that are remnants of individual meteoroids with wide range of

TABLE 9.8. Average mass for several meteorite class calculated by the author [Source: Bagnall (2000)]

Meteorite class	Average mass (kg)
I	579.0
S-I	75.6
HED	8.0
Other Chondrites	60.0
EC	17.1
OC	24.5
Carbonaceous chondrites	
CO/CV/CR	290.0
CO/CV/CR[a]	25.0
CM	10.1
CI	4.3

[a]Excluding the unusually massive Allende CV3 meteorite.

initial (pre-entry) sizes up to tens of meters in size. For example, the initial Tagish lake C-chondrite meteoroid was ~4 meters in diameter [Brown et al. (2000)].

Fireball data from the MORP survey show systematic behavior in the heights of the beginning and ending of visible trails for meteoroids from various sources (Table 9.9). These data also show an average mass of 10–25 g for high velocity cometary meteors, the Leonids (P/comet Tempel–Tuttle) and Perseids (P/comet 1862III Swift–Tuttle), and the slower Southern and Northern Taurids dust streams from P/comet Encke (Table 9.9). These masses correspond to meteoroids ~3 cm in diameter but larger meteoroids are associated with P/comet Encke, an almost extinct comet. Large meteoroids are more prevalent among 'low-velocity cometary' and asteroidal meteors that were identified by Halliday et al. (1996) on the basis of fragmentation behavior and stream association. Their initial velocities (Table 9.9) resemble the mean entry velocity of Apollo asteroids that include 2101 Adonis (α-Capricornid stream) and 2201 Oljato, displaying residual cometary activity (the parent of the Southern and Northern χ-Orionid streams). The Tagish Lake C-chondrite is probably also from an Apollo asteroid [Brown et al. (2000)].

The density of 'low-velocity cometary' and asteroidal meteors (Table 9.9) as a function of the initial velocity show that C-type meteoroids enter the atmosphere with velocities ranging from ~ 15 to ~ 30 km/s (Fig. 9.3). Halliday et al. (1996) found normal distribution for entry velocities of the asteroidal group between 12 and 25 km/s (mean = 18 km/s) while the cometary group has a broad peak at ~ 30 km/s (excluding fast-moving cometary meteors). Most C-meteors appear to be associated with sources having cometary velocities (Fig. 9.3). The entry velocities used in Fig. 9.3 were not corrected for effects due to inclination and perihelion distance. For example, the high initial velocity of fireballs #661 and #883 that penetrated deep into the atmosphere, and thus appear to be asteroidal, is caused by their short perihelion distance, 0.356 and 0.61 AU [Halliday et al. (1996)]. Using the Fireball Network classification [Ceplecha et al. (1998)] the OC meteors are a minority but should be common among surveyed meteors with lower entry velocities.

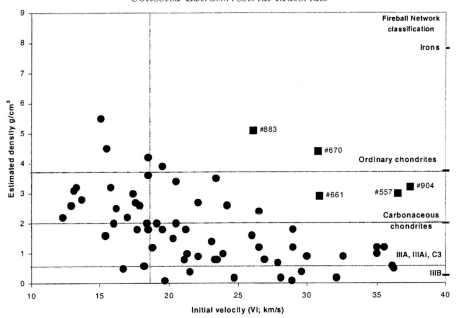

FIGURE 9.3. Meteor estimated density (g/cm^3) versus initial velocity (km/s) for low-velocity cometary meteors (dots) from the MORP survey [Halliday *et al.* (1996)]. The filled squares are selected meteors with apparently anomalous velocity as discussed in the text. For the Fireball Network classification see the caption of Fig. 9.1. The horizontal lines indicate carbonaceous chondrite and cometary meteors in this classification scheme that is based on the physical interactions (such as break-up) of individual meteors. The Halliday *et al.* (1996) classification uses slightly different criteria (see text) and found that the entry velocity of the asteroidal group is between 12 and 25 km/s (mean = 18 km/s; vertical line). The measured density for OC meteorites (see Fig. 9.1) ranges from 2.7 to 3.75 g/cm^3 while the measured (Fig. 9.1) and estimated this figure) densities for carbonaceous chondrites are similar. Several NEAs have been identified as the parents of meteor streams [see Rietmeijer (2000)].

9.3.4. *Summary*

The OC meteorites abound in the annual meteorite influx that includes smaller contributions by HED, iron- and CC-meteorites but rare iron- and common OC-meteorites dominate the mass influx. Many, or even most, OC-meteorites might be associated with NEAs with higher entry velocities than objects from the asteroid belt. The MORP data appear to indicate a sizable amount of CC-meteorites among the 'cometary meteor' group. Catastrophic meteor disruption of CC, OC and EC meteors could lead to ablation of their individual structural entities as small meteors in a denser atmosphere. A noticeable effect might be expected for meteoroids with high amounts of CAIs, but those that do (CO, CV meteoroids) are rare.

9.4. Micrometeorites

9.4.1. *Collection*

Micrometeorites, $\sim$25–500 μm (up to 1 mm) in diameter, are found in deep-sea sediments, the Antarctic ice-sheet and snow deposits, the artificial South Pole water well, glacier ice-cores, Greenland cryoconites and blue lakes, sandy beaches, desert sands, sedimentary rocks (sandstone) and swamps, e.g. Tunguska [Zolensky *et al.* (1988); Robin *et al.* (1990); Maurette *et al.* (1987); Taylor & Brownlee (1991); Maurette *et al.* (1994);

TABLE 9.9. Mean heights above seal level for the beginning and end of meteor trails (H_B, H_E), the initial atmospheric entry velocity (VI) and initial mass (MI) from the MORP survey [Halliday et al. (1996)]. Meteoroid mass estimates show multi-modal distributions with relative proportions indicated in brackets (after [Rietmeijer (2000)]). Using the ratio range/$l\sigma$ vs. N(population size) all populations are normal distributions at 95% confidence limit. Diameters are calculated using $\rho = 2.5$ g/cm^3 [Rietmeijer (1998a)] and $\rho = 3.5$ g/cm^3

	Leonids	Perseids	Southern & Northern Taurids	Low-Velocity Cometary meteors	Asteroidal meteors
H_B(km)	106.5	99	87	89; mode = 94	1. 72(H_E <50)
					2. 84(H_E >50)
H_E(km)	88	80	64–59	56.5	1. 39
					2. 65
VI(km/s)	71	60	30.5	30	1. 17.6
					2. 23.3
MI(g)	12.9	11.4	1. 23.1(30%)	1. 22.4(30%)	1. 47.5(16%)
			2. 554(70%)	2. 670(70%)	2. 428(60%)
					3. 3,010(14%)
					4. 10,740(10%)
Diameter (cm)	2.1	2.1	1. 2.6	1. 2.6	1. 3.0[a]
			2. 7.5	2. 11.7	2. 6.2
					3. 11.8
					4. 18.0

[a]From [Halliday et al. (1996)].

Beckerling & Bischoff (1995); Brownlee et al. (1997); Taylor et al. (2000)]. The relative proportions among MM types vary in different deposits. The oxygen isotope, D/H, ^{87}Sr/^{86}Sr and ^{26}Al/^{10}Be ratios, the near saturation levels of cosmogenic isotopes (^{26}Al, ^{53}Mn and ^{10}Be), as well as high carbon contents (3×CM) show that MMs were centimeter-sized, or smaller bodies, in space or they came from the surface of large parent bodies [Brownlee et al. (1997); Engrand & Maurette (1998)]. Micrometeorites are mostly spheres with variable degrees of fractional evaporation, (partial) melting and quenching. Stony MMs include unmelted and highly vesicular scoriaceous (SC) MMs that represent a stage in between unmelted stony MMs and quenched-melt stony spherules that are solidified quenched-melt droplets. Broadly speaking, spherules >50 μm in diameter probably originated in the main asteroid belt. Because millimeter-size and smaller meteoroids can reach the inner solar system by Poynting–Robertson drag, MMs are considered to be a less-biased sampling of the asteroid belt than 'conventional' meteorites [Brownlee et al. (1997)].

9.4.2. Classification

Classification (Table 9.10) is based on a combination of chemistry, texture and mineralogy [Taylor & Brownlee (1991); Beckerling & Bischoff (1995); Brownlee et al. (1997); Engrand & Maurette (1998); Taylor et al. (2000)]. Brownlee et al. (1983) described the formation of SC MMs when fine-grained, volatile-rich (H_2O, CO_2, S) micrometeoroids become initially depleted in volatile elements (K, Na) first followed by Fe and Si, and occasionally Mg. Refractory elements, Ca, Ti and Al, are concentrated in the ablation sphere [Brownlee et al. (1983)]. The morphology of the Antarctic unmelted (25–100 μm in diameter) and scoriaceous MMs and cosmic spherules (mostly >100 μm) [Kurat et al. (1994)] delineate this continuous ablation sequence. Scoriaceous IDPs that are identical

TABLE 9.10. Micrometeorite classification

Petrographic type	Petrographic sub-type	Properties
Spherules (quenched-melt MMs)		
Iron(I)		Magnetite, wüstite and hematite (silicate-free; often with a large Fe,Ni core or nugget of Pt-group elements
Iron–sulfur– nickel (FSN)		Fe-oxides and Fe,Ni-sulfides. Some I-spherules could be evolved FSN spherules after complete S-loss
Stony (S)		Chondritic composition: strong S, and moderate K and Na depletions; some spheres have a Fe,Ni core or nugget of Pt-group elements
	'common'	∼90% of S-spherules; olivine, magnetite; magnesioferrite (mostly micron-sized or less) in a glass matrix
	G-type	A dendritic network of magnetite with interstitial glass; spherical vesicles
	Vitreous (V) & cryptocrystalline (CC)	Completely glassy or cryptocrystalline, generally Si-rich matrix with continuous arrays of magnetite; relic grains present; spherical vesicles
	Barred-olivine (BO)	Lath-shape olivine and small magnetite crystals with interstitial glass
	Porphyritic (P)	Equidimensional olivine and magnetite grains in a glass matrix
	Relic-grain-bearing (RGB)	Relic grains in a glass matrix
	CAT	Enriched Ca, Al, Ti, and Mg; Fe-depleted
Scoriaceous (SC) MMs		
Stony		Chondritic composition; relic grains present in a glass matrix; irregular vesicles; highly irregularly (distended) particles at extreme vesiculation
Unmelted MMs		
Stony	*Fine-grained (FG)*	Compact C-rich chondritic matrix Fe,Ni-metal, sulfides and silicates, including fully-hydrated FG particles; rims of magnetite along the perimeter, lining irregular vesicles and disseminated in olivine
	coarse-grained (CG)	Chondritic composition; Fe-poor pyroxene and olivine grains with Fe-metal inclusions in an Fe-rich matrix; olivine can be poikilitically included in pyroxene; Ni-rich colloidal ferrihydrite occurs in COPS (Carbon-oxygen-phosphor-sulfur) inclusions up to ∼ 25 µm

to SC MMs occur [Rietmeijer (1996b)]. The FSN spheres are rare in surface deposits but these spheres < 20 µm are common IDPs. Stony MMs may contain variable proportions of relic grains that include olivines, pyroxenes, feldspar minerals, Fe,Ni-metal, Fe,Ni-sulfides, magnetite, chromite and perovskite. The relic olivine and pyroxene compositions show that only a few stony MMs can be linked to OC parent bodies [Beckerling & Bischoff (1995)] supporting the notion that they are uncommon asteroid belt objects (section 9.2.2).

9.4.3. *Chemistry*

Pre-entry compositions of stony MMs are represented by the CM bulk composition (Table 9.11). Deviations arise from post-depositional processing such as the mobilization of water-soluble elements (Mg, Ca, Na, Ni and Mn) from MMs ([Kurat *et al.* (1994)]. Selective leaching of deep-sea silicate spheres > 100 μm causing a relative enrichment of resistant Fe-spherules might explain the high Fe/Si ratio for deep-sea spherules (Table 9.11). The interpretation that CAT spheres show extreme Fe-loss and less-severe silicon loss relies on assumptions of the nature of their progenitor. For example, when the progenitor was similar to the refractory stratospheric IDPs [Rietmeijer (1998a)] or MMs [Zolensky *et al.* (1988)], the spheres survived compositionally almost intact. A similar albeit more subtle situation exists for the stony cosmic spherules. Brownlee *et al.* (1997) compared the Si-normalized element abundances in these spherules (Table 9.11) with those of unmelted chondritic IDPs. This comparison presumes that IDP-like dust was the progenitor of CM-type S-spherules. So far there is no evidence to support this genetic relationship although CM- type IDPs could be more common than currently accepted (section 9.2.5). The CI-standard normalized compositions show that continued progenitor ablation shifts the Mg/Si, Al/Si and Ca/Si ratios from <CI to >CI, while Mn/Si and Cr/Si ratios shift from CI to <CI but no change in Fe/Si ratios [Brownlee *et al.* (1997)]. The interpretation of this shift remains open to debate. Brownlee *et al.* (1997) did not accept selective volatilization of Si but Taylor *et al.* (2000) using results of the same evaporation experiments reached the opposite conclusion. If progenitor dust had a proto-CI composition, the spheres would only show enrichments in Mg and Al with general patterns in compositional changes of the spherules that are consistent with evaporative loss of silicon. The very nature of flash-evaporation favors kinetically controlled reactions that makes it difficult to identify a unique encompassing solution although trends in behavior are recognizable. Gravitational buoyancy during flash melting whereby a high-density Fe-rich core separates from a silicate-rich melt in molten MMs is another way to lose iron and siderophile elements [Brownlee *et al.* (1983); Taylor *et al.* (2000)].

9.4.4. *Summary*

Parallel I-, S-I and S-type classifications of 'conventional' meteorites and MMs is significant because the different size fractions of incoming meteoroids highlight that liberation from parent bodies and atmospheric entry survival depends on the tensile strength. High-velocity 'cometary' dust has a higher probability of severe heating than debris from the asteroid belt with the same mass and cross-section entering the atmosphere at ~11 km/s, that is, ignoring effects caused by orbital inclination and perihelion distance. The adjective 'cometary' refers to the entry velocity of incoming matter. It is not used as a genetic distinction of comets and asteroids. While cometary dust is more abundant in the Zodiacal cloud than asteroidal dust, selective heating and gravitational focusing of low-velocity dust in (near) circular orbits reverse this ratio in favor of asteroidal dust in an 80:20 ratio in the stratosphere [Brownlee (1985); Flynn (1994a); Flynn (1994b)]. The CM-like MMs are from (IR) S-class asteroids [Brownlee *et al.* (1997)].

9.5. Interplanetary Dust Particles

9.5.1. *Collection*

The near-Earth flux of IDPs >10 μm is 3×10^{-6} m^2/s. They begin deceleration at 130–100 km altitudes. Visual meteor observations, including 'nebulous' Leonid meteors, found

TABLE 9.11. Average atomic ratios for CM meteorites [Schramm *et al.* (1989)]

	CM	(1)	(2)	(3)	(4)	(5)	(6)	(7)	(8)	(9)
Mg/Si	1.04	1.06	1.06	1.06	0.785	0.95	1.04	0.91	0.97	1.42
Fe/Si	0.84	0.63	0.53	1.02	0.60	1.02	0.50	1.95	0.37	0.04
S/Si	0.23	a	a	?	?	0.05	?	?	?	?
Al/Si	0.08	0.23	0.09	0.08	0.10	0.06	0.09	0.09	0.08	0.59
Ca/Si	0.072	0.16	0.06	0.035	0.007	0.035	0.10	0.10	0.10	0.21
Na/Si	0.035	n.d.	0.002	0.002	0.03	0.06	0.015	?	?	?
Ni/Si	0.046	0.01	0.01	0.02	0.01	0.01	0.01	0.03	0.001	n.d.
Cr/Si	0.012	0.007	0.006	0.008	0.01	0.01	0.01	?	?	?
Mn/Si	0.006	0.005	0.007	0.006	0.004	0.006	0.006	0.013	0.010	0.005
Ti/Si	0.002	?	0.002	0.002	0.004	0.002	0.004	0.005	0.005	0.009
K/Si	0.004	?	?	?	0.004	0.002	n.d.	?	?	?

Columns are as follows: (1) cosmic spherules from the stratosphere (2) Antarctic ice; (3) deep-sea sediments [Brownlee *et al.* (1997)]; (4) composition of Antarctic unmelted MMs; (5) Antarctic scoriaceous MMs; (6) Antarctic stony spherules; (7) barred olivine; (8) glass; and (9) CAT spherules. (4), (5) and (6) show the effect of leaching during <3,000 years residency in ice [Kurat *et al.* (1993); Kurat *et al.* (1994)], which is not a problem in (7), (8), and (9) spherules from the South Pole Water Well that spent < 900 years in ice [Taylor *et al.* (2000)]. 'n.d.' means not detectable, and ? means no data.
[a] Antarctic spherules and stratospheric stony spheres (1.5–18 μm) are S-free and Fe-depleted.

higher initial ablation altitudes suggesting that many meteors are rich in carbonaceous materials [Murray *et al.* (1999); Spurný *et al.* (2000)]. Upon reaching a 'rest' velocity of cm/s at ∼80 km they settle towards the Earth's surface and have no longer any information on the pre-entry orbit and entry velocity. However, entry velocity of an IDP, and thus broadly speaking its source, could be inferred from a combination of its peak heating temperature, size and mass [Love & Brownlee (1991)]. Stratospheric IDPs are the only samples for which we have a constraint on the time they entered the atmosphere which can be used to explore links to annual meteor showers [Rietmeijer & Jenniskens (2000)] or other periodic events [Rietmeijer & Warren (1994)].

The slow-down from entry velocity >11 km/s to the settling velocity causes a 10^6-fold increase in particle number density for IDPs in the stratosphere between 17–19 km altitudes. The earliest collections date back to 1976 but beginning in 1981, the NASA Johnson Space Center Cosmic Dust Program has sampled dust using inertial-impact, flat-plate collectors mounted in pylons underneath the wings of high-flying U2 and W57B aircraft. Chondritic IDPs ranging from 2 μm to 41 μm have a gaussian size distribution with a mean of 12 μm. Larger aggregate particles up to ∼ 60μm in size exist before they break apart during collection [Rietmeijer (1998a), Fig. 25]. Particles that survived 'intact' typically experienced flash heating temperatures between ∼ 300 °C and ∼ 1,000 °C as indicated by chemical modification and mineral transformations [Rietmeijer (1998a)] which is still below the melting point at ∼ 1,500 °C for chondritic IDPs. The distinct ∼25 μm-sized non-aggregate CM-type IDPs survived with only melting at the surface. In general, some fraction of IDPs survived atmospheric entry with negligible thermal alteration and mass loss, others show significant dynamic pyrometamorphic alteration or partial melting while yet others became quenched-melt spheres. This behavior is possible because of a low particle-size–density product. Particles <10 μm can not efficiently dissipate the thermal energy. They will be flash-melted or evaporated. Several review papers [Brownlee

TABLE 9.12. Dimensions (micrometers) of IDPs on collector L2005. All sizes are presented in the root-mean-square size, i.e. rms $= (a^2 + b^2)^{1/2}$, where a and b are orthogonal dimensions across an IDP (Modified after [Rietmeijer (2000)])

Petrographic type	Number of IDPs in each type		Size	
	'unaltered'	'heated'	Mean	Range
Iron	0	46 spheres	8.0; mode = 7.5	2–17
Iron–sulfur–nickel	46(Fe ≈ S)		8.5; mode = 11.5	2–23
(FSN)		40 spheres(Fe ≈ S)	8.6; mode = 9.6	4–14
		18 spheres(Fe > S)	11.5; mode = 9.6	5–24
	Stony (silicate)			
Si,Mg±Fe	23		20.5	7–44
		46 spheres	10.6	2–30
Si,Mg,Ca±Al	4		34.1	28–44
		26 spheres	10.5	3–17
Si,Al,Ca±Fe	13		22.8	8–52
Si,Al,Ca		14 spheres	11.2	3–20
	Stony (chondritic)			
	47		5.8	3–9
		59	19	6–30

(1985), Sandford (1987), Mackinnon & Rietmeijer (1987), Flynn (1994a), Flynn (1996), Thomas *et al.* (1996), and Rietmeijer (1998a) (with a complete listing of original papers)] describe the collection and curation procedures, the physical, mineralogical and chemical properties, including isotopic, minor and trace element compositions and the extraterrestrial signatures. They also show images of IDPs and their constituents.

9.5.2. *Classification*

The basic IDP (Table 9.12) and MM classifications are similar. The major rock-forming element abundances in chondritic IDPs are within a factor of 2 of the CI abundances. For a number of reasons pertaining to hierarchical accretion, thermal or aqueous parent body alteration and atmospheric entry flash heating, the volatile element abundances (e.g. sulfur) can vary among individual IDPs. Chondritic dust includes 'unaltered' and 'heated' IDPs with a glazed surface, thin platy lobes or discs emerging from the surface and/or an equatorial skirt of fused lobes and discs that formed during atmospheric entry. All subtypes show a gaussian size distribution defined at a 95% significance level using range/1σ versus N (population size). The distribution of IDP types and subtypes in Table 9.12 is representative for the cosmic dust collection with the exception of the high incidence of heated type-CM IDPs on this particular collector.

Early IDPs classifications relied on a combination of IR spectroscopy, mineralogy and chemistry (mostly carbon content). However, they used both primary and secondary (alteration) properties that introduced an undesirable genetic aspect although they could made predictions on properties of unaltered chondritic aggregate IDPs. Classification (Fig. 9.4) according to composition as chondritic and non-chondritic IDPs with a subdivision into aggregate IDPs (Fig. 9.5), non-aggregate and fragment IDPs is inclusive of all collected dust types and their interrelationships. Chondritic material is often present at the surface of non-chondritic fragments and refractory aggregates. The IDP spheres

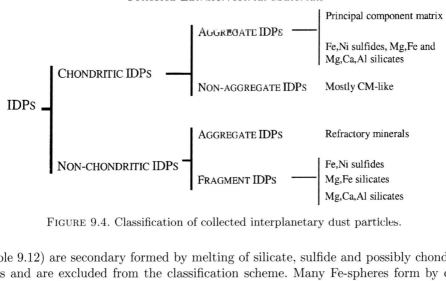

FIGURE 9.4. Classification of collected interplanetary dust particles.

(Table 9.12) are secondary formed by melting of silicate, sulfide and possibly chondritic IDPs and are excluded from the classification scheme. Many Fe-spheres form by complete sulfur loss from FSN-spheres but hollow Fe-spheres are droplets spalled off from developing meteorite fusion crusts [Rietmeijer (1998a)]. Most chondritic aggregate and cluster IDPs consist of identical building blocks that include predominantly olivines, pyroxenes and pyrrhotite that occur in randomly variable relative proportions [Rietmeijer (1998a)]. This unique feature points toward an accretion hierarchy of a limited number of non-chondritic solar nebula dusts with discrete size distributions that increase as a function of accretion time and hence increasing aggregate size. This model provides for a continuum between ever-larger aggregate IDPs that ultimately become the matrix of chondrite meteorites wherein post-accretion alteration and compaction has generally erased all textural vestiges of this accretion history.

9.5.3. *Petrography and Chemistry*

Hierarchical dust accretion began with three fundamental building blocks, or *principal components* (PCs), 90 nm to ~1000 nm in diameter that form the fractal matrix of all aggregate IDPs. These non-chondritic PCs are:
(1) carbonaceous units of refractory hydrocarbons and amorphous (often-vesicular) carbons,
(2) carbon-bearing ferromagnesiosilica units with ultrafine (2 to ~50 nm) Fe,Mg-olivines and pyroxenes and Fe,Ni-sulfides in a matrix of hydrocarbons and amorphous carbons, and
(3) pure ferromagnesiosilica PCs that include (a) coarse-grained (10–410 nm) Mg-rich PCs with trace amounts of Al and Ca that are original circumstellar dust condensates [Rietmeijer *et al.* (1999)], and (b) Fe-rich PCs with ultrafine (<50 nm) Fe,Mg-olivines and pyroxenes, Fe,Ni-sulfides and magnetite in an amorphous ferromagnesiosilica matrix.

The PCs resemble the size, mass and composition of the CHON (carbon–hydrogen–oxygen–nitrogen), 'mixed carbon-silicate', and 'silicate' dust in the coma of P/comet Halley [Rietmeijer (1998a); Rietmeijer & Nuth (2000a)]. Continued aggregation of matrix aggregates, Mg, Fe-silicates and Fe,Ni-sulfides, both ~0.1 to 3–5 µm in size, resulted in 10–15 µm-sized chondritic aggregate IDPs. They can also have small amounts of amorphous or crystalline aluminosilica ($\pm$ Ca) grains. Aggregate IDP densities range from 0.1 to 4.3 g/cm^3 (average = 2.0 g/cm^3; Rietmeijer (1998a)]. This is similar to meteoroid densities, 0.01–1.06 g/cm^3 [Verniani (1967); Verniani (1969)] and the light-element (0.1–0.3 g/cm^3) and silicate dust (0.5–3 g/cm^3) in comet P/Halley [Grün & Jessberger (1990)].

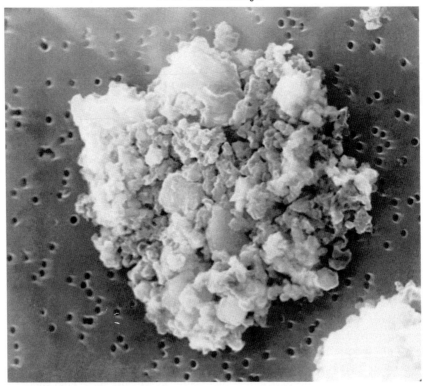

FIGURE 9.5. Scanning electron microscope image of a large porous, or fluffy, chondritic aggregate IDP collected in the lower stratosphere showing the matrix with embedded larger platy silicate grains. The image is 20 micrometers across. The background is a nucleopore-filter supporting the particle during electron microscope imaging. Courtesy of the National Aeronautics Space Administration, Particle W7029B13 (NASA number S−82 − 27575, Cosmic Dust Catalog 2(1), NASA Johnson Space Center Curatorial Branch Publication 62, 1982).

Chondritic aggregate IDPs (>10 μm) and unmelted silicate and Fe,Ni-sulfide IDPs (<25 μm) accreted to form cluster IDPs >60 μm in size [Thomas *et al.* (1995)].

Both ferromagnesiosilica PCs have lower Mg/Si and Fe/Si ratios than aggregate and cluster IDPs that had accreted Mg,Fe-silicate and Fe,Ni-sulfide particles (Table 9.13). Rietmeijer (1999) argued that in cometary dust Na/Si (at.) = 0.11 rather than 0.06 (Table 9.13). The chemical differences between aggregate matrix, aggregate and cluster IDPs reflect hierarchical accretion of evolving solar nebula dusts whereby growing aggregates gradually acquired a proto-CI bulk composition for major rock-forming elements as a function of time and thus aggregate size. A corollary of hierarchical dust accretion is that carbon is locked in early during dust accretion through accumulation of carbonaceous PCs. The resulting chondritic aggregate IDPs contain 1–46 wt.% carbon with an average bulk carbon content that is ∼2–3 times CI [Thomas *et al.* (1996)]. During flash heating the carbonaceous phases may flow around other IDP constituents and fuse into contiguous patches [Rietmeijer (1998b), fig. 9].

9.5.4. *Summary*

The IDPs ∼10 μm to ∼30 μm in size represent ∼8% of the meteoroid influx and define a peak in the terrestrial mass accretion for extraterrestrial materials smaller than the MM precursors [Rietmeijer (2000)]. The textures (low-density aggregates), chemistry (proto-CI) and mineralogy (anhydrous minerals and PCs) indicate that chondritic ag-

TABLE 9.13. Atomic ratios for ultrafine-grained (ufg) and coarse-grained (cg) PCs in aggregate IDPs, aggregate and cluster IDPs, CI chondrites and proto-CI material (from section 9.3.2) (Sources: [Rietmeijer (1998a), Schramm *et al.* (1989)])

	Aggregate matrix		Aggregate IDPs	Cluster IDPs	Proto-CI	Bulk CI
	ufg units	cg units				
C/Si	zero	zero	4.09[a]	1.1	0.52	0.7
Mg/Si	0.64	0.48	0.85	0.7	0.80	1.06
Fe/Si	0.46	0.18	0.63	0.7	0.67	0.90
S/Si	0.2	0.0	0.35	0.4	0.35	0.46
Al/Si	0.09	0.09	0.063	0.1	0.07	0.085
Ca/Si	0.027	0.04	0.048	0.07	0.04	0.07
Na/Si	no data	no data	0.049	0.2	0.05	0.06
Ni/Si	0.032	0.001	0.037	0.04	0.03	0.05
Cr/Si	0.018	0.0	0.012	0.01	0.01	0.01
Mn/Si	0.007	0.004	0.015	0.01	0.006	0.01
Ti/Si	no data	no data	0.0022	no data	0.002	0.002
K/Si	no data	no data	no data	no data	0.003	0.01

gregate IDPs are the least modified solar system materials. Ablation of these IDPs adds marginally to the major rock-forming elements in meteoric vapors produced by larger meteoroids but they might add to the C, H and N abundances as surviving aerosols (fragments) or meteoric vapor. The recent finding of initial ablation altitudes >130 km for cometary meteors might imply that a fraction of the light elements will be deposited into the thermosphere. The masses of IDPs and various solar system dusts (Table 9.14) show that former resemble dust from active comets and the Zodiacal cloud. Among the aggregate IDPs we may look for particles producing sporadic meteors and particles associated with meteor showers and storms.

The Zodiacal cloud is maintained by dust spiraling inwards from their parent bodies under the influence of Poynting–Robertson drag whereby it spends 10^4–10^5 years traveling through interplanetary space [Brownlee (1985); Mackinnon & Rietmeijer (1987)]. During this time particle aging due to space weathering will probably modify its properties. Dust in streams of active periodic comets spends less time in orbit, ranging from the latest perihelion passage, than dust released several hundreds of years previously. It will have aged less, which might favor survival of volatile pre-biotic constituents. The current Leonid meteor storm activity showed that the orbital ages of individual meteor streams from P/comet Tempel–Tuttle are known and their entry into the atmosphere can be predicted [Asher *et al.*(1999)]. There is the possibility that collected IDPs include 'old' Zodiacal and 'young' meteor shower dust.

9.6. Mesospheric Metals

9.6.1. *Chemistry*

Meteor spectra show qualitatively the same major elements as those in comets [Millman (1972)], including comet P/Halley [Jessberger *et al.* (1988)] and IDPs [Rietmeijer (2000)]. To compare mesospheric metal abundances with compositions of surviving materials, I assumed efficient non-fractional meteor evaporation. This assumption may not be en-

TABLE 9.14. Mass (g) and density (ρ in g/cm^3) of solar system dust (modified after [Rietmeijer (2000)] listing the original sources)

Sources	Mass
Comet P/Halley coma	10^{-20}–10^{-17} up to 10^{-5}
Tails of 46P/Wirtanen and 65P/Gunn; $\rho = 2$	4×10^{-4}–1.0 (upper limit)
	5×10^{-10}–1.0×10^{-6} (lower limit)
Tail comet Tabur Q1 $\rho = 2$	3.5×10^{-3}–1.0×10^1 (upper limit)
	3.5×10^{-9}–1.6×10^{-5} (lower limit)
Zodiacal cloud (sporadic meteors)	$\sim 10^{-15}$–10^3; maximum at 10^{-5}
Average meteoroid stream	10^{-15}–10^1; maximum at 10^{-1}
Irregular meteoroids (satellite impact data, Mean $\rho = 2.0$–2.5; characteristic value, $\rho = 1.5$	10^{-12}–10^{-6}
Principal components; D = 500 nm($\rho = 2.5$)	1.6×10^{-13}
Aggregate IDPs; D = 10 μm($\rho = 2$)	1.0×10^{-9}
Cluster IDPs; D = 100 μm($\rho = 1.0$)	5.2×10^{-7}

tirely valid. For example, the correlated release of Ca, Ti and Al was linked to thermal fluctuations along the Cechtice fireball trajectory [Borovička (1993); Borovička & Spurný (1996)]. These elements make up CAT spheres and refractory IDPs and CAIs and also occur in chondrite matrix. Meteor fragmentation and subsequent ablation of structurally different host materials may contribute to the phenomenon of differential ablation but without the necessity of thermal fluctuations during meteor ablation. Differential was invoked to explain temporal variations in abundances of Na and Ca during meteor ablation [McNeil et al. (1998)]. This process could cause the sudden release of Ca, Al, Ti and Mg but this interpretation presumes a texturally homogeneous meteor. There are also uncertainties in the kinetically controlled non-equilibrium evaporation of the different host phases. For example, Mg-vapor production could be affected because this element occurs in pyroxene and olivine crystals with melting temperatures of ~ 1100–$1400\,^{\circ}$C, in fine-grained amorphous material with lower melting points and refractory materials [Rietmeijer (1998a)].

Meteoric vapors from ablating meteoroids provide the metals in the mesosphere [Plane (1991)]. Geochemical constraints put the total mass accretion for meteoroids of all sizes at $(39$–$56) \times 10^6$ kg/year [Love & Brownlee (1993)]. Here I adopt a value of 4.8×10^7 kg. The mass accretion by MMs is 4×10^7 kg [Love & Brownlee (1993)] and the remainder, 8×10^6 kg, is provided by 'conventional' meteorites and IDPs that represent $\sim 8\%$ of this mass. The amount of annually produced MM ablation vapor, 2×10^6 kg [Love & Brownlee (1993)], has a CM composition with some augmentation of silicon from the formation of stony MM spheres (section 9.4.3). Fireball data and their recovered meteorite mass indicate 80–95% mass-loss from the initial meteoroid as meteoric vapor, fragments, many of which are not recovered, and dust [see Rietmeijer (2000)]. Assuming $\sim 50\%$ of the initial mass is vaporized, meteorite-producing meteoroids deposit 3.7×10^6 kg vapor mostly in the upper stratosphere and lower mesosphere [Rietmeijer (2000)].

In order to constrain the compositions of meteoric meteors we only need to consider I-, HED, CC and OC meteorites (Table 9.7). The other meteorite classes can be ignored for this purpose. The low relative mass of HED meteorites implies that they too are irrelevant in this context. Both I- and CC meteorites are relatively rare. The massive iron meteorites representing 50% of the annual incoming mass in this size range (Tables 9.7 and 9.8) are likely to produce occasional transient incursions of high amounts of iron into the mesospause region. Ablation of OC meteorites probably provides a continuous 'steady-state' supply of Mg, Al, Si, S, K, Ca, Ti, Cr, Mn, Fe and Ni, with average meteoric vapor Fe and Mg abundances close to those in OC meteorites (Table 9.4). The total annual amount of meteoric vapor is about 5.7×10^6 kg from I-, OC and CC, mostly CM, meteorites.

The Mg/Ca ratio in P/comet Halley dust, chondritic aggregate IDPs and meteorites is similar to this ratio in asteroidal and cometary meteors and is not surprisingly similar to the mesospheric metal ratio (Table 9.15). I calculated mesospheric metal ratios (Table 9.15) using data for the annual mass influx [Hughes (1994); Love & Brownlee (1993)], the OC- and I-meteorite compositions (Table 9.4) for meteoroids > 1 cm, the CM composition for incoming micrometeoroids (Table 9.11) and the average IDP composition (Table 9.13) for dust. The calculated abundance ratios have considerable errors. It is somewhat surprising then that the calculated mesospheric Mg/Fe ratio is similar to those in the collected extraterrestrial samples.

There are major differences between observed and calculated Na/Ca and Mg/Na ratios. Table 9.15 shows two calculated values for these mesospheric and proto-CI or aggregate IDP ratios. The first value is based on the proto-CI sodium value (Table 9.13). The second one used Na/Si = 0.11 [Rietmeijer (1999)], which is similar to Na/Si ≈ 0.17 for 27 IDPs [Flynn *et al.* (1996b)]. The observed Na/Ca and Mg/Na values for mesospheric metals match those of meteors and the higher cometary Na/Si ratio. All ratios for the other extraterrestrial materials show severe sodium depletion due to mobilization of sodium in the parent bodies or their terrestrial deposits. Organogenic elements, C, H, O, N and P do not appear to be provided by 'conventional' meteorites but hydrocarbon and carbonaceous materials in MMs and IDPs might import significant amounts of these elements.

9.6.2. *Summary*

Observed mesospheric abundances for Mg, Fe and Ca match well with calculated abundances placing boundary conditions on the compositions of ablating meteoroids. Sodium and possibly potassium are mostly present as highly volatile alkali phases in incoming meteoroids. Residues of these phases might still be present in IDPs. The presence of volatile alkali phases would also increase the beginning altitudes of visible and radar meteor ablation in a manner similar to what was proposed for volatile organic species. High velocity cometary meteors such as the Leonids and Perseids will add their ablation products to the thermosphere while 'low velocity cometary' and asteroidal meteors contribute less directly to these metal abundances at these high altitudes. The present data support that loss of alkali elements might occur at the very onset of ablation. Ablation of CM MMs [Engrand & Maurette (1998)] and C-rich aggregate and cluster IDPs could deliver significant amounts of carbon, H and N into the middle and upper atmosphere. For example, the carbon abundance in the middle an upper atmosphere is calculated to be C/Fe (at.) = 0.33 and C/Mg (at.) = 0.26.

9.7. Meteoric Dust

9.7.1. *Collection*

Hunten *et al.* (1980) suggested that meteoric vapors condense at the mesopause as large molecules or small particles (i.e. Aitken molecules) that form into small clusters or smoke

TABLE 9.15. Mean and range (parenthesis) of metal abundances (ionic for meteors) and atomic ratios for mesospheric metals, comet P/Halley dust, IDPs, proto-CI and bulk CI, CM, and OC meteorites. The average meteor abundance includes the Perseids, β-Taurids Giacobinids, Geminids, Leonids, and Sporadics (after ([Rietmeijer (2000)] and references cited therein).

	Na/Ca	Mg/Ca	Mg/Na	Mg/Fe
Mesosphere(observed)	3	21	8	n.a.
Mesosphere(calculated)	0.7/0.7	16	23.5/21.8	1.3
12/17–12/18/1999 meteors	2.5	n.a.	n.a.	n.a.
Average meteors	2.7(0.08–7.6)	21.3(4.8–47.5)	6.4(0.95–10.7)	1.7(0.3–5.7)
Draconids	1.7	18.4	10.8	1.1
Comet P/Halley(bulk)	1.6	15.9	10.0	1.9
PCs [in IDPs]	n.a.	24.0	n.a.	1.4
Aggregate IDPs	1.5	15.2	10.0	1.2
Cluster IDP L2008 # 5	2.9	10.0(0.6–22.9)	3.5(0.2–8.0)	1.0(0.06–2.3)
Average IDPs	1–2.4	22–39	16–18	1.1–1.4
Proto-CI	12/2.75	20.0	16.0/7.3	1.2
Bulk CI meteorites	0.8	14.9	17.7	1.2
Bulk CM meteorites	0.5	14.4	29.7	1.2
OC meteorites	1.0	18.8	18.8	1.3

n.a., not available.

particles <20 nm in diameter. Rietmeijer (2001) re-assessed the notion that ice particles in noctilucent clouds may have a catalytic role in the formation of meteoric dust by providing a support that facilitates the transformation of clusters to massive grains. Hunten *et al.* (1980) suggested that meteoric dust is chemically homogenous and that coagulation should cause further homogenization. When meteoric dust of different compositions could co-exist, this suggestion is not obvious. The ultimate composition might be a function of mixing compositionally different metal-oxide nanograins. The resulting aggregates collapse in the presence of a whetting agent such as water vapor or sulfuric acid at the surface of the meteoric dust.

The upper stratosphere is a convenient location to collect meteoric dust but success depends on a compromise between environmental and technical factors. Environmental factors include the presence of dust that originated at the Earth's surface. Technical factors include the availability of collection and laboratory techniques to determine the crystallographic and chemical properties of nanoparticles. In early attempts it was impossible to obtain quantitative identification of collected dust and aggregates of condensed matter [Bigg *et al.* (1970); Bigg *et al.* (1971)]. More than 90% of dust < 8 μm at ∼ 35 km altitudes was linked to the 1982 El Chichón volcanic eruption [Rietmeijer (1993)]. The remainder included silica and Fe_3O_4 grains that may be meteoric dust [Rietmeijer (1998a)].

9.7.2. *Petrology and Chemistry*

Kinetically controlled gas-to-solid condensation of M-SiO-H_2-O_2 (M: Fe, Mg or Al) vapors favored formation a single crystals with simple compositions such as SiO_2, Al_2O_3, MgO, Fe_3O_4 and γ-Fe_2O_3 particles <100 nm among copious amounts of amorphous 'AlSiO', 'MgSiO' and 'FeSiO' grains [Rietmeijer & Karner (1999); Rietmeijer *et al.* (1999)]. The finding of 40 nm-sized hematite (Table 9.1) at 60 km altitude [Bohren & Olivero (1984)] is consistent with these results although Plane *et al.* (1999) showed that FeO_2 and FeO_3 will be the mostly likely reservoirs of iron in the mesopause and lower thermosphere region. The data suggest that condensed meteoric particles are simple oxides that can be modified by interactions with the ambient atmosphere [Rosinski & Pierrard (1964)]. Rietmeijer (2001) suggested that Fe-oxides were hydrated to goethite in noctilucent cloud particles, while Rosinski & Pierrard (1964) argued that condensed meteoric silica particles ≪ 10 nm in diameter become hydrated or react with atmospheric nitrogen.

Jessberger *et al.* (1992) suggested that a narrow Fe-oxide rim on IDPs is due to precipitation of iron from the mesospheric Fe-layer. This mechanism remains a matter of contention [Flynn (1994b)]. Using the chondritic Fe,Ni ratio = 18.8, Rietmeijer & Flynn (2000) submitted that condensed meteoric Fe,Ni-oxide(?) aerosols (Fe,Ni = 20.5; range = 6.1–54.6) was present at the surface of volcanic particles in the upper stratosphere. Using an airborne time-of-flight mass spectrometer Murphy *et al.* (1998) determined the compositions of aerosols mostly between 200 nm and 2 μm in size at 19 km altitudes at the Northern Hemisphere during a period of low volcanic activity. The aerosols included a group with a high Fe-ion peak plus minor peaks for Mg, Na, Al, K, Ca, Cr, and Ni, and $^{39}K/^{23}Na$ ratios similar to those determined in meteors. There is no quantitative information on these elements. Murphy *et al.* (1998) interpreted the aerosols as meteoritic dust. This origin seems unlikely because kinetically controlled condensation yields chemically simple solids. Still, supersaturation of meteoric vapors at the mesopause, the coldest place on Earth, might be conducive to metastable eutectic condensation [Rietmeijer & Nuth (2000b)] whereby meteoric dust has a quenched-vapor composition. It seems more likely that the aerosols are IDP fragments smaller than the cut-off value (∼2 μm) for the NASA cosmic dust collectors [Rietmeijer & Warren (1994)].

9.7.3. *Summary*

Meteoric dust remains elusive despite progress in collection and analyses techniques, and in our understanding of atmospheric chemistry, IDPs and kinetically controlled vapor phase condensation. Fe-rich aerosols, e.g. nm-sized FeOOH grains, are the likely candidates for meteoric dust. The identification of middle and upper atmospheric aerosols as either meteoric dust or IDP fragments is important because it might lead to better constraints on a steady-state, i.e. annual average, composition of meteoric vapors.

9.8. General Summary

From the vast literature on meteorites, micrometeorites and interplanetary dust particles I have extracted the information I believe to be relevant and of interest to appreciate the compositions of fireballs and meteors, of meteoric vapors and their condensates and of the mesospheric metal abundances. As such there are no conclusions, only issues for future research as an open invitation to a dialog that by the very nature of its subject matter has to be interdisciplinary. Accepting that the collected surviving extraterrestrial materials represent the compositions and physical properties of incoming meteoroids that (almost) completely vaporized in the Earth's upper and middle atmosphere, it is possible to derive some remarkable simplifications to be made on the steady state, annual average, composition of meteoric vapors, and on spatially and temporally confined augmentations of individual elements, such as iron. The contributing meteoroids are (1) ordinary chondrites with lesser amounts of iron meteoroids, (2) CM-type micrometeoroids, and (3) carbon-rich chondritic interplanetary dust particles. Ablation of OC meteoroids was efficient when they are associated with near-Earth asteroids rather than being from the asteroid belt. The CM-type materials are from parent bodies in the asteroid belt with unique physical properties that explain the enormous abundances of this material among the MMs. The IDPs represent the least altered icy protoplanets, which include comet nuclei and carbonaceous and 'ultra-carbonaceous' outer asteroids. Recent laboratory experiments provided a tantalizing hint that the chemical composition of CI meteorites may have been depleted in silica during aqueous alteration on their parent body. The more silica-rich proto-CI material might be preserved in icy protoplanets of which fragments could produce fireballs and meteors. The chemical difference between proto-CI and CI meteors is probably large enough to be detectable by meteor spectroscopy.

Using relative mass abundances of incoming meteoroids, it is possible to calculate average mesospheric metal abundances that are in good agreement with observations of co-occurring Fe, Ca and Na [Plane (1991); Gardner et $al.$ (1993); von Zahn et $al.$ (1999)]. Apart from iron, these metals are minor elements in meteoroids. Silicon and carbon could be abundant metals in the middle and upper atmosphere. The low calculated atmospheric ratios C/Fe (at.) = 0.33 and C/Mg (at.) = 0.26, and the fact that CHON-like cometary dust is yet to be collected appears to be consistent with the very high beginning heights of meteor ablation. Quantitative determination of C/Fe and C/Mg ratios will constrain the atmospheric interactions and compositions of less-than-centimeter-sized cometary meteors that include chondritic aggregate and cluster IDP materials. Fireball compositions are poorly known because meteorites are only available for six fireballs to calibrate the physical models and quantitative data for meteors are scarce. This situation is changing for meteor compositions [Borovička (1999), among others] while meteor light curves now provide constraints on the physical properties of meteors and the possible sizes of their constituent grains [e.g., Murray et $al.$ (1999)]. Meteor properties might be calibrated using the physical, chemical and petrographic properties of collected extraterrestrial materials. Differential meteor ablation might also be determined by textural different meteor constituents.

Acknowledgements: I am grateful to George Flynn for a thoughtful review and to Rhian Jones for constructive discussions and providing Fig. 9.2. This work was supported by the National Aeronautics and Space Administration grant, NAG5-4441.

REFERENCES

ANDERS E. & GREVESSE N. 1989. Abundance of the elements: Meteoritic and solar, *Geochim. Cosmochim. Acta*, **53**, 194–214.

ASHER D. J., BAILEY M. E. & EMEL'YANENKO V. V. 1999. Resonant meteoroids from Comet Tempel–Tuttle in 1333: the cause of the unexpected Leonid outburst in 1998, *Mon. Not. R. Astron. Soc.*, **304**, L53–L56.

BAGNALL P. M. 2000. StarFall 2000: Non-Antarctic meteorite database, *www.ticetboo.demon.co.uk/starfall.htm*.

BECKERLING W. & BISCHOFF A. 1995. Occurrence and composition of relict minerals in micrometeorites from Greenland and Antarctica – Implications for their origins, *Planet. Space Sci.*, **43**, 435–449.

BELL J. F., DAVIS D. R., HARTMANN W. K. & GAFFEY M. J. 1989. Asteroids: The big picture, in *Asteroids II*, edited by R. P. Binzel, T. Gehrels, and M. S. Matthews, University of Arizona Press, Tucson, AZ, pp. 921–945.

BIGG E. K., ONO A. & THOMPSON W. J. 1970. Aerosols at altitudes between 20 and 37 km., *Tellus*, **22**, 550–563.

BIGG E. K., KVIZ Z. & THOMPSON W. J. 1971. Electron microscope photographs of extraterrestrial particles, *Tellus*, **23**, 247–260.

BLANCHARD M. B. & CUNNINGHAM G. G. 1974. Artificial meteor ablation studies: Olivine, *J. Geophys. Res.*, **79**, 3973–3980.

BOHREN C. F. & OLIVERO J. J. 1984. Evidence for haematite particles at 60 km altitude, *Nature*, **310**, 216–218.

BOROVIČKA J. 1993. A fireball spectrum analysis, *Astron. Astrophys.*, **279**, 627–645.

BOROVIČKA J. & SPURNÝ P. 1996. Radiation study of two very bright terrestrial bolides, *Astron. Instit. Czech.*, pp. 1–25.

BOROVIČKA J. 1999. Meteoroid properties from meteor spectroscopy, in *Meteoroids 1998*, edited by W. J. Baggaley, & V. Porubčan, Slovak Academy of Sciences, Bratislava, Slovak Republic, 355–362.

BRADLEY J. P., KELLER L. P., BROWNLEE D. E. & THOMAS K. L. 1996. Reflectance spectroscopy of interplanetary dust particles, *Meteoritics Planet. Sci.*, **31**, 987–994.

BREARLEY A. J. & JONES R. H. 1998. Chondritic meteorites, in *Planetary Materials*, edited by J. J. Papike, Revs. Mineral. **36**, Chapter 3, Mineral. Soc. Am., Washington, DC, 1–398.

BROWN P. G., HILDEBRAND A. R., ZOLENSKY M. E., GRADY M., CLAYTON R. N., MAYEDA T. K., TAGLIAFERRI E., SPALDING R., MACRAE N. D., HOFFMAN E. L., MITTLEFEHLDT D. W., WACKER J. F., BIRD J. A., CAMPBELL M. D., CARPENTER R., GINGERICH H., GLATIOTIS M., GREINER E., MAZUR M. J., MCCAUSLAND P. J. A., PLOTKIN H. & MAZUR T. R. 2000. The fall, recovery, orbit, and composition of the Tagish Lake meteorite: A new type of carbonaceous chondrite, *Science*, **290**, 320–325.

BROWNLEE D. E. 1978. Microparticle studies by sampling techniques, in *Cosmic Dust*, edited by J. A. M. McDonnell, Wiley, New York, 295–336.

BROWNLEE D. E. 1985. Cosmic dust: Collection and research, *Ann. Rev. Earth Planetary Sci.*, **13**, 147–173.

BROWNLEE D. E., BATES B. & BEAUCHAMP R. H. 1983. Meteor ablation spherules as chondrule analogs, in *Chondrules and Their Origins*, edited by A. E. King, Lunar and Planetary Institute, Houston, TX, pp. 10–25.

BROWNLEE D. E., JOSWIAK D. J., SCHLUTTER D. J., PEPPIN R. O., BRADLEY J. P. & LOVE S. G. 1995. Identification of individual cometary IDP's by thermally stepped He release (Abstract), in *Lunar Planet. Sci.*, **XXVI** 183–184.

BROWNLEE D. E., BATES B. & SCHRAMM L. S. 1997. The elemental composition of stony cosmic spherules, *Meteoritics Planet. Sci.*, **32**, 157–175.

CEPLECHA Z., BOROVIČKA J., ELFORD W. G., REVELLE D. O., HAWKES R. L., PORUBČAN V. & ŠIMEK M. 1998. Meteor Phenomena and Bodies, *Space Sci. Rev.*, **84**, 327–471.

CEPLECHA Z., JACOBS C. & ZAFFERY C. 1997. Correlation of ground- and space-based bolides, in *Near Earth Objects*, edited by J. L. Remo, New York Academy of Science, New York, NY, 145–154.

CHYBA C. F. 1993. Explosions of small Earthwatch objects in the Earth's atmosphere, *Nature*, **363**, 701–703.

DERMOTT S., JAYARAMAN S., XU Y. L., GUSTAFSON B. Å. & LIOU J. C. 1994. A circumsolar ring of asteroidal dust in resonant lock with the Earth, *Nature*, **369**, 719–723.

DODD R. T. 1981. *Meteorites – A Petrologic-Chemical Synthesis*, Cambridge University Press, New York, 368 pp.

ELFORD W. G., STEEL D. I. & TAYLOR A. D. 1997. Implications for meteoroid chemistry from height distribution of radar meteors, *Adv. Space Res.*, **20** (8), 1501–1504.

ENGRAND C. & MAURETTE M. 1998. Carbonaceous micrometeorites from Antarctica, *Meteoritics Planet. Sci.*, **33**, 565–580.

FLYNN G. J. 1994a. Interplanetary dust particles collected from the stratosphere: physical, chemical, and mineralogical properties and implications for their sources, *Planet. Space Sci.*, **42**, 1151–1161.

FLYNN G. J. 1994b. Changes to the composition and mineralogy of interplanetary dust particles by terrestrial encounters, in *Analysis of Interplanetary Dust*, edited by M. E. Zolensky, T. L. Wilson, F. J. M. Rietmeijer, and G. J. Flynn, American Institute of Physics, 255–275.

FLYNN G. J., THOMAS K. L., BAJT S., SUTTON S. R., KLÖCK W. & CLARK L. 1995. The chemical composition of Semarkona matrix: Implications for formation and aqueous alteration and a comparison to hydrated IDPs (abstract). *Lunar Planet. Sci.*, **XXVI**, 409–410.

FLYNN G. J. 1996. Sources of 10 micron interplanetary dust: The contribution from the Kuiper belt, in *Physics, Chemistry, and Dynamics of Interplanetary Dust*, edited by B. Å. S. Gustafson and M. S. Hanner, *Astron. Soc. Pacific Conf. Series* **104**, 171–175.

FLYNN G. J., BAJT S., SUTTON S. R., ZOLENSKY M., THOMAS K. L. & KELLER L. P. 1996. The abundance pattern of elements having low nebular condensation temperatures in interplanetary dust particles: Evidence for a new type of chondritic material, in *Physics, Chemistry, and Dynamics of Interplanetary Dust*, edited by B. Å. S. Gustafson and M. S. Hanner, Astron. Soc. Pacific Conf. Series. **104**, 291–294.

FLYNN G. J., MOORE L. B. & KLÖCK W. 1999. Density and porosity of stone meteorites: Implication for the density, porosity, cratering, and collisional disruption of asteroids, *Icarus*, **142**, 97–105.

GARDNER C. S., KANE T. J., SENFT D. C., QIAN J. & PAPEN G. C. 1993. Simultaneous observations of Sporadic E, Na, Fe, and Ca^+ layers in Urbana, Illinois: Three case studies, *J. Geophys. Res.*, **98(D9)**, 16865–16873.

GOMBOSI T. I. & HOUPIS H. L. F. 1986. An icy-glue model of cometary nuclei, *Nature*, **324**, 43–44.

GROSSMAN J. N., RUBIN A. L., NAGAHARA H. & KING E. A. 1988. Properties of chondrules, in *Meteorites and the Early Solar System*, edited by J. F. Kerridge and M. S. Matthews, University of Arizona Press, 619–659.

GRÜN E. & JESSBERGER E. K. 1990. Dust, in *Physics and Chemistry of Comets*, edited by W. F. Huebner, Springer-Verlag, 113–176.

GRÜN E., ZOOK H. A., BAGUHL M., BALOGH A., BAME S. J., FECHTIG H., FORSYTH R., HANNER M. S., HORÁNYI M., KISSEL J., LINDBLAD B.-A., LINKERT D., LINKERT G., MANN I., MCDONNELL J. A. M., MORFILL G. E., PHILLIPS J. L., POLANSKEY C., SCHWEHM G., SIDDIQUE N., STAUBACH P., SVETSKA J. & TAYLOR A. 1993. Discovery of Jovian dust streams and interstellar grains by Ulysses spacecraft, *Nature*, **362**, 428–430.

HALLIDAY I., GRIFFIN A. A. & BLACKWELL A. T. 1996. Detailed data for 259 fireballs from the Canadian camera network and inferences concerning the influx of large meteoroids, *Meteoritics Planet. Sci.*, **31**, 185–217.

HARTMANN W. K., THOLEN D. J. & CRUIKSHANK D. P. 1987. The relationship of active comets, extinct comets, and dark asteroids, *Icarus*, **69**, 33–50.

HUDSON R. S. & OSTRO S. J. 1994. Shape of asteroid 4769 Castalia (1989 PB) from inversion of radar images, *Science*, **263**, 940–943.

HUGHES D. W. 1994. Comets and asteroids, *Contemp. Phys.*, **35**, 75–93.

HUNTEN D. M., TURCO R. P. & TOON O. B. 1980. Smoke and dust particles of meteoric origin in the mesosphere and stratosphere, *J. Atm. Sci.*, **32**, 1342–1357.

HUSS G. R., KEIL K. & TAYLOR G. J. 1981. The matrices of unequilibrated ordinary chondrites: Implications for the origin and history of chondrites, *Geochim. Cosmochim. Acta*, **45**, 33–51.

JENNISKENS P., BETLEM H., BETLEM J., BARIFAIJO E., SCHLÜTTER T., HAMPTON G., LAUBENSTEIN M., KUNZ J. & HEUSSER G. 1994. The Mbale meteorite shower, *Meteoritics*, **29**, 246–254.

JENNISKENS P., BOROVIČKA J., BETLEM H., TER KUILE C., BETTONVIL F. & HEINLEIN D. 1992. Orbits of meteorite producing fireballs, The Glanerbrug – a case study, *Astron. Astrophys.*, **255**, 373–376.

JESSBERGER E. K., CHRISTOFORIDIS A. & KISSEL J. 1988. Aspects of major element composition of Halley's dust, *Nature*, **332**, 691–695.

JESSBERGER E. K., BOHSUNG J., CHAKAVEH S. & TRAXEL K. 1992. The volatile element enrichment of chondritic interplanetary dust particles, *Earth Planet. Sci. Lett.*, **112**, 91–99.

JONES R. H., LEE T., CONNOLLY, JR. H. C., LOVE S. G. & SHANG H. 2000. Formation of chondrules and CAI's: Theory versus observation, in *Protostars and Planets IV*, edited by V. G. Mannings, A. P. Boss & S. S. Russell, University of Arizona Press, 927–962.

KEIL K., HAACK H. & SCOTT E. R. D. 1994. Catastrophic fragmentation of asteroids: Evidence from meteorites, *Planet. Space Sci.*, **42**, 1109–1122.

KERRIDGE J. F. & MATTHEWS M. S. (EDITORS) 1988. *Meteorites and the Early Solar System*, University of Arizona Press. 1269 pp.

KETTRUP D., STEMMERMANN P., DEUTSCH A. & GOTTLICHER J. 2000. Micrometeorites in sandotone – A new successful separation method (Abstract), in *Lunar Planet. Sci.*, **XXXI**, CD-ROM No. 1374.

KURAT G., BRANDSTÄTTER F., PREPER T., KOEBERL C. & MAURETTE M. 1993. Micrometeorites, *Russian Geol. Geophys.*, **34**, 132–147.

KURAT G., KOEBERL C., PREPER T., BRANDSTÄTTER F. & MAURETTE M. 1994. Petrology and geochemistry of Antarctic micrometeorites, *Geochim. Cosmochim. Acta*, **58**, 3879–3904.

LOVE S. G. & BROWNLEE D. E. 1991. Heating and thermal transformation of micrometeoroids entering the Earth's atmosphere, *Icarus*, **89**, 26–43.

LOVE S. G. & BROWNLEE D. E. 1993. A direct measurement of the terrestrial mass accretion rate of cosmic dust, *Science*, **262**, 550–553.

MACKINNON I. D. R. & RIETMEIJER F. J. M. 1987. Mineralogy of chondritic interplanetary dust particles, *Rev. Geophys.*, **25**, 1527–1553.

MAKJANIC J. 1990. Carbon in chondrites: Distribution and structure. *Ph.D. Thesis*, Free University of Amsterdam, The Netherlands, 109 pp.

MAURETTE M., JÉHANO C., ROBIN E. & HAMMER C. 1987. Characteristics and mass distribution of extraterrestrial dust from the Greenland ice cap, *Nature*, **328**, 699–702.

MAURETTE M., IMMEL G., HAMMER C., HARVEY R., KURAT G. & TAYLOR S. 1994. Collection and curation of IDP's from the Greenland and Antarctic ice sheets, in *Analysis of Interplanetary Dust*, edited by M. E. Zolensky, T. L. Wilson, F. J. M. Rietmeijer, and G. J. Flynn, Am. Inst. Phys. Conf. Proc., **310**, 277–289.

McFADDEN L. A., COCHRAN A. L., BARKER E. S., CRUIKSHANK D. P. & HARTMANN W. K. 1993 The enigmatic object 2201 Oljato: Is it an asteroid or an evolved comet?, *J. Geophys. Res.*, **98**, 3031–3041.

McNEIL W. J., LAI S. T. & MURAD E. 1998. Differential ablation of cosmic dust and implications for the relative abundances of atmospheric metals, *J. Geophys. Res.*, **103**, 10899–10911.

McSWEEN, JR. H. Y. & TREIMAN A. H. 1998. Martian meteorites, in *Planetary Materials*, edited by J. J. Papike, Revs. Mineralogy, 36, Chapter 6, Mineral. Soc. Am., 1–53.

McSWEEN, JR. H. Y. 1999. Meteorites, in *The New Solar System* (4th Edition), edited by J. K. Beatty, C. C. Petersen, and A. Chaikin, Cambridge University Press, Cambridge, pp. 351–363.

MEIBOM A. & CLARK B. E. 1999. Evidence for the insignificance of ordinary chondritic material in the asteroid belt, *Meteoritics Planet. Sci.*, **34**, 7–24.

MILLMAN P. M. 1972. Cometary meteoroids, in *From Plasma to Planet*, edited by A. Elvius, Almqvist and Wiksell, Stockholm, 157–167.

MITTLEFELDT D. W., McCOY T. J., GOODRICH C. A. & KRACHER A. 1998. Non-chondritic meteorites from asteroidal bodies, in *Planetary Materials*, edited by J. J. Papike, Revs. Mineralogy, 36, Chapter 4, Mineral. Soc. Am., pp. 1–195.

MUENOW D. W., KEIL K. & McCOY T. J. 1995. Volatiles in unequilibrated ordinary chondrites: Abundances, sources and implications for explosive volcanism on differentiated asteroids, *Meteoritics*, **30**, 639–645

MURPHY D. M., THOMSON D. S. & MAHONEY M. J. 1998. In situ measurements of organics, meteoritic material, mercury, and other elements in aerosols at 5 to 19 kilometers, *Science*, **282**, 1664–1669.

MURRAY I. S., HAWKES R. L. & JENNISKENS P. 1999. Airborne intensified charge-coupled device observations of the 1998 Leonid shower, *Meteorit. Planet. Sci.*, **34**, 949–958.

NYSTROM J. O., LINDSTROM M. & WICKMAN E. F. 1988. Discovery of a second Ordovician meteorite using chromite as a tracer, *Nature*, **336**, 572–574.

ODDERSCHEDE L., MEIBOM A. & BOHR J. 1998. Scaling analysis of meteorite shower mass disctributions, *Europhys. Lett.*, **43**, 598–604.

PIETERS C. M., TAYLOR L. A., NOBLE S. K., KELLER L. P., HAPKE B., MORRIS R. V., ALLEN C. C., McKAY D. S. & WENTWORTH S. 2000. Space weathering on airless bodies: Resolving a mystery with lunar samples, *Meteoritics Planet. Sci.*, **35**, 1101–1107.

PLANE J. M. C. 1991. The chemistry of meteoritic metals in the Earth's upper atmosphere, *Int. Rev. Phys. Chem.*, **10**, 55–106.

PLANE J. M. C., COX R. M. & ROLLASON R. J. 1999. Metallic layers in the mesopause and lower thermosphere region, *Adv. Space Res.*, **24**, 1559–1570.

RABINOWITZ D. L., GEHRELS T., SCOTTI J. V., MCMILLAN R. S., PERRY M. L., WISNIEWSKI W., LARSON S. M., HOWELL E. S. & MUELLER B. E. A. 1993. Evidence for a near-Earth asteroid belt, *Nature*, **363**, 704–706.

RAMDOHR P. 1967. Die Schmlzkrüste der Meteoriten, *Earth Planet. Sci. Lett.*, **2**, 197–209.

RIETMEIJER F. J. M. 1993. Volcanic dust in the stratosphere between 34 and 36 km altitude during May, 1985, *J. Volc. Geotherm. Res.*, **55**, 69–83.

RIETMEIJER F. J. M. & WARREN J. L. 1994. Windows of Opportunity in the NASA Johnson Space Center Cosmic Dust Collection, *Analysis of Interplanetary Dust*, edited by M. E. Zolensky, T. L. Wilson, F. J. M. Rietmeijer, and G. J. Flynn, Am. Inst. Phys. Conf. Proc., **310**, 255–275.

RIETMEIJER F. J. M. 1996a. CM-like interplanetary dust particles in the lower stratosphere during 1989 October and 1991 June/July, *Meteoritics Planet. Sci.*, **31**, 278–288.

RIETMEIJER F. J. M. 1996b. The ultrafine mineralogy of a molten interplanetary dust particle as an example of the quench regime of atmospheric entry heating, *Meteoritics Planet. Sci.*, **31**, 237–242.

RIETMEIJER F. J. M. 1998a. Interplanetary dust particles, in *Planetary Materials,* edited by J. J. Papike, Revs. Mineralogy, 36, Chapter 2, Mineral. Soc. Am., pp. 1–95.

RIETMEIJER F. J. M. 1998b. Interplanetary dust, in *Advanced Mineralogy*, edited by A. S. Marfunin, Volume 3, Springer-Verlag, Berlin, pp. 22–28.

RIETMEIJER F. J. M. 1999. Sodium tails of comets: Na/O and Na/Si abundances in interplanetary dust particles, *Ap. J.*, **514**, L125–L127.

RIETMEIJER F. J. M. & KARNER J. M. 1999. Metastable eutectics in the $Al_2O_3-SiO_2$ system, *J. Chem. Phys.*, **110**, 4554–4558.

RIETMEIJER F. J. M., NUTH, III J. A. & KARNER J. M. 1999. Metastable eutectic condensation in a $Mg-Fe-SiO-H_2-O_2$ vapor: Analogs to circumstellar dust, *Ap. J.*, **527**, 395–404.

RIETMEIJER F. J. M. 2000. Interrelationships among meteoric metals, meteors, interplanetary dust, micrometeorites, and meteorites, *Meteoritics Planet. Sci.*, **35**, 1025–1041.

RIETMEIJER F. J. M. & FLYNN G. J. 2000. A cosmic Fe,Ni signature associated with dust between 34–36 km altitude during May 1985 (Abstract), *Meteoritics Planet. Sci.*, **35** Suppl., A136.

RIETMEIJER F. J. M. & JENNISKENS P. 2000. Recognizing Leonid meteoroids among the collected stratospheric dust, *Earth Moon Planets*, **82/83**, 505–524.

RIETMEIJER F. J. M. & NUTH, III J. A. 2000a. Collected extraterrestrial materials: Constraints on meteor and fireball composition, *Earth Moon Planets*, **82/83**, 325–350.

RIETMEIJER F. J. M. & NUTH, III J. A. 2000b. Metastable eutectic equilibrium brought down to Earth, *EOS, Trans. Am. Geophys. Union*, **81(36)**, 409, 414–415.

RIETMEIJER F. J. M. 2001. Identification of Fe-rich meteoric dust, *Planet. Space Sci.*, **49**, 71–77.

ROBIN E., CHRISTOPHE MICHEL-LEVY N., BOUROT-DENISE M. & JÉHANNO C. 1990. Crystalline micrometeorites from Greenland blue lakes: Their chemical composition, mineralogy, and possible origin, *Earth Planet. Sci. Lett.*, **97**, 162–176.

ROSINSKI J. & PIERRARD J. M. 1964. Condensation products of meteor vapors and their connection with noctiluscent clouds and rainfall anomalies, *J. Atm. Terr. Phys.*, **26**, 51–66.

SANFORD S. A. 1987. The collection and analysis of extraterrestrial dust particles, *Fund. Cosmic Phys.*, **12**, 1–73.

SCHRAMM L. S., BROWNLEE D. E. & WHEELOCK M. M. 1989. Major element composition of stratospheric micrometeorites, *Meteoritics*, **24**, 99–112.

SHERARER C. K., PAPIKE J. J. & RIETMEIJER F. J. M. 1998. The planetary sample suite and environments of origin , in *Planetary Materials*, edited by J. J. Papike, Revs. Mineralogy, 36, Chapter 1, Mineral. Soc. Am., pp. 1–28.

SPURNÝ P. & BOROVIČKA J. 1999. Detection of a high density meteoroid on cometary orbit, in *Evolution and Source Regions of Asteroids and Comets*, edited by J. Svoren, E. M. Pittich, and H. Rickman, *Proc. IAU Coll.* **173**, Astron. Inst. Slovak Acad. Sci., Tatranska Lomnica, pp. 163–168.

SPURNÝ P., BETLEM H., JOBSE K., KOTEN P. & VAN'T LEVEN J. 2000. New type of radiation of bright Leonid meteors above 130 km., *Meteoritics Planet. Sci.*, **35**, 1109–1115.

STEEL D. 1997. Cometary impacts on the biosphere, in *Comets and the Origin and Evolution of Life*, edited by P. J. Thomas, C. F. Chyba, and C. P. McKay. Springer-Verlag, pp. 209–242.

TAYLOR S. & BROWNLEE D. E. 1991. Cosmic spherules in the geological record, *Meteoritics*, **26**, 203–211.

TAYLOR S., LEVER J. H. & HARVEY R. P. 2000. Numbers, types, and composition of an unbiased collection of cosmic spherules, *Meteoritics Planet. Sci.*, **35**, 651–666.

THOMAS K. L., KELLER L. P. & MCKAY D. S. 1996. A comprehensive study of major, minor, and light element abundances in over 100 interplanetary dust particles, in *Physics, Chemistry, and Dynamics of Interplanetary Dust*, edited by B. Å. S. Gustafson and M. S. Hanner, Astron. Soc. Pacific Conf. Series **104**, 283–286.

THOMAS K. L., BLANFORD G. E., CLEMETT S. J., FLYNN G. J., KELLER L. P., KLÖCK W., MAECHLING C. R., MCKAY D. S., MESSENGER S., NIER A. O., SCHLUTTER D. J., SUTTON S. R., WARREN J. L. & ZARE R. N. 1995. An asteroidal breccia: The anaatomy of a cluster IDP, *Geochim. Cosmochim. Acta*, **59**, 2797–2815.

THORSLUND P. & WICKMAN F. E. 1981. Middle Ordovician chondrite in fossiliferous limestone from Brunflo, Central Sweden, *Nature*, **289**, 285–286.

VAISBERG O. L., SMIRNOV V. N., GORN L. S., IOVLEV M. V., BALIKCHIN M. A., KLIMOV S. I., SAVIN S. P., SHAPIRO V. D. & SHEVCHENKO V. I. 1986. Dust coma structure of Comet Halley from SP-1 detector measurements, *Nature*, **321**, 274–276.

VERNIANI F. 1967. Meteor masses and luminosity, *Smithsonian Contrib. Astrophys.*, **10**, 181–195.

VERNIANI F. 1969. Structure and fragmentation of meteoroids, *Space Sci. Revs.*, **10**, 230–261.

VON ZAHN U., GERDING M., HÖFFNER J., MCNEIL W. J. & MURAD E. 1999. Iron, calcium, and potassium atom densities in the trails of Leonids and other sources, *Meteoritics Planet. Sci.*, **34**, 1017–1027.

WEISSMAN P. R., A'HEARN M. F., MCFADDEN L. A. & RICKMAN A. 1989. Evolution of comets into asteroids, in *Asteroids II*, edited by R. P. Binzel, T. Gehrels and M. S. Matthews, University of Arizona Press, Tucson, AZ, pp. 880–920.

WILSON L., KEIL K. & LOVE S. J. 1999a. The internal structure and densities of asteroids, *Meteoritics Planet. Sci.*, **34**, 479–483.

WILSON L., KEIL K., BROWNING L. B., KROT A. N. & BOURCIER W. 1999b. Early aqueous alteration, explosive disruption, and reprocessing of asteroids, *Meteoritics Planet. Sci.*, **34**, 541–557.

WISDOM J. 1985. Meteorites may follow a chaotic route to Earth *Nature*, **315**, 731–733.

ZOLENSKY M. E. 1997. Structural water in the Bench crater Chondrite returned from the Moon, *Meteoritics Planet. Sci.*, **32**, 15–18.

ZOLENSKY M. E., WEBB S. J. & THOMAS K. L. 1988. The search for refractory interplanetary dust particles from preindustrial aged Antarctic ice, in *Proceedings of 18th Lunar Planetary Science Conference*, Cambridge University Press, Cambridge, 599–605.

Part 3
MODELING AND ANALYSIS

10
METEOROID IMPACTS ON SPACECRAFT

By **LUIGI FOSCHINI**[†]

Institute TeSRE-CNR, Via Gobetti 101, I-40129 Bologna, Italy

In the space age, information about the near-Earth environment is becoming more and more important, because of the potential danger to human exploration and use of space. In recent years there have been a number of *in situ* space experiments experiments, such as LDEF and EURECA, that have demonstrated the threats to satellites, space station, and astronauts from high-kinetic-energy impacts of meteoroids and space debris. Post-flight analyses of data from these satellites have revealed that, the catastrophic impact to be a rare event; however, the main danger comes from the impact-generated plasma, which can produce several types of electromagnetic interferences that can disturb or even destroy on-board electronics.

10.1. Introduction

The increasing activity of mankind in space due to the use of satellites for scientific, commercial, and military purposes, space stations, and interplanetary probes has lead to an increase in safety-related problems. The main space hazards are energetic charged particles and radiations, which are particularly troublesome for scientific satellites used in the study of high-energy astrophysics (for example Chandra, XMM), plasmas, dust and meteoroids, and space debris. See Bedingfield *et al.* (1996), Belk *et al.* (1997), Daly *et al.* (1996) for some reviews.

The meteoroid hazard has often been underestimated. Until some years ago, it was a common idea that space debris, rather than meteoroids, was the main threat to spacecraft. This was mainly due to the paper by Laurance & Brownlee (1986): they analysed data from the *Solar Max Satellite* and found that the space debris flux was several hundred times higher than the natural meteoroid flux. However, during last few years other events and post-flight data analyses have indicated a need for a revision of this hypothesis.

During the last maximum of the Perseids, Beech & Brown (1993) warned of impact hazard for the *Hubble Space Telescope* (HST) and, indeed, on the night of August 12th, 1993, astronauts aboard the *MIR* Space Station reported audible meteoroid impacts. Later, it was revealed that *MIR* experienced about 2,000 hits during 24 hours and that the solar panels were severely damaged (Beech *et al.*, 1995). In the same night, the ESA (*European Space Agency*) lost control of the *Olympus* telecommunication satellite: the failure was probably caused by the impact of a Perseid meteoroid (Caswell *et al.*, 1995).

There were also reports before 1993: as early as 1977, the *International Sun Earth Explorer* (ISEE-1) was hit by a meteoroid and damages resulted in a 25% data loss (Bedingfield *et al.*, 1996). The *MIR* Space Station has reported power shortage due to environmental effects, particularly meteoroid impact on solar arrays (Bedingfield *et al.*, 1996). The satellites LDEF (*Long Duration Exposure Facility*) and EURECA (*EUropean REtrievable CArrier*) have provided the lion's share of meteoroid–spacecraft impact data. Post-flight analyses on these satellites drastically rescaled the theories of Laurance and

[†] Email: `foschini@tesre.bo.cnr.it`

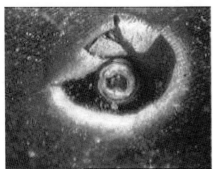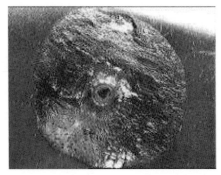

FIGURE 10.1. Different morphologies shown by impacts on *Hubble Space Telescope*. (a) Diameter 460 μm; (b) diameter 5 mm. Courtesy ESA.

Brownlee on the space debris danger. McDonnell *et al.* (1997) showed that the error in the paper of Laurance and Brownlee was due to the use of an incorrect formula for transformation from crater dimension to particle mass. This group then derived a new formula based on data from laboratory simulations (Gardner *et al.*, 1997). New flux estimates, based on data collected on EURECA and LDEF, have shown that at 500 km altitude the micrometre-sized debris population is not dominant as previously thought (McDonnell *et al.*, 1997).

In fact, above the 30 μm ballistic limit, which is the maximum thickness of material that can be penetrated for a given set of impact parameters, meteoroids dominate, while, for thickness between 4 and 5 μm, only 18% of impacts are due to interplanetary matter. Moreover, owing to atmospheric drag, the debris flux did not change appreciably in the 30 μm size regime over the period 1980–1994.

Later on, McDonnell & Gardner (1998) refined these values: the flux above 100 μm is compatible with the natural background, while an additional component (probably space debris) is needed for values of ballistic limit below 40 μm. It is worth noting an enhancement of this additional component at 2 μm.

It is worth noting that hypervelocity impact studies are not only related to catastrophes. Indeed, several probes are equipped with dust detectors, which are sensitive to plasma generated during the impact, and they can be used to determine the flux, in certain mass ranges (generally lower than 10^{-9} kg), all over the Solar System (see discussion by Grün in Chapter 3).

In this chapter we shall focus our attention on the meteoroid impact, with particular attention to physical effects related to plasma generation.

10.2. Impact Probabilities

Meteoroids are concentrated in streams (see discussion by Williams in Chapter 2) where the particle number per volume unit can reach very high values, with geocentric speeds up to 72 km/s. During the 1966 Leonid meteor storm a ZHR (*Zenithal Hourly Rate*) of 150,000 meteors per hour was recorded.[†]

The impact probability P_n of n meteoroids with an artificial satellite with area A [m²]

[†] Jenniskens (1995) has raised some doubts on this ZHR value because of lack of corroborating radar data. According to Jenniskens, the correct value should be ZHR = 15,000 ± 3,000.

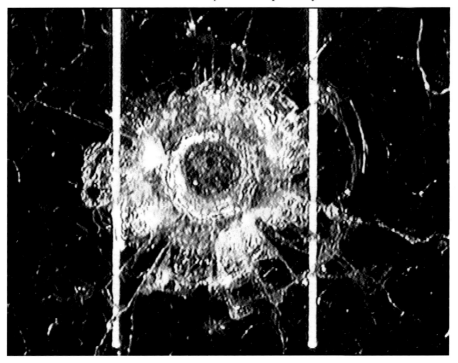

FIGURE 10.2. Micro-impact observed on the EURECA panel f7, Cell F7 I39. Diameter 1.4 mm. Courtesy ESA.

exposed for a time t [s], is given by Poisson statistics:

$$P_n = \frac{N^n}{n!} \cdot \exp(-N) \qquad (10.2.1)$$

where N is the number of impacts, given by:

$$N = k_{\mathrm{ef}} \Phi A t \qquad (10.2.2)$$

where Φ [m^{-2}·s^{-1}] is the flux of meteoroids intercepting the satellite under normal conditions (see Chapter 3).

When a storm or outburst is present, it is necessary to introduce an enhancement factor k_{ef}, which ranges from 300 to 10,000 for the Leonids (Beech & Brown, 1994).

During last few years, several studies were carried out in order to assess specifically the hazard from the Leonids, the most dangerous stream (for a review see Jenniskens ,1999; McBride & McDonnell, 1999; and references therein). Catastrophic mechanical impact does not seem to be a risk, as shown by several cases: for example, the *Space Shuttle* program, until STS68, replaced 46 windshields because of impacts; the *Hubble Space Telescope* suffered about 6,000 impacts during the first 4 years of its life (Bedingfield *et al.*, 1996). Fig. 10.1 shows a couple of impacts on *Hubble Space Telescope*, while Fig. 10.2 shows an impact on a solar cell of EURECA satellite.

The *Olympus* failure is a paradigmatic example of the danger of natural debris: in that case, the impact by a Perseid meteoroid is presumed to have caused electrical failures, leading to a chain reaction that culminated with an early end of the mission. According to Caswell *et al.* (1995), a gyro motor stopped, probably owing to a lack of power, and the satellite lost the reference. Ensuing manoeuvres to acquire a new reference (*Emergency Sun Acquisition*) failed, probably owing to a short circuit in a capacitor of the emergency

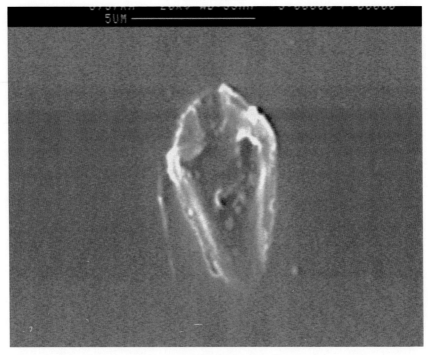

FIGURE 10.3. An example of an oblique impact on a solar cell of *Hubble Space Telescope* (size 7×4 μm). Courtesy ESA.

network. Even though it is not certain, it is plausible that the impact of a small meteoroid, entering the grounded spacecraft via the umbilical, may have generated a plasma that triggered a discharge of charged surfaces.

After the *Olympus* end-of-life anomaly, other authors examined impact-produced plasma, rather than the impact itself (McBride & McDonnell, 1999). In terms of plasma generation, meteoroid streams, particularly the Leonids, can be very dangerous even during normal conditions (i.e. not during an outburst).

The impact-produced plasma can disturb a satellite in several ways: thermal forces can magnetize the neighbourhood of craters (Cerroni & Martelli, 1982); electromagnetic radiations emitted from the plasma can disturb several resources or scientific experiments on the satellite (Foschini, 1998). But the main problem is that the plasma-generated charge can deposit on nearby surfaces, and subsequently lead to discharges (see Lai *et al.*, 2000 and references therein). *ElectroStatic Discharge* (ESD) can be directly channelled into a circuit or can hit a nearby surface and disturb a circuit by a secondary discharge. The pulse shape will depend on electrical characteristics (resistance, inductance, capacitance) of the subject materials. In addition, there can be capacitive coupling between the discharge electric field and the circuit, or inductive coupling between the discharge magnetic field and the circuit. If the first two modes are localized, and then depend on the impact place, the third and fourth modes can disturb distant components. However, these coupling effects are strongly non-linear and depend on the particular circuit layout. Thus, more detailed studies must be made on specific satellites; from analysis of past events, however, we can say that these disturbances can result in telemetry glitches, logic upsets, component failures, and spurious commands (Leach & Alexander, 1995). In addition to electrical effects, direct discharge can physically damage or degrade the spacecraft surfaces, changing the thermal and electric properties of materials.

10.3. Hypervelocity Impact Processes

Since the geocentric speed of meteoroids can be in the range 12–72 km/s, the impact of meteoroids on the walls of spacecraft exceeds the speed of sound in both bodies. The effect is that shock waves are generated and propagate along colliding bodies, compressing and heating both target and meteoroid.

The basic processes of shock waves in solids are well described in the classic book by Zel'dovich & Raizer (1966–67) and we refer to it in the following.

The main difference in the behaviour of compressed gases and condensed matter is the degree of interaction among atoms or molecules in the medium. Indeed, while the pressure in a gas is determined by particle collisions (thermal pressure), the presence of binding forces in a solid generate another type of pressure (nonthermal or elastic). In order to compress a metal by a factor of 2 requires pressures up to several hundreds GPa. Hypervelocity impacts can cause such tremendous pressures, even though we are still below the limit of about 10^{14} Pa, above which the average energy of an electron is much greater than the binding energy of the electrons in atoms and the electrons behave as a degenerate gas (see Eliezer, 1991).

Under such extreme conditions, almost the entire projectile and a fraction of target materials are evaporated and even ionized. A plasma cloud is then created almost instantaneously after the impact and expands into the surrounding vacuum, emitting electromagnetic radiation in a wide spectral range.

10.3.1. Partition of Initial Kinetic Energy

The partition of initial kinetic energy is surely one of the most important problems in hypervelocity impacts. The Rankine–Hugoniot relationships for the conservation of mass, momentum, and energy across a shock are of general validity, regardless of the medium, and therefore can be applied also in hypervelocity impacts. Let us consider V the specific volume ($\equiv \rho^{-1}$, where ρ is the density of the medium), P the pressure, U the propagation speed of the shock wave, and u the mass speed of the compressed material with respect to an undisturbed one. The conservation of mass, momentum, and energy are respectively given by:

$$\frac{V_0}{V} = \frac{U}{U-u} \tag{10.3.1}$$

$$P = \frac{Uu}{V_0} \tag{10.3.2}$$

$$E - E_0 = \frac{1}{2}P(V_0 - V) \tag{10.3.3}$$

where the subscript 0 refers to the undisturbed material. The last equation states that the total energy acquired by unit mass of material, as result of shock compression, is partitioned equally between the kinetic energy $u^2/2$ and the change in internal energy $(E - E_0)$. The latter is in turn composed of changes in elastic and thermal energies.

10.3.2. Cratering, Melting, and Ejecta

The high temperature and pressure developed in hypervelocity impact result mainly in the melting, fragmentation, and ejection of target material. Because of the very high speed, the projectile loses about 99% of its mass during the impact processes, so that it is very difficult to determine unambiguously the identification of the impactor (meteoroids or space debris), as shown in Table 10.1.

TABLE 10.1. Comparison of impact type from various data sets

Impact type	HST data	MIR data	LDEF data
Meteoroids	18–27%	23%	17–23%
Space Debris	11–40%	9%	4–11%
Unknown	42–62%	68%	$\approx 50\%$

From McDonnell & Griffiths (1998).

The main parameter to be considered in the crater formation is the ballistic limit $F_{\max}$ [m], which is the maximum thickness of material that can be penetrated for a given set of impact parameters (McBride & McDonnell, 1999):

$$F_{\max} = 1.064 d_{\mathrm{p}}^{1.056} v^{0.806} \left(\frac{\rho_{\mathrm{p}} \rho_{\mathrm{Al}}}{\rho_{\mathrm{Fe}} \rho_{\mathrm{t}}} \right)^{0.476} \left(\frac{\sigma_{\mathrm{Al}}}{\sigma_{\mathrm{t}}} \right)^{0.134} \qquad (10.3.4)$$

where d_{p} [m] is the particle diameter, v is the projectile speed in km/s, ρ is the density, and σ is the yield stress. The subscripts p, t, Al and Fe refer respectively to projectile, target, aluminium, and iron. The equation was derived from fitting of experimental data with speeds up to 16 km/s, although it is consistent with speeds as high as 100 km/s. For example, McBride & McDonnell (1999) calculated that the *Space Shuttle* can intercept, during a typical mission of two weeks, a meteoroid up to $5.4 \cdot 10^{-8}$ kg, which has a ballistic limit of 2 mm. These calculations were made by using the standard model by Grün *et al.* (1985) of meteoroid flux at 1 AU.

An impact results not only in crater formation, but also in ejection of material. According to recent studies by McDonnell & Griffiths (1998), the total ejected mass can be divided into three parts: jet, cone, and spalls. This is true for a brittle target, while for a ductile one only the first two processes are present. The total of ejected mass can be given by Gault's empirical formula (cf. McDonnell & Griffiths, 1998):

$$M = 7.41 \cdot 10^{-6} \sqrt{\frac{\rho_{\mathrm{p}}}{\rho_{\mathrm{t}}}} E^{1.133} \cos^2 \theta \qquad (10.3.5)$$

where E is the kinetic energy and θ is the angle of impact.

Jetting is the generation of jets of target/projectile material during the penetration of the meteoroid. This process accounts for the release of most of the projectile material, except in cases of grazing or low-kinetic-energy impact.

Another process is the fragmentation of material close to the crater, because of tensile and compressive break up (cone). Almost all these fragments are ejected at the same elevation angle, giving the characteristic conical shape to the debris cloud.

If the target is made of brittle material, there is also the rupture of material near free surfaces around the primary impact site (spallation).

It is worth noting that all these processes are strongly dependent on the impact angle, and that vertical impact are quite rare. Indeed, McDonnell & Griffiths (1998) found that about 80% of impacts with a diameter less than 10 μm are oblique (for example, see Fig. 10.3).

In oblique impact, the shock effects (particularly the temperature) decrease with decreasing of impact angle, but vaporization increases, showing that shear heating must play an important role (Schultz, 1996).

The decreasing of shock effects leads to a reduction of the crater efficiency, which is accompanied by an increase of the fraction of energy carried away by the ricocheted meteoroid. The experimental results by Shultz (1996) are recently confirmed by numerical studies by Pierazzo & Melosh (2000), even though they studied mainly planetary impacts.

10.3.3. *Light Flash*

Another evident phenomenon related to hypervelocity impact is the emission of light. The internal energy of the plasma cloud is partially lost by radiation of light. Laboratory investigations have shown that the spectral distribution of the emitted light can be due to black body radiation (Eichhorn, 1978; Kadono & Fujiwara, 1996). Line emissions can also occur, but the black body approximation gives a reasonable estimation of the temperature of the expanding plasma.

Therefore, let τ be the optical thickness of the cloud, then the energy loss by radiation, dL, can be approximately written:

$$dL = 4\pi R^2 \tau \sigma T^4 \, dt \tag{10.3.6}$$

where R is the cloud radius, σ is the Stefan–Boltzmann constant, and T is the temperature.

According to results by Kadono & Fujiwara (1996), the light energy radiating from the plasma cloud accounts for about 10^{-4}–10^{-5} of the initial kinetic energy of the projectile. These results are in agreement with those obtained previously by Eichhorn (1976, 1978), who found a dependence of radiated light on the kinetic energy of the impacting projectile. Eichhorn (1976) found that the fraction of impact energy that appears as light was in the range from $2 \cdot 10^{-6}$, for iron impacting gold at 4 km/s, to 10^{-2} for carbon impacting tungsten or gold at 7.5 km/s.

However, laboratory studies cannot simulate realistically the speed of natural meteoroids, since velocities as high as 72 km/s cannot be reached. Therefore, it is useful to study the light flashes of impacts on the Moon. Recent studies by Bellot Rubio *et al.* (2000) found a value of $2 \cdot 10^{-3}$ from observations of the 1999 Leonids impacting the Moon.

10.3.4. *Plasma Generation*

Among the different processes occurring during a hypervelocity impact, vaporization and plasma generation are surely the most important for the spacecraft safety. The fraction of kinetic energy partitioned to produce plasma depends mainly on the speed, but the impact angle also plays an important role (Schultz, 1996).

As mentioned above, Ratcliff & Allahdadi (1996) studied the impact of a particle on a surface at a speed comparable with meteoroid speed (94 km/s); however, the particle was very small ($4 \cdot 10^{-19}$ kg; 70 nm diameter). The projectile was made of boron carbide, while the target was silver-doped aluminium. The configuration of the experiment was similar to the dust detector aboard the Cassini–Huygens probe and it allowed direct measurement of ion energies, so that it was possible to estimate the partition of impact energy into various modes.

Ratcliff & Allahdadi (1996) recorded three time-of-flight spectra: one for an impact at 6.9 km/s, one with speed 15.4 km/s, and the last one at the speed of 94 km/s. The features of the three spectra are quite different: in the two low speed spectra, alkaline metals (Na, K) are the most intense lines, while the high-speed spectra show clearly the carbon line, and also the hydrogen one. In addition, at 94 km/s there is the development of leading flanks. The best fitting of the lines was with a gaussian ion-energy distribution, meaning that ion energies are randomly distributed around a mean value and that the plasma is not in thermal equilibrium. Some of the results of Ratcliff & Allahdadi's experiment are

TABLE 10.2. Energies involved in the impact of a 70 nm boron carbide particle on Ag-doped aluminium at 94 km/s. All energies are expressed in Joule and the percentages in parentheses are referred to the kinetic energy of the incoming particle

	Mean ion energy	Total ion energy	Total energy of ionization	Total energy of vaporization	Total measured energy
H	$2.6 \cdot 10^{-17}$	$3.0 \cdot 10^{-11}$ (1.7%)	$2.6 \cdot 10^{-12}$ (0.15%)	Undefined	$> 3.4 \cdot 10^{-11}$ (2.1%)
C	$6.4 \cdot 10^{-18}$	$1.9 \cdot 10^{-11}$ (1.1%)	$5.6 \cdot 10^{-12}$ (0.32%)	$> 4.6 \cdot 10^{-12}$ (0.3%)	$> 2.9 \cdot 10^{-11}$ (1.7%)
Al	$1.5 \cdot 10^{-18}$	$1.1 \cdot 10^{-12}$ (0.06%)	$6.7 \cdot 10^{-13}$ (0.04%)	$4.6 \cdot 10^{-13}$ (0.026%)	$2.2 \cdot 10^{-12}$ (0.13%)
Total		$5.1 \cdot 10^{-11}$ (2.9%)	$8.8 \cdot 10^{-12}$ (0.5%)	$> 5.1 \cdot 10^{-12}$ (0.3%)	$> 6.4 \cdot 10^{-11}$ (3.7%)

From the experiment of Ratcliff & Allahdadi (1996).

shown in Table 10.2. It is worth noting that plasma production accounts for about 4% of kinetic energy of the projectile. By contrast, at lower speeds, e.g. 6 km/s, less than 1% of the kinetic energy goes into plasma production.

McBride & McDonnell (1999) found an empirical formula for evaluation of total charge production during a hypervelocity impact. This equation, rearranged in order to emphasize projectiles dimensions and densities of the projectile, can be written as:

$$Q \simeq 3.04 \delta^{1.02} r^{3.06} v^{3.48} \; [\text{C}] \tag{10.3.7}$$

where δ is the meteoroid density [kg/m^3], r is the meteoroid radius [m] and v its speed [km/s]. The plasma is still globally neutral, even though some phenomena could generate space charges (electron drift, pre-existing electric fields). If there are pre-existing surface charges on the spacecraft, the plasma can neutralize such surface charges in a few microseconds, generating current pulses that can reach very high values. McBride & McDonnell (1999) showed that for each square metre exposed to meteoroid flux in low Earth orbits, a plasma event with current of about 0.8 A would occur each year.

Once the generated charge is known, it is possible to evaluate the characteristic plasma parameters, i.e. the Debye length, λ_D, and the Langmuir frequency, ν. The first parameter gives the scale length over which there can be departures from global charge neutrality. If plasma quantities change on length scales greater than λ_D, collective properties are important; if not, then we have a simple ionized gas. The Debye length is:

$$\lambda_D = \sqrt{\frac{\epsilon_0 k T}{e^2 n_e}} \; [\text{m}] \tag{10.3.8}$$

where ϵ_0 is the electric permittivity of vacuum, k the Boltzmann constant, T the temperature,[†] e the electronic charge, and n_e the electron volume density.

The second parameter arises from the fact that small charge imbalances generate an electric field which acts to restore the global neutrality, but the recalled particles overshoot, so that oscillations with frequency ν are generated. This frequency is called the

---

[†] Generally, the electron temperature should be considered, because electrons move faster than ions.

TABLE 10.3. Examples of charge generation and plasma parameters for Leonids; for explanation of symbols, see the text

r [m]	Q [C]	n_e [m^{-3}]	λ_D [m]	ν [Hz]
10^{-4}	$2.8 \cdot 10^{-3}$	$5.0 \cdot 10^{17}$	$1.7 \cdot 10^{-5}$	$6.3 \cdot 10^{9}$
10^{-3}	3.2	$5.7 \cdot 10^{20}$	$5.0 \cdot 10^{-7}$	$2.1 \cdot 10^{11}$
10^{-2}	$3.7 \cdot 10^{3}$	$6.6 \cdot 10^{23}$	$1.5 \cdot 10^{-8}$	$7.3 \cdot 10^{12}$

From Foschini (1998).

Langmuir or plasma frequency:

$$\omega_0 = 2\pi\nu = \sqrt{\frac{e^2 n_e}{\epsilon_0 m_e}} \ [\text{rad/s}] \qquad (10.3.9)$$

where m_e is the electron rest mass.

It is possible to evaluate the parameters of the plasma generated during a hypervelocity impact of a meteoroid if some simplifying arguments, namely charge uniformity, are made. In Table 10.3 some order-of-magnitude values are shown for the Leonids (Foschini, 1998).

Because of the energy range, plasma production is related to the chemical composition of the meteoroid. Cometary streams, richer in lower-ionization potential elements, will be more dangerous than other streams. The Leonid meteoroid stream turns out to be the most dangerous, even during normal condition (McBride & McDonnell, 1999).

For low speed particles, it is worth noting that meteoroids in space are electrically charged due to cosmic rays, solar UV, and other effects (solar wind and ion and electron impacts). The electric charge changes according to environmental conditions. In interplanetary space, the dust grain are positively charged, because of photoelectron emission. However, meteoroids in plasma (magnetosphere) can be negatively charged: for example, from studies by Horányi (1996, 2000) on dust streams nearly Jupiter and Saturn, grains in a cold plasma are negatively charged, while those in a hot plasma switch their charge to positive values. Horányi (2000) found average values for surface potentials from -30 V in the cold plasma to $+3$ V elsewhere.

At low speed (5–10 km/s), the presence of surface charge can increase the degree of impact ionization by field electron emission (Burchell et al., 1998). This can be explained by the effect of the electric field generated by surface charges of the incoming meteoroid, which increases the electron emission (Sysoev et al., 1997). This, in turn, stimulates the desorption and ionization of atoms or molecules of the surface involved (Sysoev et al., 1997). In addition, the electric field can induce currents in the target, which result in joule heating of the material. These processes require the projectile to have a low speed; at high speeds there is not sufficient time for the field interaction.

10.3.5. *Generation of Electric and Magnetic Fields*

The presence of plasma in hypervelocity impacts leads to the generation of electric and magnetic fields. There are several mechanisms and probably some of these processes will be explained in the framework of a unique theory in the near future, but for the moment we have to consider a mixture of processes, itemized below:

(*a*) Thermally driven electric currents and their associated magnetic fields can be generated in plasma clouds produced during early stages of impacts (Crawford & Schultz,

1991, and references therein). This mechanism can also lead to the magnification of pre-existing magnetic fields (Cerroni & Martelli, 1982).

(b) Radiofrequency emission (about 100 kHz (Bianchi *et al.*, 1984)) – these authors suggested a mechanism similar to electromagnetic emissions before earthquakes (electric double layer), even though it is only a guess that authors offered to explain the electromagnetic radiation. Probably, authors wanted to refer to the mechanism of the following point.

(c) Quasistatic fields generated by space charge separation (Oberc, 1996, and references therein) – these fields were recorded during Vega-2 flyby of comet Halley: it turned out that the electric field response was dependent on the antenna geometry and the arrival direction of larger dust particle (Oberc, 1993).

There are also other effects, mainly electrostatic induction, which result in charging of the antenna and/or the spacecraft (Oberc, 1996).

10.4. Plasma Oscillations

Among different mechanisms that generate an electromagnetic field in hypervelocity impact, plasma oscillations were often neglected. The main reason, as pointed out by M. Ryle in 1949, is that the identity of the plasma frequency with the critical frequency for total reflection would prevent the escape of any radiation produced inside the plasma. However, G.B. Field (1956) later suggested that under certain conditions the electromagnetic radiation can escape.

Plasma oscillations were discovered by Tonks and Langmuir in 1929, and the theory thereof was developed subsequently by Vlasov and Landau, among others. The general approach to the theoretical study of a plasma is based on the use of the kinetic equations, defining the distribution function of its particles (cf. Jackson, 1975). Let us consider the electronic distribution function in equilibrium $f_0(v)$ (Maxwell distribution) and a small deviation from it:

$$F = f_0(v) + f(\vec{v}, \vec{r}, t) \qquad (10.4.1)$$

where $\vec{v}$ is the speed of particles, $\vec{r}$ their position, and t is the time. F is the resulting distribution. The kinetic equation is:

$$\frac{\partial f}{\partial t} + (\vec{v} \cdot \nabla)f - \frac{e}{m}\left(\nabla\psi \cdot \frac{\partial f_0}{\partial \vec{v}}\right) = 0 \qquad (10.4.2)$$

where ψ is the electric field potential given by the Poisson equation:

$$\nabla^2\psi = -\frac{e}{\epsilon_0}\int f \, d\sigma \qquad (10.4.3)$$

where $d\sigma = dv_x \, dv_y \, dv_z$. The equilibrium electronic charge:

$$e\int f_0 \, d\sigma \qquad (10.4.4)$$

is compensated by positive charge of ions.

Irrespective of the initial particle distribution, after a time τ, the particle distribution becomes Maxwellian. The question is then how a certain particle distribution, which differs from a Maxwellian,[†] changes in a time interval $t < \tau$. In hypervelocity impact, the time of expansion of a plasma cloud is of the order of tens of microseconds; during this

---

[†] Experimental investigations by Ratcliff & Allahdadi (1996) showed that the energy distribution at early stages is gaussian.

time we can consider that we are far from the equilibrium time, τ. In this case, the effect of self-consistent fields plays the dominant role and binary collision are not important; consequently, we can neglect the collision integral in the kinetic equation above.

Before continuing, it is necessary to stress that in a plasma there are two types of collisions: close and long-range collisions. Close collisions occur when the impact parameter b_0 takes a value such that the kinetic and potential energies of the particles are comparable at closest approach. In a neutral gas, b_0 is of the order of the size of the gas particles and long-range interactions are very weak. In a plasma, however, the presence of the Coulomb force leads to scattering at substantially longer range due to the strong Coulomb force.

Therefore, during the first instants of the expansion of the plasma cloud, where self-consistent fields (particle electrostatic fields) dominate the dynamics, we have only long-range collisions, i.e. small deflection angles, and we can consider the plasma to be collisionless. In addition, because of the action of Coulomb fields, the distribution functions of the particles, and, therefore, the fields, are subjected to oscillations which can be damped or undamped, longitudinal or transverse.

The longitudinal oscillations are irrotational (i.e. the curl is equal to zero), while the transverse are solenoidal (in this case, the divergence is zero). In a homogeneous unmagnetized plasma, these two are independent, but the presence of density gradients or magnetic fields allow a coupling, so that independent plasma modes are a mixture of these two types.

Physically, the longitudinal and transverse oscillations correspond to two different ways to generate and maintain an electric field: charge motion and changing of the magnetic field, respectively.

Longitudinal plasma oscillations do not propagate, because we have $\nabla \times \vec{E} = 0$ (irrotational field). This means that $\vec{E}$ is a longitudinal field coming from a scalar potential, $\vec{B} = 0$ and we have *no radiation*. But, as shown by Field (1956), if we include the effects of pressure, there is propagation.[†] The dispersion relation for longitudinal oscillations was calculated by Field (1956), to which we refer for more details:

$$\omega^2 = \omega_0^2 + v_T^2 k^2 \tag{10.4.5}$$

where $v_T = \sqrt{3kT/m_e}$ is the root mean square electron speed. Eq. (10.4.5) states that if v_T is zero, the phase speed of any plasma wave is infinite and any disturbance simply oscillates with the Langmuir frequency, but with the same phase everywhere: i.e., the disturbance does not propagate. But, if v_T is different from zero, we have propagation. For historical reasons, effects for $v_T = 0$ are called simply "plasma oscillations", while those effects for $v_T \neq 0$ are "plasma waves", although they are the same physical phenomenon, but in different situations.

There are also plasma oscillation modes which are transverse and, therefore, they can radiate. If we consider that the oscillations behave in time as $\exp(i\vec{k} \cdot \vec{r} - i\omega t)$, the dispersion relation is:

$$\omega^2 = \omega_0^2 + c^2 k^2 \tag{10.4.6}$$

where ω_0 is the Langmuir frequency or the proper angular frequency of the plasma, as given in Eq. (10.3.9). The above equation is valid for the electrons, but also ions can

---

[†] Plasma oscillations including pressure effects are often referred as *plasma waves*.

oscillate with their proper angular frequency:

$$\omega_0 = \sqrt{\frac{(Ze)^2 n_i}{\epsilon_0 m_i}} \qquad (10.4.7)$$

It is worth noting that the angular frequency of these oscillations is determined only by the particle density and not by the velocity distributions of the particles. Since the charge density is dependent on the meteoroid speed and mass (see McBride & McDonnell, 1999), in principle we should know the species in the plasma if we measure the frequency spectrum.

As stated above, in presence of density gradients or static magnetic fields, the two modes (longitudinal and transverse) are coupled. In Field's equations the inhomogeneity is represented by the gradient of $(1 - \omega_0^2/\omega^2)$, which in turn implies a density gradient according to Eq. (10.3.9). Physically, the coupling is due to plasma waves which are refracted by the density gradient, producing a curl. It is straightforward to note that if the waves propagate along the gradient, there is no refraction and hence no coupling. The proof is quite complex and we refer to the paper by Field (1956) for more details.

10.5. The Radiation Field

The plasma cloud generated during a hypervelocity impact and expanding into vacuum creates the conditions of density gradient and non-zero temperature which are necessary to convert the plasma wave energy into electromagnetic energy. To avoid the Landau damping, the wavelength of the plasma must be longer than the Debye length and the boundary of the expanding cloud can be considered a discontinuity for such waves; this generally occurs in hypervelocity impacts.

The conversion occurs with a certain efficiency, defined as the ratio of the normal transmitted flux to that of the incident wave, depending on plasma properties. Field (1956) gave some equations under certain conditions, but they are quite complicated. A sufficiently accurate expression for the mean conversion coefficient can be (Field, 1956):

$$\langle C \rangle = 4k_{\mathrm{L}} \frac{3kT}{m_e c^2} \qquad (10.5.1)$$

where k_{L} is the longitudinal wave number.

The evaluation of the radiated field in hypervelocity impact of meteoroids from the Leonid shower was done by Foschini (1998). The calculation was made using several simplifying assumptions, although numerical simulations are needed to study the problem in more depth. Foschini's calculations are an *upper limit*, but are sufficient to suggest that for a Leonid impact it is possible to reach the microwave frequency range and the radiated power may be two orders of magnitude greater than the electronic noise in circuits on board satellites.

An order-of-magnitude calculation can be done with the electric dipole approximation and assuming a full conversion, i.e. $C = 1$. In this case, the electric and magnetic field generated by an oscillating electric dipole can be easily calculated by the classical formulas in spherical coordinates (see, for example the book by Lorrain *et al.*, 1988):

$$E_r = \frac{p \cos\theta \exp \imath\omega t}{2\pi\epsilon_0 \ell^2} \left\{ \frac{\ell^2}{r^3} + \imath\frac{\ell}{r^2} \right\} \qquad (10.5.2)$$

$$E_\theta = \frac{p \sin\theta \exp \imath\omega t}{4\pi\epsilon_0 \ell^2} \left\{ \frac{\ell^2}{r^3} - \frac{1}{r} + \imath\frac{\ell}{r^2} \right\} \qquad (10.5.3)$$

$$H_\phi = \frac{cp \sin\theta \exp \imath\omega t}{4\pi\ell^2} \left\{ -\frac{1}{r} + \imath \frac{\ell}{r^2} \right\} \qquad (10.5.4)$$

where $\ell = \lambda/2\pi$, r is the distance from the source, θ is the colatitude, ϕ is the azimuth, p is the module of the dipole moment, ω is the angular frequency of the oscillating dipole (cf. Eqs. 10.4.5 and 10.4.6), and c is the speed of the light in vacuum.

When $r \gg \ell$ we have to consider the radiation terms, i.e., those falling off as r^{-1}. In our case, with a Langmuir frequency (see Eq. 10.3.9) of $\nu = 3 \cdot 10^{11}$ Hz, i.e. the lower limit for microwaves, we have $\ell \approx 0.2$ mm. This very small value means that, in hypervelocity impacts, almost all electronic devices aboard the satellite are exposed to the radiation field, because it quite difficult to have intact circuits so close to the expanding plasma vapour cloud.

Moreover, because of the small value of the transition from induction to radiation field, direct measurement of the induction fields, but also of the radiation field, is quite complex under the extreme conditions of hypervelocity impact. Until now, several types of experiments were set up in order to measure the different electric and magnetic emissions from impact-generated plasmas, with particular reference to the magnetic field in studies on craters and to the electric field for dust detectors on board space probes, but there are still several problems and open questions. Specifically, whether or not the used lines of experimental attack are still valid for meteoroid impacts.

For example, after the above discussion, we can expect a broad band noise, with a cutoff limit given by the Langmuir frequency for electrons. In order to measure an electric or magnetic field of this kind of radiation, it is necessary to set up a proper antenna and receiver.

For frequencies lower than 30 MHz, it is necessary to use a vertical rod antenna, generally 1 m long, and therefore not tuned. For frequencies in the range 30 MHz–1 GHz, it is necessary to use a dipole antenna, resonant at half wavelength of the field to be measured. For higher frequencies there are other proper antennas (type horn) to be used. However, it is very difficult to place this type of antenna near a hypervelocity facility.

The proper choice of the transducer is very important, particularly in these studies where measurements are strongly dependent on the frequency response of the antenna. Often a loop antenna for magnetic field measurement was used; however, this type of antenna, *under ideal conditions*, has a proper behaviour only to tens of MHz. In other words, it is not the proper transductor for measuring a signal with GHz frequency.

Another important factor is the choice of the signal receiver and analyser, where the frequency response should be selected with care. The best choice is an EMI (*ElectroMagnetic Interferences*) analyser with a wide dynamic response, so as to be able to measure a broad band spectrum of impulsive signals. Also a spectrum analyser is a good choice, even though it has a limited dynamic band. Often a preamplifier is used in addition to a spectrum analyser to give a good dynamic behaviour.

It is not our intention to examine in deepest detail or to be overly critical with respect to past and actual experiments, but we would like to stress some sides often underestimated or even neglected, that, nevertheless, could determine the goodness of the experiment.

10.6. Concluding Remarks

During recent years, meteoroid hazards to spacecraft have manifested themselves in all their complexity. Because of the high geocentric speed (up to 72 km/s), the kinetic energies in impacts generates high pressures and temperatures. After analyses on satellites, it has become obvious that even though the mechanical damage is not serious, the plasma cloud generated during the impact can be a serious threat. Specifically, the release of electric charge and of electromagnetic energy can disturb or even destroy on-board electronics. Studies on electromagnetic effects are of great importance for future missions: indeed, forthcoming engines for spacecraft will be based on electromagnetic principles (magnetohydrodynamic propulsion, field effect thrusters, ion propulsion, and so on).

Last, but not least, and apart from practical applications, meteoroid impacts on spacecraft present very interesting scientific questions that challenge our theoretical understanding.

I wish to thank the European Office of Aerospace Research and Development for partially supporting this work. I also thank G. Drolshagen and the European Space Agency (ESA) for information and data on Eureca and HST. Last, but not least, I acknowledge the helpful critical reading of this manuscript by Rainer Dressler and the suggestions by Ed Murad.

REFERENCES

BEDINGFIELD K. L., LEACH R. D. & ALEXANDER M. B. 1996 Spacecraft system failures and anomalies attributed to the natural space environment. NASA RP-1390.

BEECH M. & BROWN P. 1993 Impact probabilities on artificial satellites for the 1993 Perseid meteoroid stream. *Mon. Not. R. Astron. Soc.*, **262**, L35–L36.

BEECH M. & BROWN P. 1994 Space-Platform impact probabilities – The threat of the Leonids. *ESA Journal*, **18**, 63–72.

BEECH M., BROWN P. & JONES J. 1995 The potential danger to space platforms from meteor storm activity. *Quart. J. R. Astron. Soc.*, **36**, 127–152.

BELK C. A., ROBINSON J. H., ALEXANDER M. B., COOKE W. J. & PAVELITZ S. D. 1997 Meteoroids and orbital debris: effects on spacecraft. NASA RP-1408.

BELLOT RUBIO L. R., ORTIZ J. L. & SADA P. V. 2000 Luminous efficiency in hypervelocity impacts from the 1999 Lunar Leonids. *Ap. J.*, **542**, L65–L68.

BIANCHI R., CAPACCIONI F., CERRONI P., CORADINI M., FLAMINI E., HURREN P., MARTELLI G. & SMITH P. N. 1984 Radiofrequency emissions observed during macroscopic hypervelocity impacts experiments. *Nature* **308**, 830–832.

BURCHELL M. J., COLE M. J. & McDONNELL J. A. M. 1998 Role of particle charge in impact ionization by charged microparticles. *Nuclear Instruments and Methods in Physics Research B* **143**, 311–318.

CASWELL R. D., McBRIDE N. & TAYLOR A. 1995 Olympus end of life anomaly – A Perseid meteoroid impact event? *Int. J. Impact Eng.* **17**, 139–150.

CERRONI P. & MARTELLI G. 1982 Magnification of pre-existing magnetic fields in impact-produced plasmas, with reference to impact craters. *Planet. Space Sci.* **30**, 395–398.

CRAWFORD D. A. & SCHULTZ P. H. 1991 Laboratory investigations of impact generated plasma. *J. Geophys. Res.* **96**, 18807–18817.

DALY E. J., HILGERS A., DROLSHAGEN G. & EVANS H. D. R. 1996 Space environment analysis: experience and trends. In *Proceedings of Symposium on Environment Modelling for Space-Based Applications*, ESA SP-392.

EICHHORN G. 1976 Analysis of the hypervelocity impact process from impact flash measurements. *Planet. Space Sci.* **24**, 771–781.

EICHHORN G. 1978 Heating and vaporization during hypervelocity particle impact. *Planet. Space Sci.* **26**, 463–467.

ELIEZER S. 1991 High-pressure equation of state – A perspective. In *High-pressure equation of state: theory and applications* (eds. S. Eliezer & R. A. Ricci). p. 1–11, North-Holland, Amsterdam.

FECHTIG H., GRÜN E. & KISSEL J. 1978 Laboratory simulations. In *Cosmic Dust* (ed. J. A. M. McDonnell). pp. 607–669, Wiley, Chichester.

FIELD G. B. 1956 Radiation by plasma oscillations. *Ap. J.* **124**, 555–570.

FOSCHINI L. 1998 Electromagnetic interference from plasmas generated in meteoroids impacts. *Europhys. Lett.* **43**, 226–229.

GARDNER D. J., SHRINE N. R. G. & McDONNELL J. A. M. 1997 Determination of hypervelocity impactor size from thin target spacecraft prenetrations. In *Proceedings 2nd European Conference on Space Debris*, pp. 493–496, ESA-SP393.

GRÜN E., ZOOK H. A., FECHTIG H. & GIESE R. H. 1985 Collisional balance of the meteoritic complex. *Icarus* **62**, 244–272.

HORÁNYI M. 1996 Charged dust dynamics in the Solar System. *Annu. Rev. Astron. Astrophys.* **34**, 383–418.

HORÁNYI M. 2000 Dust streams from Jupiter and Saturn. *Phys. Plasmas* **7**, 3847–3850.

KADONO T. & FUJIWARA A. 1996 Observation of expanding vapor cloud generated by hypervelocity impact. *J. Geophys. Res.* **101E**, 26097–26109.

JACKSON J. D. 1975 *Classical Electrodynamics*. Wiley, New York.

JENNISKENS P. 1995 Meteor stream activity II – Meteor outbursts. *Astron. Astrophys.* **295**, 206–235.

JENNISKENS P. 1999 Update on the Leonids. *Adv. Space Res.* **23**, 137–147.

LAI S. T., MURAD E. & McNEIL W. J. 2000 Spacecraft interactions with hypervelocity particulate environment. *AIAA 38th Aerospace Science Meeting & Exhibit*, 10–13 January 2000, Reno NV.

LAURANCE M. R. & BROWNLEE D. E. 1986 The flux of meteoroids and orbital space debris striking satellites in low earth orbit. *Nature*, **323**, 136–138.

LEACH R. D. & ALEXANDER M. B. 1995 Failures and anomalies attributed to spacecraft charging NASA RP-1375.

LORRAIN P., CORSON D. R. & LORRAIN F. 1988 *Electromagnetic fields and waves*. W. H. Freeman and Co., New York.

McBRIDE N. & McDONNELL J. A. M. 1999 Meteoroid impacts on spacecraft: sporadics, streams, and the 1999 Leonids. *Planet. Space Sci.* **47**, 1005–1013.

McDONNELL J. A. M. & GARDNER D. J. 1998 Meteoroid morphology and densities: decoding satellite impact data. *Icarus* **133**, 25–35.

McDONNELL J. A. M. & GRIFFITHS A. D. (EDS) 1998 Meteoroid and debris flux and ejecta models – Summary Report. ESA Contract No. 11887/96/NL/JG. Unispace Kent.

McDONNELL J. A. M., RATCLIFFE P. R., GREEN S. F., McBRIDE N. & COLLIER I. 1997 Microparticle populations at LEO altitudes: recent spacecraft measurements. *Icarus* **127**, 55–64.

OBERC P. 1993 Simultaneous observations of quasistatic electric fields and large dust particles during the Vega-2 flyby of comet Halley. *Planet. Space Sci.* **41**, 609–617.

OBERC P. 1996 Electric antenna as a dust detector. *Adv. Space Res.* **17**, (12)105–110.

PIERAZZO E. & MELOSH H. J. 2000 Understanding oblique impacts from experiments, observations, and modeling. *Ann. Rev. Earth Planet. Sci.* **28**, 141–167.

RATCLIFF P. R. & ALLAHDADI F. 1996 Characteristics of the plasma from a 94 km/s microparticle impact. *Adv. Space Res.* **17**, (12)87–91.

SCHULTZ P. H. 1996 Effect of impact angle on vaporization. *J. Geophys. Res.* **101E**, 21117–21136.

SYSOEV A. A., IVANOV V. P., BARINOVA T. V., SURKOV Y. A. & VYSOCHKIN V. V. 1997 Mass spectra formation from charged microparticles. *Nucl. Instr. Methods Phys. Res. B* **122**, 79–83.

ZEL'DOVICH Y. B. & RAIZER Y. P. 1966–67 *Physics of Shock Waves and High Temperature Hydrodynamic Phenomena*. 2 vols. Academic Press, New York.

11
MODELS OF METEORIC METALS IN THE ATMOSPHERE

By WILLIAM J. McNEIL[1†], EDMOND MURAD[2‡], AND JOHN M. C. PLANE[3¶]

[1]Radex, Inc., 3 Preston Court, Bedford, MA 01731, USA

[2]Space Weather Center of Excellence, Space Vehicles Directorate, Air Force Research Laboratory, Hanscom AFB, MA 01731, USA

[3]School of Environmental Sciences, University of East Anglia, Norwich NR4 7TJ, UK

Meteoroids entering the Earth's atmosphere are heated by collisions with the atmospheric species and, eventually, reach temperatures where components of the meteoroids evaporate. Modeling in detail this process requires a knowledge of entry speeds, size distribution, entry angle, and composition of the meteoroids. Not all this information is currently available, although several assumptions, grounded in sound physical principles, can be made and reasonable quantities derived for the vapor input into the atmosphere. An as-yet unsettled question is whether the meteoroids can reach a thermodynamic equilibrium during the transit through atmosphere or not. Once the vapor input, consisting of metal atoms, is derived, chemical processes take place that result in the transformation of the metal atoms into complex chemical compounds. Model development then involves the use of reliable laboratory measurements of reaction cross sections. Following these components of modeling, methodology is summarized for integrating them into geophysical parameters, such as diffusion, gravity waves, and electric fields.

11.1. Historical Survey

It has been accepted for many years that meteoric influx is the direct cause of the presence of several metallic species in the Earth's atmosphere [Megie & Blamont (1969)]. The atomic metals, concentrated in layers near 90 km, have been studied extensively by the lidar technique since the first of these experiments more than 30 years ago [Bowman (1969)]. Figure 11.1 shows the profiles of the annual mean layers of Na, Fe, K and Ca, observed by lidar at several mid-latitude locations.

The Na layer has been studied in much greater detail than the other metals because historically it was the easiest metal to observe. The height of the peak of the layer varies between 88 and 92 km, with the highest peak heights occurring during summer. The full width at half maximum of the layer is about 10 km, and it is usually characterized by strikingly small scale-heights of 2–3 km on the top side of the layer, and less than 2 km on the bottom side (the scale height is the distance over which the concentration changes by a factor of e). Fe, K and Ca are all depleted relative to Na, when compared with their relative abundances in chondritic meteorites recovered from the Earth's surface. It is worth noting here and throughout the discussion of composition, that the issue is somewhat complicated [see Rietmeijer (2000) and Chapter 9 of this book] due to the fact that whereas the phenomenon we are studying and describing in this chapter refers to dust expelled from comets, composition data come for the most part from meteorites that are

[†] Email: William.McNeil@hanscom.af.mil
[‡] Email: ed.murad@hanscom.af.mil
[¶] Email: J.Plane@uea.ac.uk

Annual Mean Metal Layers (at mid-latitudes)

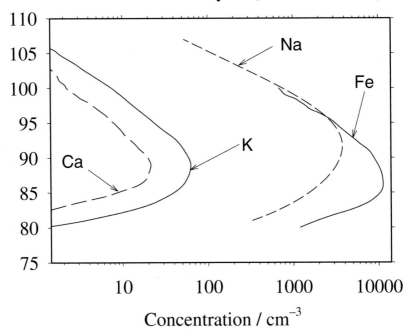

FIGURE 11.1. Vertical profiles of the annual mean concentrations of Fe, Na, K and Ca measured by lidar at a number of mid-latitude locations in the U.S. and Europe [Helmer *et al.* (1998); Eska *et al.* (1999); Plane *et al.* (1999a); Gerding *et al.* (2000)]

collected on the ground. These come from asteroids. Thus, the appropriate composition becomes problematic. An important contribution towards resolving this issue has come from recent spectral observations of Leonid meteors [Borovička *et al.* (1999)], as will be discussed in more detail below. With this cautionary remark in mind, it merits note that, compared with chondrite composition, Ca is depleted by the enormous factor of 120–360 depending on season [Gerding *et al.* (2000)]. The K, Ca and Fe peaks are also several kilometers lower and have considerably smaller bottom-side scale heights, as shown in Figure 11.1.

It has also been established that ionized metal layers exist in the E and F regions of the ionosphere [Istomin & Pokunkov (1963); Narcisi (1968); Hanson & Sanatani (1970); Hanson *et al.* (1970); Hanson & Sanatani (1971); Hanson *et al.* (1972); Kopp & Herrmann (1984)]. These metallic species can be transported as ions into the thermosphere by winds and electric fields, leading to substantial densities at altitudes at and above 1000 km [Gérard & Monfils (1974)].

The historical path that led to our present state of knowledge has been summarized, to some extent, in the Introduction. Hence, we will limit the scope of this chapter to the current state of modeling of meteoric metals in the atmosphere. The development of these models has been driven by the need to understand the many curious features of the metals outlined above – indeed, to explain how a highly reactive species such as atomic Na can exist in an essentially oxidizing atmosphere.

The modeling of atmospheric metals requires knowledge of several key parameters describing the process of meteor ablation, the chemical kinetics of the metals once deposited, and the processes that transport these metals from the lower E region, where they are deposited, to higher and lower altitudes. Modeling, in conjunction with observation, is of intrinsic importance because the results shed light on several very different but interdependent processes in near-space physics. There are essentially two general approaches to the modeling of atmospheric metals. A steady-state approach, which assumes that the chemical kinetics are fast compared with deposition of the metal and with transport, has been used extensively to study the seasonal variability in the neutral layers of several of the metals [Helmer *et al.* (1998); McNeil *et al.* (1995); Plane *et al.* (1999a); Eska *et al.* (1999); Gerding *et al.* (2000)]. A second type of model is time-dependent, solving the sequential problem of metal deposition, transport, and chemical transformation and removal of the metal self-consistently. This second type of model is required for study of metal transport into the thermosphere [Fesen *et al.* (1983); Carter & Forbes (1999); McNeil *et al.* (1996)]. Time dependent models are also required for the modeling of the effects of meteor showers on the background metal layers [McNeil *et al.* (2000)], and have been applied to studying the formation of sporadic (or sudden) metal layers [Cox & Plane (1998)] and the evolution of long-enduring meteor trains [Jenniskens *et al.* (2000a)]. First, however, we examine the modeling of meteor ablation in the atmosphere, which is the starting point of all models of atmospheric metals.

11.2. Ablation of Meteoroids

Meteoroids heat as they enter the Earth's atmosphere, and the meteoroid components subsequently evaporate into the atmosphere. This problem, treated in detail by several investigators [Öpik (1936a); Öpik (1936b); Öpik (1937); Whipple (1943); Whipple (1950); Whipple (1951); Öpik (1958); Jones & Kaiser (1966); Hawkes & Jones (1975); Nicol *et al.* (1985); Fyfe & Hawkes (1986); Flynn (1989); Kalashnikova *et al.* (2000)], becomes manageable for the very small meteoroids because heat conductivity can be neglected. Recent analysis suggests that for particles of diameter ≤ 1 mm, the neglect of conductivity is justified [Love & Brownlee (1991)]. For the cosmic dust background, most of the mass is contributed by particles having masses $\leq 10^{-4}$ grams [Love & Brownlee (1993)]. Assuming a density of 1 g cm^{-3}, a 1 mm diameter meteoroid has a mass of $\sim 5 \times 10^{-4}$ g. Therefore the neglect of heat conductivity seems justified. The heating of the meteoroid by collisions with the ambient atmosphere is counterbalanced by radiative losses and heat capacity (melting, sublimation, phase transitions). Calculation of the sum total of these terms depends on a number of parameters that are only approximately known: meteoroid shape, density, and composition. Fortunately, recent studies of the spectra of Leonid meteors [Borovička *et al.* (1999)] have shown that the composition of the Leonid meteors agrees with that of chondrites. This study has also shown that Na is evaporated earlier in the flight of the meteoroid through the atmosphere. Most models treat the total amount of a particular metal deposited in the atmosphere as a variable parameter; an exception is recent models [McNeil *et al.* (1998); McNeil *et al.* (2000); Gardner *et al.* (1999)], and, even here, the composition is introduced only in an average sense. Even for dustball meteoroids, the initial heating leads to melting of the "glue" or binding material [Hawkes & Jones (1975)], and the remaining part is likely to consist of dustballs which are then heated by momentum transfer in collisions with the atmosphere. The evaporation then is from these dustballs. The problem is tractable for a spherical particle; following the treatment given by Jones & Kaiser (1966), the balance between the heating and cooling per unit surface area is given by:

$$\frac{1}{2}\Lambda\rho_a v^3 = 4\sigma\epsilon(T_s^4 - T_a^4) + \frac{4}{3}R\rho_m C\frac{dT_m}{dt} \qquad (11.2.1)$$

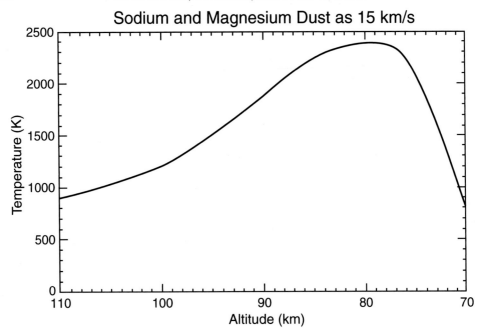

FIGURE 11.2. Calculated meteoroid temperature as a function of altitude.

The left side of the equation represents the collisional heating term, while the first term on the right side of the equation represents radiative losses, and the second term on the right side represents energy losses due to heat capacity (vaporization, phase transitions, and heating). In the above equation, Λ is the heat transfer coefficient or fraction of the total kinetic energy of the air molecules that is transferred to the meteoroid, ρ_a the atmospheric density, v the meteoroid velocity, ϵ the emissivity of the meteoroid, σ Stefan's constant, T_s the surface temperature of the meteoroid, T_a the ambient atmosphere temperature, R the initial meteoroid radius, ρ_m the meteoroid density, C the meteoroid specific heat, T_m the mean temperature, and t the time.

There are two limiting cases for Equation 11.2.1: (1) radiative cooling is $\gg$ than the heat capacity term (i.e. for the case where R is very small), in which case the meteoroid does not get hot enough to ablate; and (2) radiative cooling is $\ll$ the heat capacity (large R), in which case the meteoroid will evaporate. The limiting meteoroid radius was calculated to be ~ 100 μm [Jones & Kaiser (1966)]. Ignoring phase transitions, the second term on the right side of Equation 11.2.1 may be rewritten to include the vapor pressure explicitly:

$$H_v C p_v(T) \sqrt{\frac{m}{T_s}} \tag{11.2.2}$$

where the term $p_v(T)$ is the vapor pressure at a given temperature, m the atomic weight, and H_v the heat of vaporization. Figure 11.2 shows the temperature profile of a particle assuming relatively non-volatile components. Generally, cosmic dust is irregularly shaped (fractal); hence its surface area to volume ratio is larger than for a sphere. This means that radiative cooling is greater than for a sphere; however, such a particle cannot be treated analytically. Simultaneous with the collisional heating of the meteoroids, a deceleration

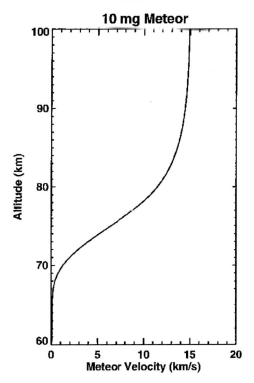

FIGURE 11.3. Calculated meteoroid velocity as a function of altitude. The calculation assumes a spherical particle with $\rho = 3.2$ g cm^{-3}, and an entry angle of 45°. Vapor pressures are taken from Love & Brownlee (1991).

takes place according to

$$\frac{dv}{dt} = \frac{\Gamma \rho(h) \pi r^2 v^2}{m} \qquad (11.2.3)$$

where v is the velocity of the meteoroid, Γ the atmospheric drag parameter, $\rho(h)$ the atmospheric density as a function of altitude, r the meteoroid radius, and m its mass. In Figure 11.3 we show a typical profile of the velocity as a function of altitude for a 10 mg meteoroid.

For realistic modeling, a great deal of information about the meteoroid is needed: composition, solid structure, heat capacity, shape, and velocity. This information is not available, and, hence, guesses have to be made. Some of these issues are discussed in Chapter 9 by Rietmeijer and in a separate publication [Rietmeijer (2000)].

11.2.1. *Vaporization*

All the models of the vaporization of meteoroids and subsequent deposition of metals in the Earth's atmosphere have at some point or another to consider the composition of the meteoroids. Here, the subject is quite uncertain due to the great variability in composition of the different types of meteorites, as discussed in detail by Rietmeijer in Chapter 9 and elsewhere [Rietmeijer (2000)]. Furthermore, it may be that the composition of the meteoroids that ablate in the upper atmosphere is different from that of the meteorites that survive transit through the atmosphere [Rietmeijer (2000)]. In Table 11.1 we show the composition given by Mason (1971), by Sears & Dodd (1988), and by Lodders & Fegley (1998).

TABLE 11.1. CI–Chondrite elemental abundances from three references. Si is taken to be 1.00 $\times 10^6$

Element	Abundances		
	Mason[a]	Sears & Dodd[b]	Lodders & Fegley[c]
Na	6.3×10^4	5.7×10^4	5.7×10^4
Mg	1.04×10^6	1.075×10^6	1.07×10^6
Al	8.6×10^4	8.49×10^4	8.4×10^4
Si	10^6	10^6	10^6
S	—	5.15×10^5	4.4×10^5
K	—	3770	3700
Ca	5.8×10^4	6.11×10^4	6.1×10^4
Fe	8.73×10^5	9.00×10^5	8.60×10^5

[a]Mason (1971).
[b]Sears & Dodd (1988).
[c]Lodders & Fegley (1998) – wt % figures were converted into abundances relative to Si for the purpose of comparison.

The agreement between these compilations is quite good, although two points should be stressed: (1) these compilations rely heavily on the measurements of solar system abundances by Anders and his colleagues over a number of years [Anders & Ebihara (1982); Grevesse & Anders (1991)], and (2) this agreement is limited to the CI chondrites. Most of the models developed to date for meteor metal input into the atmosphere [Plane & Helmer (1994); McNeil et al. (1998)] have relied on the data given by Mason [Mason (1971)] or by Sears and Dodd [Sears & Dodd (1988)]. Once the meteoroid has reached a sufficiently high temperature (> 1000 K) evaporation (boiling or sublimation) begins. There is some dispute at present as to the nature of the evaporation.

Based on modeling of planetary formation [Fegley & Cameron (1987)] and on thermodynamic arguments it has been suggested [McNeil et al. (1998)] that fractionation of the meteoroid takes place, with the low boiling material (such as Na) evaporating first and the high boiling material (such as Ca) evaporating last. This model, called *differential ablation* by McNeil et al. (1998), is based on the fractionation model of Fegley & Cameron (1987). This model explained the lidar observations of meteor trails, particularly the 100-fold depletion of Ca relative to Na in the atmospheric layers [von Zahn et al. (1999)]. Using the *MAGMA* model provided by Profs. Fegley and Cameron, we show in Figure 11.4 the results of calculation of thermodynamic distributions of the metals as a function of evaporated material at 3000 K. While this temperature is too high for meteoroid heating in the atmosphere, it illustrates that type of calculations that are necessary for the development of an a priori model for metal deposition in the atmosphere from meteoroid ablation. What this figure shows is that at 3000 K, when 80% is left in the condensed phase, the vapor consists of Si, Mg, Fe, smaller amounts of K, and very little Ca; Na will have completely evaporated. When 50% is left in the condensed phase, Na and K will have completely disappeared, and the vapor will consist of Si, Mg, Fe, and Ca. In the next subsection we will discuss this subject in more detail and will compare the deposition of different metals at more realistic meteoroid temperatures, namely 1500 K and 2000 K.

It is worth noting that this is an entirely thermodynamic picture. It assumes that a particle entering the atmosphere remains molten for long enough, and that diffusion of

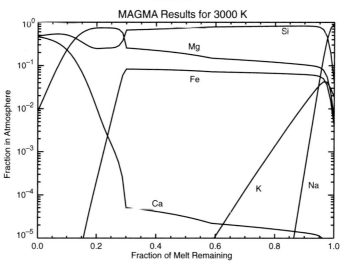

FIGURE 11.4. Thermodynamic calculation of fraction of different metals remaining in the condensed phase after reaching a meteoroid temperature of 3000 K [Fegley & Cameron (1987)].

volatile metallic species through the melt and their evaporation is rapid enough, for a thermodynamic model based on equilibrium conditions to apply. Although these assumptions have not been investigated in detail, the success of the model of differential ablation described below suggests that thermodynamics largely controls the ablation process.

11.2.2. *Fractionation or Differential Ablation*

A model has been developed to investigate the meteor ablation process, specifically focusing on the overall depletion of calcium in the atmosphere relative to sodium [McNeil *et al.* (1998)]. Sodium and calcium are of approximately equal abundance in meteoric material [Mason (1971); Lodders & Fegley (1998)] while in the atmospheric neutral layers, sodium abundance exceeds calcium abundance by a factor of 50–100. This model made the assumption that sodium ablates first from meteorites, at a lower temperature, while calcium begins to be released lower in the atmosphere when the temperature of the meteorite is substantially higher. This process was termed differential ablation. The ablation of sodium and calcium were represented by differing vapor pressure laws computed from simplified thermodynamic models following Fegley and Cameron [Fegley & Cameron (1987)]. These were then used in the standard equations for meteor flight and heating, Equation (11.2.1) through Equation (11.2.3), resulting in deposition profiles for the two metals which differed in peak deposition rates, since a smaller fraction of the Ca than Na ablated from the smaller meteorites. More importantly, the deposition profiles differed by some 10 km in the altitude of peak deposition. This led to calcium being deposited at substantially lower altitudes, where the atom was converted much more rapidly into sink species and removed from the system. The results showed a depletion of calcium relative to sodium in good agreement with experiment, provided that one assumed that the majority of the incoming cosmic dust particles were slow, < 20 km/s. This makes sense if one considers that the majority of the influx to the Earth arises from decayed cometary streams which have become circularized and co-orbital with the Earth [Hughes (1978)]. A similar model [McNeil *et al.* (2000)] was later applied to the four-fold depletion of Mg^+ relative to Fe^+ in the ionosphere, which has been noted in several rocket-borne mass spectrometer experiments [Kopp (1997)]. It was found that a 50-fold

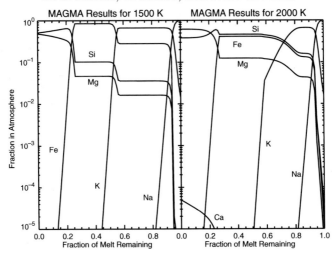

FIGURE 11.5. Vapor pressures of the different metals as derived from the *MAGMA* model
[Fegley & Cameron (1987)].

increase in Fe volatility over Mg gave good agreement for the ion depletion, while reproducing a neutral Fe layer also agreeing well with observations. Borovička *et al.* (1999)
have obtained spectra of bright meteors during the 1998 Leonid meteor shower. Their
data show earlier vaporization of Na, in agreement with the suggestion that differential
ablation occurs for meteoroids.

As mentioned above in Section 11.2.1, meteoroids reach temperatures of 1500–2000 K
during their entry into the Earth's atmosphere. These lower temperatures lead to different
mixes of vapor composition. In Figure 11.5 we show calculations using the *MAGMA*
model for 1500 K and 2000 K.

What we notice now is that at 1500 K Ca does not even show up because it is completely
condensed at that temperature. Na is completely released in the first 5% of ablation.
When 80% of the meteoroid is left, the gaseous atmosphere consists mostly of Na and
a little of K. When 60% of the meteoroid is left the vapor consists mostly of K, Na
having evaporated completely by then; the melt consists of Fe, Si, and Mg. When 20%
is left, the vapor consists mostly of Fe and smaller amounts of Si and Mg; the condensed
phase contains mostly Si, Mg, and all the Ca. By contrast, at 2000 K, when 80% of the
meteoroid is left, most of the Na will have disappeared and the vapor consists mostly of
K and smaller amounts of the other metals. When 60% remains the vapor consists of Si,
Fe, and Mg; Ca is still retained in the condensed phase.

11.3. Steady-state Models

All models of meteoric metals require as input an altitude profile of the deposition of
the metal in question. An early calculation of the deposition was carried out by Hunton
and co-workers [Hunton *et al.* (1980)] and has been used in several studies [Helmer *et
al.* (1998); Plane & Helmer (1994)]. Other studies have started with refined versions of
this calculation examining various effects such as a bi-modal velocity distribution of the
incoming meteors [Eska *et al.* (1999)] and varying volatility according to the particular
species of metal, leading to an altitude separation in the deposition of one metal from
another [McNeil *et al.* (1998)]. The mechanics needed to compute such a deposition profile
were outlined in Section 11.2. For steady-state models, the next step is the calculation of

thc altitude profile of total metal containing species, which is to say the sum of all species containing the metal M. Since these models are generally limited to the mesosphere where transport is assumed to be by eddy diffusion only, the total flux of all metal containing species, $n(M)$ is given by [Banks & Kockarts (1973)]:

$$\Phi(M) = -K_E \left\{ \frac{dn(M)}{dz} + n(M)\left(\frac{1}{H} + \frac{1}{T_a}\frac{dt}{dz}\right) \right\} \tag{11.3.1}$$

where K_E is the eddy diffusion coefficient, H the scale height of the atmosphere, and T_a the atmospheric temperature. This equation expresses the total downward flux of metal species at any altitude which, in the steady-state approximation, is equal to the total metal deposited at all altitudes above z. Therefore

$$\Phi(M) - \int_z^\infty q(z)\,dz \tag{11.3.2}$$

with $q(z)$ the altitude-dependent deposition function. With all other quantities known, Equation (11.3.1) can then be solved for the altitude profile $n(M)$. The solution requires specification of the total meteoric influx of the particular metal, equivalent to a scaling of the deposition function, and the specification of $n(M)$ at a chosen reference altitude.

The final step is the development and application of a chemical kinetics scheme which governs the partitioning of a metal into various forms. Note that Chapter 12 describes the major techniques that have been used to measure the rate coefficients for reactions of metallic species, and also provides a compilation of recommended rate coefficients. Figure 11.6 is a schematic diagram of the important gas-phase cycles of sodium in the upper atmosphere. Note that the rate coefficient for each of these reactions has been measured. The partitioning of sodium between the different Na-containing species changes as a function of altitude, because the ambient atmospheric constituents, e.g. O, H, O_3, NO^+, vary with altitude.

Suppose there are N different metal containing species in the kinetic model, e.g. Na, NaO, NaO_2, NaOH, ... Then the set of N equations for species S_i are solved simultaneously:

$$\frac{dS_i}{dt} = 0 = \sum_{j=1, j\neq 1}^{N} (P_j S_j - L_i S_i) \tag{11.3.3}$$

where the P_j are the rates at which species j are converted into species i and L_i is the rate at which species i is removed. This yields the density profile for each of the N metal containing species. Figure 11.7 illustrates such a calculation for Na.

The most common application of steady-state models has been the investigation of the seasonal variability of the abundances of various atmospheric metals. Annual variations in the column density of neutral atomic metals have been measured by lidar for all the predominant atmospheric metals [Gerding *et al.* (2000)], with the exception of magnesium, for which no appropriate resonance line exists. The variation of the layer density with season is especially pronounced for iron [Kane & Gardner (1993); Helmer *et al.* (1998)] and sodium [Plane *et al.* (1999a)], where the average column density in winter is nearly three times that in mid-summer at mid-latitudes. Steady-state models have been very successful in explaining this variation, as shown in Figure 11.8. The primary effect of season on metal abundance arises from temperature changes in the mesosphere and the resulting changes in the rates of various reactions. The increase in the winter temperature around the metal layer peak causes an increase in the rate of the recycling reaction (e.g. Equation 11.3.4) and an increase in the overall metal abundance. Since the recycling reaction is the one most strongly effected, the overall change is most pronounced on the

W. J. McNeil, E. Murad, and J. M. C. Plane

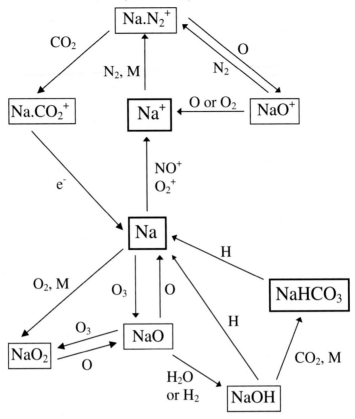

FIGURE 11.6. A schematic diagram of the major chemical cycles of sodium in the upper mesosphere/lower thermosphere region [Plane *et al.* (1999a)]

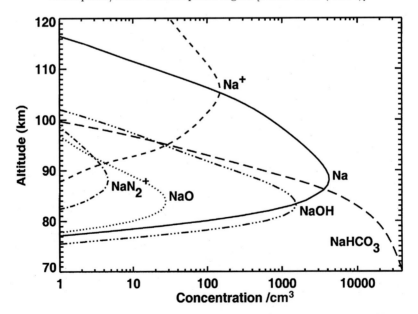

FIGURE 11.7. An example of a steady-state model calculations for Na.

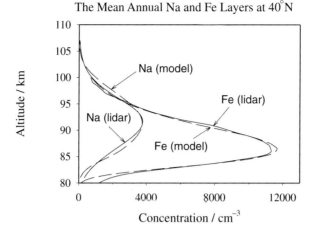

The Mean Annual Na and Fe Layers at 40°N

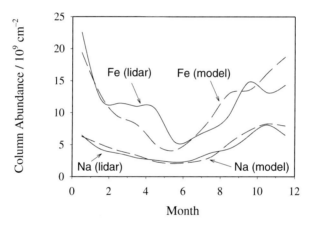

Column Abundances of Na and Fe at 40°N

FIGURE 11.8. Comparison between lidar measurements and steady-state model predictions for the Na and Fe layers at 40° N. Top panel: mean annual profiles; Bottom panel: seasonal variations in layer column abundances [Helmer *et al.* (1998); Plane *et al.* (1999a)].

bottom side of the layer. The layer peak therefore moves down in winter and the layer becomes broader. Temperature variations of the rate coefficients are not the only source of seasonal variability, however. Changes in the eddy diffusion profile [Helmer *et al.* (1998)] and variations in the meteoric influx rates [Eska *et al.* (1999); Gerding *et al.* (2000)] have also been called upon in various models to explain the observations. Monthly changes in the ambient atmosphere are also important. These tend to be incorporated off-line from the output of two- or three-dimensional atmospheric models. Seasonal temperature changes affect the partitioning of odd oxygen (O and O_3) and odd hydrogen (H, OH, HO_2) species, with more O and H at higher temperatures. These atomic species convert the metallic compounds to free metal atoms (Figure 11.6).

Plane & Helmer (1994) were the first to present a model for sodium for summer and winter conditions showing a substantial increase in the wintertime layer. A later version [McNeil *et al.* (1995)] included also the seasonal variation in the sodium night glow, driven primarily by variation in sodium density. The most comprehensive model for sodium at mid-latitudes has been published recently [Plane *et al.* (1999a)]. This combines extensive

observations of both Na and temperature with the latest measurements of the kinetic
rate coefficients of relevant Na reactions.

These techniques have also been applied to the seasonal variation in iron [Helmer
et al. (1998)], resulting in reasonable agreement with the data (Figure 11.8) by invok-
ing changes in temperature and in the eddy diffusion profile. A steady-state model of
potassium has also recently been published [Eska et al. (1999)] which attempts to model
observed variations in the potassium density with latitude, in addition to the seasonal
variation. For the latitude variation, it was found that change in the eddy diffusion coefficient with latitude was necessary for good agreement. In order to achieve satisfactory
agreement with the observed seasonal variation, it was also necessary to assume that
the wintertime deposition of meteoric K was reduced by 30% compared with that in the
summertime. A model for the seasonal variability of the Ca layer has also been recently
advanced [Gerding et al. (2000)]. The calcium layer shows three strong peaks, two during
the summer and one in November, which can roughly be identified with meteor showers.
In this model, the authors invoked a nearly three-fold increase in meteoric deposition
during these peak months, obtaining corresponding peaks in the modeled calcium layer.

The implication of these modeling results is that variation in the meteoric influx is
necessary to explain the seasonal variation in the Ca and to a lesser extent the K layers.
For Na and Fe, on the other hand, temperature and eddy diffusion changes appear to
be sufficient. This apparent contradiction can be understood by consideration of the
depletion in Ca and K relative to Na and Fe in relation to the meteoric and cosmic
abundances. It has been shown [McNeil et al. (1998)] that the underabundance of Ca
relative to Na can be explained by assuming a substantially higher volatility of meteoric
Na and by assuming that the majority of the background meteoric influx is comprised of
slow meteors. Fast meteors, on the other hand, ablate both Ca and Na completely and
at altitudes high enough so that the differences in volatility do not lead to significant
differences in atmospheric abundance. Therefore when meteor showers occur that consist
primarily of fast meteors, the relative increases strongly favor those elements which are
naturally depleted due to low volatility. The steady-state modeling results to date have
provided important clues both for the chemical kinetics and for the ablation process.
What is perhaps necessary at this point is a comprehensive model for all four metals,
treating seasonal variations in temperature, eddy diffusion and changes in meteor ablation
due to the annual showers.

The kinetic schemes contained in the steady-state models contain one or more chemical
reservoir species for the metal. For example, in the sodium scheme in Figure 11.6, Na^+ is
the reservoir above the atomic layer, and $NaHCO_3$ is the reservoir beneath. A requirement
of this approach is that the chemical system must be "closed", meaning that each species
must have at least one path to one other species. Otherwise, the partitioning would result
in only the sink species at all altitudes. It is worth noting that the rate of recycling of
the sink species, e.g.

$$NaHCO_3 + H \rightarrow Na + H_2CO_3 \qquad (11.3.4)$$

is treated as a variable parameter in the models, controlling most strongly the bottom
side of the neutral atomic metal layer. This is necessary because these recycling reactions
are extremely difficult to measure. Recently, however, true "closure" has been obtained
for sodium through the measurement of the rate coefficient of Equation (11.3.4) [Cox et
al. (1998)].

Of course, the $NaHCO_3$ shown in Figure 11.7 below 80 km as the major Na-containing
species is unlikely to remain in the form of isolated molecules. $NaHCO_3$ possesses an
enormous dipole moment (6.8 Debyes [Cox et al. (1998)]), so that it will polymerize with

itself and other metallic molecules, forming meteoric "smoke" particles that probably provide a permanent sink for the metals [Hunton *et al.* (1980)]. One important piece of evidence for this removal of $NaHCO_3$ below 80 km is that the photolysis cross-section has recently been measured [D. E. Self and J. M. C. Plane, University of East Anglia, pers. comm.]. Although the resulting photodissociation coefficient is small (10^{-4} s^{-1}), it is still large enough that photolysis of $NaHCO_3$ would produce a sub-layer of Na around 80 km during daytime. This has not been observed in a series of detailed diurnal observations of the Na layer [States & Gardner(2000)], indicating that $NaHCO_3$ is converted to a more stable (condensed) form.

11.4. Time-dependent Models

There are applications for which the assumptions of the steady-state approach cannot be made, in which case a fully time-dependent modeling scheme must be employed. These include instances where transport of metallic species by forces other than eddy diffusion become important and instances where time scales for meteor injection or other processes are comparable to or faster than those required to reach chemical equilibrium between the various metallic species. An important application is to the transport of metal ions into the thermosphere, where metal ions act essentially as passive tracers of transport phenomena. Modeling can therefore test our understanding of these processes. Another application is in the modeling of meteor showers, where injection time scales can be fast relative to the chemistry.

All time-dependent models involve solution of one form or another of the general continuity equation

$$\frac{\partial n_i}{\partial t} = P - L - \nabla \cdot (n_i \mathbf{V}) \tag{11.4.1}$$

which expresses the time rate of change of component i. P is the rate of production of the component, including chemistry and deposition by meteors, L is the rate of loss of the component, usually through chemistry alone, and the gradient term expresses the bulk motion of the component due to diffusion, winds, and electric fields in the case of ionized species, with V the bulk velocity of the species at any particular point and time.

11.5. Models of Metals in the Thermosphere

The presence of metallic ions in the Earth's F-region was first demonstrated by Hanson and Sanatani [Hanson & Sanatani (1970)]. Shortly thereafter, Hanson and co-workers presented model results showing that these ions could be transported upward from the E region by the equatorial electrojet in a narrow region around the magnetic equator [Hanson *et al.* (1972)]. Diffusion and winds would thereafter transport these ions along field lines, eventually returning them to the E region at higher latitudes. Fesen and co-workers [Fesen *et al.* (1983)] developed a similar but more rigorous model which reproduced well many aspects of the global morphology of thermospheric metals. Among the important findings of these workers was the fact that a source of metal ions at the magnetic equator was not necessary, but rather the assumption of a constant layer of metal ions at 125 km and all latitudes was sufficient to provide reasonable agreement with observations. They also showed that longitudinal differences in metal morphology could be explained by differences in the offset between the geographic and magnetic equator. Finally, they were able to model correctly the more frequent observation of metal ions at the sunset terminator and the observation that metal ions attain higher altitudes at solar maximum.

McNeil *et al.* (1996) have developed a one-dimensional model for thermospheric magnesium, applicable at the geomagnetic equator. They later extended this model to sodium and calcium, reproducing quite well equatorial measurements of the airglow intensities of the three metals [Gardner *et al.* (1999)]. The model included transport via diffusion and the equatorial electric field alone. An important result was that the equatorial electric field was sufficient to produce the observed metal densities in the thermosphere, in the absence of winds, at least at the equator. Another advance was the correct prediction of the existence and abundance of neutral Mg and Na in the thermosphere, these arising from the radiative recombination of the ionized species after transport to high altitudes. Carter and Forbes have developed both one-dimensional and two-dimensional models of thermospheric iron applicable to tropical latitudes [Carter & Forbes (1999)]. These models included both comprehensive chemistry and zonal and meridional winds, in addition to the electric field. Using the one-dimensional model, the authors examine in detail the effects of electric field and wind components, independently and in combination. They demonstrate the importance of winds in determining the detailed structure of thermospheric metal densities. With the two-dimensional model, solving for Fe^+ density as a function of altitude and magnetic latitude, the model produces a redistribution of metal ions from the equator, up and down the field lines, demonstrating the equatorial fountain effect.

Recently, a new model has been developed to explain the morphology of sporadic E layers (E_s) within the central polar cap [MacDougall *et al.* (2000)]. The challenge in explaining these layers is that, with almost vertical magnetic field lines, it is difficult to obtain significant vertical motion of the ionization from convective or neutral wind driven movement of ionization. Nevertheless, central polar cap E_s are frequently observed that are not auroral in origin, and the layers usually occur close to the height where there are strong negative gradients in the vertical ion motion. This vertical motion is due to the combined effects of neutral winds *and* drift due to electric fields, with the effects of the electric field usually dominant [e.g. Kirkwood & von Zahn (1991); Bristow & Watkins (1997); Parkinson *et al.* (1998)]. E_s behavior is also strongly influenced by the distribution of metallic ions in the lower thermosphere [Bedey & Watkins (1997)], since E_s layers have been shown to be composed mostly of Fe^+ and Mg^+ ions, with smaller concentrations of Na^+ [e.g. Heinselman *et al.* (1998)]. Central polar cap E_s appear to start as transient layers that are associated with gravity waves, and so the recent one-dimensional model begins by simulating the action of gravity waves on metallic ions in the lower F region [MacDougall *et al.* (2000)]. The ions form into downward propagating E_s layers, which in winter are transient phenomena that disappear around 120 km. In contrast, during summer these layers often become thin intense E_s layers at about 100 km, persisting for several hours. This is because the metal ion component of the E_s is maintained by charge exchange with the ambient ionization in the sunlit polar thermosphere [MacDougall *et al.* (2000)].

11.6. Modeling of Meteor Shower Effects

One situation in which time-dependent models are absolutely required is in the simulation of the effects of meteor showers on the background metal layers and on the ionosphere in general. There has been considerable debate on whether the annual meteor showers have a measurable impact on either the neutral metal layers or the ionosphere [Kopp (1997); Grebowsky *et al.* (1998)]. Recently, a model has been developed to assess the impact of both annual showers and the rarer intense meteor storms [McNeil *et al.* (2000)]. This model includes for the first time realistic representation of the meteor shower mass

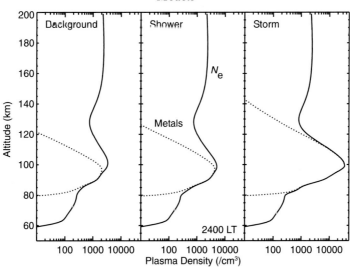

FIGURE 11.9. An example of a time-dependent calculation of metal abundances in the background atmosphere, following a shower, and following a major Leonid storm – see text for details.

influx both in terms of the relative distribution of meteoric particles and the absolute influx. Both these features are derived from parameters available from visual observations of meteor showers. Using these, the resulting metal abundances can be directly compared with those using only the background influx.

Figure 11.9 shows one example of ionospheric modification of total metal density along with the total plasma density, including the ambient ionospheric species. The panels from left to right show the result without meteor showers, the densities after the annual Perseids meteor shower, and the densities after a major Leonid meteor storm of magnitude comparable to that in 1966. The results are for local midnight, when the ambient E region ionosphere is sparse. For the latter calculation, hyperthermal collisions between the meteoroid species and the atmosphere were assumed to give rise to collisional ionization, with the net effect being that the total electron content of the ionosphere is greatly increased. The conclusion of the calculation is that the annual showers raise the E region density by a factor of 2–3 while a major storm would create an anomalously large peak in total electron density at the altitudes of the metal ion layer.

Time-resolved models have also been used to study the evolution of long-enduring meteor trains [Jenniskens *et al.* (2000b); Kruschwitz *et al.* (2001)]. Leonid meteors in particular seem to produce this striking phenomenon. Long-enduring trains sometimes persist for tens of minutes, and often separate into two well-defined parallel tracks. Models have focused on the evolution of the train intensity with time. A significant fraction of the visible emission is produced by the chemiluminescent reaction between O and NaO (formed from ablated Na reacting with O_3). However, comparing a model with calibrating intensity measurements [Kruschwitz *et al.* (2001)] has shown that this reaction is not the only source of light. Indeed, chemiluminescence from excited states of FeO, formed by the reaction between Fe and O_3, was identified recently in the spectrum from a train [Jenniskens *et al.* (2000a)]. Other emissions involving atomic O and N produced by the impact of the meteoroid on the atmosphere may also be involved. The separation of the train into two tracks has historically been explained by the train, which is roughly cylindrical along the path of the meteoroid, becoming dark in the center: when viewed from the ground, the resulting hollow "tube" of light would appear as parallel tracks. However, recent models shows that this explanation does not account for the observed variation in intensity across the train, nor with the evolution of the tracks with time [Jenniskens et al. (2000b); Kruschwitz et al. (2001)]. Another explanation, perhaps involving the shedding of vortex eddy pairs as a large meteoroid descends through the upper mesosphere, or the role of small-scale turbulence mixing fresh O_3 from the ambient atmosphere into the metal-rich train, appears to be required.

11.7. Gravity Waves

Gravity waves originate in the troposphere from a variety of sources, including orographic forcing, cumulo-nimbus storms and cyclonic fronts. Much of the energy and momentum flux is filtered out in the stratosphere, but a significant portion of the shorter period waves propagate to the upper mesosphere [Fritts (1995)]. As a wave travels upwards through the mesosphere, the wave amplitude increases with falling pressure until the wave becomes unstable and breaks, depositing both energy and momentum. This has a profound effect on the thermal structure of the mesopause, because gravity waves force the meridional wind circulation [Walterscheid (1995)]. Indeed, if the summer mesopause at high latitudes were in thermal equilibrium it would have a temperature around 220 K, compared with temperatures below 140 K [Garcia & Solomon (1985)]. Observing the spectrum of gravity waves that reach the MLT, and being able to parameterize the effects of gravity wave breaking, is therefore crucial for understanding and modeling the general circulation of the entire atmosphere.

There are two major techniques for observing gravity waves in the upper mesosphere. Wave-driven perturbations to an airglow layer, such as the Meinel emission between 85 and 89 km from vibrationally excited OH, or $Na(^2P)$ emission around 89 km, can be observed using an all-sky camera. This type of instrument, which employs a charge-coupled device (CCD) detector behind an appropriate lens and interference filter [Swenson et al. (1998)], is suitable for observing waves with horizontal wavelengths less than about 50 km, since the field-of-view in the upper mesosphere is about 300 km. The other important technique for studying mesospheric gravity waves is the metal resonance lidar, which has the necessary height and time resolution to observe waves propagating through one of the meteoric metal layers. Note that the vertical wavelength must be less than about 20 km because the metallic layers extend vertically over about this distance, and the horizontal wavelength as large as possible because the lidar observes in a Eulerian framework [Gardner & Taylor (1998)].

One question that has arisen is whether the metals are suitable inert tracers of atmospheric motion, or whether the apparent dynamical perturbation has been amplified by a fast chemical response to changes in temperature or the concentrations of minor species [Hickey & Plane (1995)]. The vertical displacement of a parcel of air causes heating/cooling because of adiabatic compression/rarefaction as the parcel moves down/up. As discussed earlier, the reactions controlling the partitioning of a metal between its atomic and reservoir forms are often quite temperature dependent. Furthermore, the metal atom layer and the species such as O, H, and O_3 that control its formation have vertical concentration profiles that are quite different from the atmospheric density, i.e. their mixing ratios change significantly with height, so that a vertical displacement produces a change in the local concentration.

Wave-induced perturbations can be modeled either by combining a steady-state approach to the metal chemistry with a simplified analytical treatment of a monochromatic gravity wave [Swenson et al. (1998); Plane et al. (1999b)], or by a full time-resolved wave

Chemical Amplification of the Fe Layer

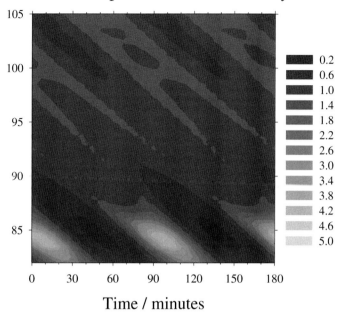

FIGURE 11.10. Chemical amplification factor for the Fe layer during June at $40°$ N, during the passage of an undamped monochromatic gravity wave with the following parameters: wave amplitude ge at 90 km = 0.01; vertical wavelength gl_z = 10 km; period gt = 90 min.

packet treatment [Hickey & Plane (1995)]. Both approaches show that the Na layer is effectively a conservative tracer of short-period gravity waves. By contrast, the Fe layer is predicted to exhibit significant chemical amplification below 85 km [Plane *et al.* (1999b)]. Figure 11.10 illustrates the chemical amplification factor, which is unity where the tracer is conserved, for two cycles of a gravity wave passing through the Fe layer – the amplitude and vertical wavelength are in the optimal range for lidar observations. The model predicts that between 82 and 87 km there should be a significant chemical amplification, by a factor of 5. This prediction, which is a test of the current understanding of iron chemistry in the upper mesosphere, should be tested in the future. Note that metal lidars can also be used to measure the temperature and wind profile (using the hyperfine structure within one of the Na D lines, or the relative populations of the spin-orbit multiplets of ground state $Fe(^5D_J)$). This provides important additional information for wave analysis [Gardner & Taylor (1998)].

11.8. Sporadic Metal Layers

Sporadic (or sudden) neutral metal layers (in the case of sodium termed Na_s) are thin layers of metal atoms 1–3 km wide) that occur at altitudes between 90 and 105 km [Clemesha (1995)]. They usually appear quite suddenly (in a matter of minutes), can last for several hours, and then disappear rapidly. These layers were discovered using ground-based lidars [e.g., Beatty *et al.* (1989); Batista *et al.* (1989); Hanson & von Zahn (1990)], and are defined as an abrupt increase in the metal atom density over the level of the background mesospheric metal layer, sometimes by more than an order of magnitude. Although sporadic layers have been observed for all of the metals that can be measured by ground-based lidar, most observations and models have been concerned with Na_s.

Currently the most promising explanation for Na_s formation is the neutralization of Na^+ concentrated in a sporadic E_s layer. The evidence for this is the high correlation in time and space between Na_s and E_s [e.g. Kwon et al. (1988); von Zahn & Hansen (1988); Batista et al. (1989); Hanson & von Zahn (1990); Clemesha (1995)].

A recent laboratory and modeling study [Cox & Plane (1998)] has shown that Na^+ ions are neutralized in the upper atmosphere by first forming ion clusters which then undergo dissociative recombination with electrons. The process of ion cluster formation is extremely sensitive to altitude. Figure 11.6 shows how a sporadic Na layer can form from a layer of Na^+ ions. The first step is recombination of Na^+ with N_2 to form the weakly bound $Na^+ \cdot N_2$ cluster ion. Above 100 km, however, this ion will generally be broken down to Na^+ again by atomic O. Below 100 km, switching reactions with CO_2 (or less frequently with H_2O, which is much less abundant), cause stable clusters to form that are immune to attack by O and will eventually undergo dissociative electron attachment to produce Na. If a sporadic E layer forms and then descends, the release of Na can turn on very rapidly because the initial formation of $Na^+ \cdot N_2$ depends on $[N_2]^2$, and the ratio of $[CO_2]/[O]$ increases markedly below 100 km. Figure 11.11 compares an observation of a sporadic Na layer with a time-dependent model prediction using this ion–molecule chemistry. Note the sudden release of Na when the sporadic E layer reaches 96 km, in very good agreement with the observation. This mechanism has also been applied successfully to model high-latitude Na_s [Heinselman et al. (1998)]. However, it should be noted that even though this mechanism appears to provide a convincing explanation for sporadic neutral layers in many cases, there may be other production mechanisms that can operate, such as auroral precipitation acting on meteoric dust particles [Hanson & von Zahn (1990)].

11.9. Conclusions

An important conclusion to be drawn from this review is that the modeling of meteoroid entry into the atmosphere has progressed to the point where several qualitative and a few quantitative features of meteor entry into the atmosphere can be predicted. Important fundamental questions have still to be answered before a fully quantitative model can be developed for predicting meteor impact on atmospheric phenomena. The missing information can be summarized below: (a) Laboratory measurements of the Langmuir vapor pressures of the elements over their respective silicate phases; (b) The composition of meteoroids before breakup upon re-entry (this question may be answered by sample collections at high altitudes; (c) Concerted field data about the metal ion composition from lidar and mass spectrometric studies; and (d) Common-volume observations by lidar and radar, made simultaneously of the same trails.

In terms of the modeling of the neutral metal atom layers around 90 km, substantial progress has been made in the last decade with laboratory measurements of the pertinent reaction rate coefficients. This has enabled seasonal models of the Na, Fe, K and Ca layers to be developed that agree quite satisfactorily with observations, although the unexpected difference in seasonal behavior between the Na and K layers has still to be fully explained. However, all of these models were developed to simulate the *nighttime* metal layers. Progress in lidar technology has now enabled full diurnal observations of several metal layers to be made, and it is already clear that full diurnal models will need to provide additional sinks for the metals below 80 km, that are most likely heterogeneous in nature.

It is worth noting here that an important, recent, study of the auroral effects on meteoric sodium [Heinselman (2000)] was not included in this review due to limitations

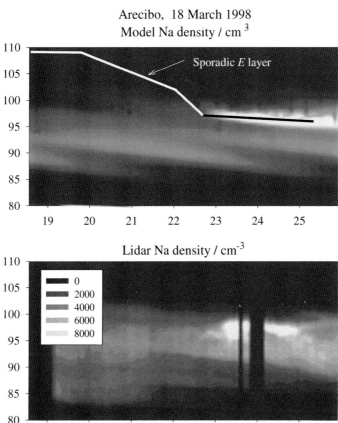

Arecibo, 18 March 1998
Model Na density / cm^3

Lidar Na density / cm^{-3}

Local time / hours

FIGURE 11.11. An example of a sporadic Na layer that formed at 96 km over the Arecibo Observatory, Puerto Rico at about 2315 hrs and lasted for about 2 hours. Top panel: a model prediction using the ion–molecule chemistry in Figure 11.6; the solid white/black line shows the height of a sporadic E layer measured simultaneously using an incoherent scatter radar. Bottom panel: observations of the Na density using a resonance lidar. Data provided courtesy of M. C. Kelley and S. C. Collins, Cornell University.

of scope and to due to uncertainty about its application to meteoric metals other than sodium. Undoubtedly future work will clarify the application of auroral to the other metals.

Acknowledgements: We thank Profs. Fegley and Cameron for providing a copy of their *MAGMA* model and for permission to reproduce the results of the calculations in this chapter. WJMc acknowledges support under contract F19628-00-C-0010. JMCP's work at the University of East Anglia is supported by grant GR3/11754 from the Natural Environment Research Council. We also thank R. L. Hawkes and J. M. Grebowsky for helpful comments.

REFERENCES

ANDERS E. & EBIHARA M. 1982. Solar system abundances of the elements, *Geochim. Cosmochim. Acta,* **46**, 166–173.

BANKS P. M. & KOCKARTS G. 1973. *Aeronomy*, Academic Press, New York, 300 pp.

BATISTA P. P., CLEMESHA B. R., BATISTA I. S. & SIMONICH D. M. 1989. Characteristics of the sporadic sodium layers observed at 23° S, *J. Geophys. Res.*, **94**, 15349–15358.

BEATTY T. J., COLLINS R. L., GARDNER C. S., HOSTETLER C. A., SECHRIST C. F. & TEPLEY C. A. 1989. Simultaneous radar and lidar observations of sporadic E and Na layers at Arecibo, *Geophys. Res. Lett.*, **16**, 1019–1022.

BEDEY D. F. & WATKINS B. J. 1997. Large scale transport of metallic ions and the occurrence of thin ion layers in the polar ionosphere, *J. Geophys. Res.*, **102**, 9675–9681.

BRISTOW W. A. & WATKINS B. J. 1994. Effect of the large-scale convection electric field structure on the formation of thin ionization layers at high latitudes, *J. Atmos. Terr. Phys.*, **56**, 401–415.

BOROVIČKA J., STORK R. & BOČEK J. 1999. First results from video spectroscopy of 1998 Leonid meteors, *Meteorit. Planet. Sci.*, **34**, 987–994.

BOWMAN M. R., GIBSON A. J. & SANDFORD M. C. W. 1969. Atmospheric sodium measured by a tuned laser radar, *Nature*, **221**, 456–457

CARTER L. N. & FORBES J. M. 1999. Global transport and localized layering of metallic ions in the upper atmosphere, *Ann. Geophysicae*, **17**, 190–209.

CLEMESHA B. R. 1995. Sporadic neutral metal layers in the mesosphere and lower thermosphere, *J. Atmos. Terr. Phys.*, **57**, 725–736.

COX R. M. & PLANE J. M. C. 1998. An ion–molecule mechanism for the formation of neutral sporadic Na layers, *J. Geophys. Res.*, **103**, 6349–6360.

COX R. M., SELF D. E. & PLANE J. M. C. 2001. A study of the reaction between $NaHCO_3$ and H: apparent closure on the neutral chemistry of sodium in the upper mesosphere, *J. Geophys. Res.*, **106**, 1733–1739.

ESKA V., VON ZAHN U. & PLANE J. M. C. 1999. The terrestrial potassium layer (75–110 km) between 71° S and 54° N: Observations and modeling, *J. Geophys. Res.*, **104**, 17173–17186.

FEGLEY, JR. B. & CAMERON A. G. W. 1987. A vaporization model for iron/silicate fractionation in the Mercury protoplanet, *Earth Planet. Sci. Lett.*, **82**, 207–222.

FESEN C. G., HAYS P. B. & ANDERSON D. N. 1983. Theoretical modeling of low-latitude Mg^+, *J. Geophys. Res.*, **88**, 3211–3223.

FLYNN G. J. 1989. Atmospheric entry heating of micrometeorites, *19th Lunar and Planetary Science Conference, Houston, TX*, pp. 673–682.

FRITTS D. C. 1995. Gravity wave forcing and effects in the mesosphere and lower thermosphere, in *The Upper Mesosphere and Lower Thermosphere: A Review of Experiment and Theory*, R. M. Johnson and T. L. Killeen (Editors), American Geophysical Union Geophysical Monograph, **87**, 89–100.

FYFE J. D. D. & HAWKES R. L. 1986. Residual mass from ablation of meteoroid grains detached during atmospheric flight, *Planet. Space Sci.*, **34**, 1201–1212.

GARCIA R. R. & SOLOMON S. 1985. The effect of breaking gravity waves on the dynamics and chemical composition of the mesosphere and lower thermosphere, *J. Geophys. Res.*, **90**, 3850–3868.

GARDNER C. S. & TAYLOR M. J., 1998. Observational limiets for lidar, radar, and airglow imager measurements of gravity wave parameters, *J. Geophys. Res.*, **103**, 6427–6437.

GARDNER J. A., BROADFOOT A. L., MCNEIL W. J., LAI S. T. & MURAD E. 1999. Analysis and modeling of the GLO-1 observations of meteoric metals in the thermosphere, *J. Atm. Sol. Terr. Phys.*, **61**, 545–562.

GÉRARD J.-C., & MONFILS A. 1974. Satellite observations of the equatorial Mg II equatorial dayglow intensity distribution, *J. Geophys. Res.*, **79**, 2544–2550.

GERDING M., ALPERS M., VON ZAHN U., ROLLASON R. J. & PLANE J. M. C. 2000. The atmospheric Ca and Ca^+ layers: Midlatitude observations and modeling, *J. Geophys. Res.*, **105**, 27131–27146.

GREBOWSKY J. M., GOLDBERG R. A. & PESNELL W. D. 1998. Do meteor showers significantly perturb the ionosphere? *J. Atm. Sol. Terr. Phys.*, **60**, 607–615.

GREVESSE N. & ANDERS E. 1991. Solar Element Abundances, in *Solar Interior amd Atmosphere*, A. N. Cox, W. Livingston, and M. S. Matthews (Editors), University of Arizona Press, Tucson, AZ, 1416 pp.

HANSON W. B. & SANATANI S. 1970. Meteoric ions above the F2 Peak, *J. Geophys. Res.*, **75**, 5503–5509.

HANSON W. B., SANATANI S., ZUCCARO D. & FLOWERDAY T. W. 1970. Plasma measurements with the retarding potential analyzer on Ogo-6, *J. Geophys. Res.*, **75**, 5483–5501.

HANSON W. B. & SANATANI S. 1971. Relationship between Fe^+ ions and equatorial spread F, *J. Geophys. Res.*, **76**, 7761–7768.

HANSON W. B., STERLING D. L. & WOODMAN R. F. 1972. Source and identification of heavy ions in the equatorial F layer, *J. Geophys. Res.*, **77**, 5530–5541.

HANSEN G. & VON ZAHN U. 1990. Sudden sodium layers in polar latitudes, *J. Atmos. Terr. Phys.*, **52**, 585–608.

HAWKES R. L. & JONES J. 1975. A quantitative model for the ablation of dustball meteors, *Mon. Not. R. Astr. Soc.*, **173**, 339–356.

HEINSELMAN C. J., THAYER J. P. & WATKINS B. J. 1998. A high-latitude observation of sporadic sodium and sporadic E-layer formation, *Geophys. Res. Lett.*, **25**, 3059–3062.

HEINSELMAN, C. J. 2000. Auroral effects on the gas phase chemistry of meteoric sodium, *J Geophys. Res.*, **105**, 12181–12192.

HELMER M., PLANE J. M. C., QIAN J. & GARDNER C. S. 1998. A model of meteoric iron in the upper atmosphere, *J. Geophys. Res.*, **103**, 10913–10925.

HICKEY M. & PLANE J. M. C. 1995. A chemical-dynamical model of wave-driven sodium fluctuations, *Geophys. Res. Lett.*, **22**, 2861–2864.

HUGHES D. W. 1978. Meteors, in *Cosmic Dust*, J. A. M. McDonnell (Editor), Wiley, London pp. 123–185.

HUNTON D. M., TURCO R. P. & TOON O. B. 1980. Smoke and dust particles of meteoric origin in the mesosphere and stratosphere, *J. Atmos. Sci.*, **37**, 1342–1357.

ISTOMIN V. G. & POKHUNKOV A. Z. 1963. Mass spectrometer measurements of atmospheric composition in the USSR, *Space Research*, **3**, 117–131.

JENNISKENS P., LACEY M., ALLAN B. J., SELF D. E. & PLANE J. M. C. 2000. FeO "Orange arc" emission detected in optical spectrum of Leonid persistent train, *Earth, Moon and Planets*, **82**, 429–438.

JENNISKENS P., NUGENT D. & PLANE J. M. C. 2000. The dynamical evolution of a tubular Leonid persistent train, *Earth, Moon and Planets*, **82**, 471–488.

JONES J. & KAISER T. R. 1966. The effects of thermal radiation, conduction, and meteoroid heat capacity on meteoroid ablation, *Mon. Not. R. Astr. Soc.*, **133**, 411–420.

KALASHNIKOVA O., HORÁNYI M., THOMAS G. E. & TOON O. B. 2000. Meteoric smoke production in the atmosphere, *Geophys. Res. Lett.*, **27**, 3293–3296.

KANE T. J. & GARDNER C. S. 1993. Structure and seasonal variability of the nighttime mesospheric Fe layer at midlatitudes, *J. Geophys. Res.*, **98**, 16875–16886.

KIRKWOOD S. & VON ZAHN U. 1991. On the role of auroral electric fields in the formation of low altitude-E and sudden sodium layers, *J. Atmos. Terr. Phys.*, **53**, 389–407.

KOPP E. & HERRMANN U. 1984. Ion composition of the lower ionosphere, *Ann. Geophys.*, **2**, 83–94.

KOPP E. 1997. On the abundance of metal ions in the lower ionosphere, *J. Geophys. Res.*, **102**, 9667–9674.

KRUSCHWITZ C. A., KELLEY M. C., GARDNER C. S., SWENSON G., LUI A. Z., CHU X., DRUMMOND J. D., GRIME B. W., ARMSTRONG W. T., PLANE J. M. C. & JENNISKENS P. 2001. Observations of persistent Leonid meteor Trails II: Photometry and numerical modeling, *J. Geophys. Res.*, **106**, 21525–21541.

KWON K. H., SENFT D. C. & GARDNER C. S. 1988. Lidar observations of sporadic sodium layers at Mauna Kea Observatory, Hawaii, *J. Geophys. Res.*, **93**, 14199–14208.

LODDERS K. & FEGLEY, JR. B. 1998. *The Planetary Scientist's Companion*, Oxford University Press, New York, 371 pp.

LOVE S. G., & BROWNLEE D. E. 1991. Heating and thermal transformation of micrometeoroids entering the Earth's atmosphere, *Icarus*, **89**, 26–43.

LOVE S. G. & BROWNLEE D. E. 1993. A direct measurement of the terrestrial mass accretion rate of cosmic dust, *Science*, **262**, 550–553.

MacDOUGALL J. W., PLANE J. M. C. & JAYACHANDRAN P. T. 2000. Polar cap Sporadic-E: Part 2, Modeling, *J. Atmos. Terr. Solar Phys.*, **62**, 1169–1176.

MASON B. (EDITOR) 1971. *Handbook of Elemental Abundances of the Elements in Meteorites*, Gordon and Breach, Newark, NJ, 555 pp.

McNEIL W. J., MURAD E. & LAI S. T. 1995. Comprehensive model for the atmospheric sodium layer, *J. Geophys. Res.*, **100**, 16847–16855.

McNEIL W. J., LAI S. T. & MURAD E. 1996. A model for meteoric magnesium in the ionosphere, *J. Geophys. Res.*, **101**, 5251–5259.

McNEIL W. J., LAI S. T. & MURAD E. 1998. Differential ablation of cosmic dust and implications for the relative abundances of atmospheric metals, *J. Geophys. Res.*, **103**, 10899–10911.

McNEIL W. J., DRESSLER R. & MURAD E. 2001. The impact of a major meteor shower on the Earth's ionosphere: A modeling study, *J. Geophys. Res.*, **106**, 10447–10465.

MEGIE G. & BLAMONT J. E. 1969. Laser sounding of atmospheric sodium: Interpretation in terms of global atmospheric parameters, *Planet. Space Sci.*, **25**, 1093–1109.

NARCISI R. S. 1968. Processes associated with metal-ion layers in the E-region of the ionosphere, *Space Res.*, **8**, 360–369.

NICOL E. J., MACFARLANE J. & HAWKES R. L. 1985. Residual mass from atmospheric ablation of small meteoroids, *Planet. Space Sci.*, **33**, 315–320.

ÖPIK E. J. 1936a. Researches on the physical theory of meteor phenomena. I. Theory of the formation of meteor craters, *Publ. de l'Observatoire astronomique de l'Universté de Tartu*, **28**, 1–12.

ÖPIK E. J. 1936b. Researches on the physical theory of meteor phenomena. II. The possible consequences of the collisions of meteors in space, *Publ. de l'Observatoire astronomique de l'Universté de Tartu*, **28**, 13–27.

ÖPIK E. J. 1937. Researches on the physical theory of meteor phenomena. III. Basis of the physical theory of meteor phenomena, *Publ. de l'Observatoire astronomique de l'Universté de Tartu*, **33**, 1–67.

ÖPIK E. J. 1958. *Physics of Meteor Flight in the Atmosphere*, Interscience Publishers, New York, 174 pp.

PARKINSON M. L., DYSON P. L., MONSELESAN D. P. & MORRIS R. J. 1998. On the role of electric field direction in the formation of sporadic-E layers in the southern polar cap ionosphere, *J. Atmos. Terr. Phys.*, **60**, 471–491.

PLANE J. M. C. & HELMER M. 1994. Laboratory studies of the chemistry of meteoric metals, *Research in Chemical Kinetics*, **2**, 313–365.

PLANE J. M. C., GARDNER C. S., YU J., SHE C. Y., GARCIA R. R. & PUMPHREY H. C. 1999. Mesospheric Na layer at 40-deg N: Modeling and observations, *J. Geophys. Res.*, **104**, 3773–3788.

PLANE J. M. C., COX R. M. & ROLLASON R. J. 1999. Metallic layers in the mesopause and lower thermosphere region, *Adv. Space Res.*, **24**, 1559–1570.

RIETMEIJER F. J. M. 2000. Interrelationships among meteoric metals, meteors, interplanetary dust, micrometeorites, and meteorites, *Meteor. Planet. Sci.*, **35**, 1025–1042.

SEARS D. W. & DODD R. T. 1988. Overview and Classification of Meteorites, in *Meteorites and the Early Solar System*, J. F. Kerridge & M. S. Matthews (Editors), University of Arizona Press, Tucson, AZ. 1260 pp.

STATES R. J. & GARDNER C. S. 2000. Thermal structure of the mesopause region (80–105 km) at 40 degrees N latitude. Part II: diurnal variations, *J. Atmos. Sci.*, **57**, 78–92.

SWENSON G. R., QIAN J., PLANE J. M. C., ESPY P. J., TAYLOR M. J., TURNBULL D. N. & LOWE R. P. 1998. Dynamical and chemical aspects of the mesospheric Na "wall" event on 9 October 1993 during the ALOHA campaign, *J. Geophys. Res.*, **103**, 6361–6380.

VON ZAHN U. & HANSEN T. L. 1988. Sudden neutral sodium layers: a strong link to sporadic E layers, *J. Atmos. Terr. Phys.*, **50**, 93–104.

VON ZAHN U., GERDING M., HÖFFNER H., McNEIL W. J. & MURAD E. 1999. Iron, calcium, and potassium atom densities in the trails of Leonids and aother meteors: Strong evidence for differential ablation, *Meteor. Planet. Sci.*, **34**, 1017–1027.

WALTERSCHEID R. L. 1995. Gravity wave mean state interactions in the upper mesosphere and lower thermosphere, in *The Upper Mesosphere and Lower Thermosphere: A Review of Experiment and Theory*, R. M. Johnson and T. L. Killeen (Editors), American Geophysical Union Geophysical Monograph, **87**, 133–144.

WHIPPLE F. L. 1943. Meteors and the Earth's upper atmosphere, *Rev. Mod. Phys.*, **15**, 246–264.

WHIPPLE F. L. 1950. The theory of micro-meteorites. Part I. In an isothermal atmosphere, *Proc. Nat'l. Acad. Sci.*, **36**, 687–695.

WHIPPLE F. L. 1951. The theory of micro-meteorites. Part I. In heterothermal atmospheres, *Proc. Nat'l. Acad. Sci.*, **37**, 19–30.

12
LABORATORY STUDIES OF METEORIC METAL CHEMISTRY

By JOHN M. C. PLANE[†]

School of Environmental Sciences, University of East Anglia, Norwich NR4 7TJ, UK

The ablation of meteoroids in the Earth's atmosphere gives rise to layers of metal atoms between 80 and 110 km that are global in extent. In order to understand the geophysical significance of these layers, models have to be constructed using fundamental physico-chemical data such as rate coefficients and photolysis cross sections. This chapter describes the laboratory techniques that are employed to study the reactions of metal atoms and metal-containing molecules with atmospheric species such as O_3, O_2, O, H, H_2, H_2O and CO_2. This is followed by a discussion of the use of quantum theory calculations to extrapolate laboratory data to the low temperatures and pressures of the upper atmosphere. The chapter ends with a compilation of recommended rate coefficients for the neutral and ion–molecule reactions of species containing sodium, iron, calcium, magnesium, potassium and lithium.

12.1. Introduction

The presence of free metal atoms in the Earth's upper atmosphere was discovered over 60 years ago, when it was established that the source of the Na radiation observed in the nightglow spectrum at 589 nm was located within the atmosphere [Slipher (1929); Bernard (1938)]. Since then, the techniques of twilight photometry [Hunten (1967)] and lidar [Bowman *et al.* (1969)] have been developed to study the layers of a number of different metal atoms that exist at altitudes between 80 and 105 km. Rocket-borne mass spectrometry has been employed to observe a range of metallic ions [Narcisi (1968); Kopp (1997)].

Historically, a number of questions arose from these observations. For instance, is the source of the metals interplanetary or terrestrial? Why do highly reactive species such as Na and K exist as free metal atoms in an essentially oxidising atmosphere? Why do the metal layers appear around 90 km, and why are they only a few kilometres wide? Why are the relative abundances of the metal atoms quite different from their relative abundances in common minerals found in the solar system, and how can their different seasonal behaviours be explained? Finally, what is the impact of these metals on the general chemistry of the lower atmosphere, particularly on the removal of ozone in the stratosphere? Clearly, most of these questions can only be answered through a detailed understanding of the atmospheric chemistry of the relevant metals. However, because only the metal atoms and their atomic ions can actually be observed (above 80 km), knowledge of this chemistry has had to be acquired through laboratory kinetic studies of the pertinent reactions, combined with atmospheric modelling [McNeil *et al.* (1995); Helmer *et al.* (1998); Plane *et al.* (1999a); Eska *et al.* (1999); Gerding *et al.* (2000)].

Laboratory studies under mesospheric conditions are particularly challenging, because of the extremes of temperature and pressure and the high photon energies involved. For instance, the mesosphere is the coldest part of the atmosphere, with summertime mesopause temperatures falling below 120 K at high latitudes [Lübken (1999)]. There is

[†] Email: J.Plane@uea.ac.uk

also a substantial seasonal variation in the *absolute* temperature: the winter mesopause is warmer by about a factor of 2 at high latitudes. At mid-latitudes, summer mesopause temperatures of about 170 K are typical, which then increase to between 200 and 230 K in winter [Plane *et al.* (1999a)]. The atmospheric pressure varies from about 10 mTorr at 80 km to less than 1 mTorr (1 Torr = 133 Pa) in the lower thermosphere above 96 km, so that termolecular reactions are essentially at their low-pressure limits. Above 105 km, the pressure falls to the point where molecular diffusion becomes more important than bulk eddy diffusion as a mode of molecular transport. Solar radiation at wavelengths longer than 190 nm, as well as a significant fraction of Lyman-α at 121.6 nm (which is not effectively absorbed by the Schumann-Runge bands of O_2), penetrates to the upper mesosphere. Thus, photochemistry with photon energies in excess of 5 eV takes place, producing ionic and neutral species that are often in highly excited electronic and vibrational states [Plane (2000)].

This chapter will provide a review of the substantial progress that has been made over the last decade in laboratory studies. It should be noted that the chemistry of meteoric metals has been reviewed by the author up to 1994 [Plane (1991); Plane & Helmer (1994)], so that the laboratory techniques and measurements presented here will focus on work during the last six years. However, results from studies before 1995 are included in the compilation of recommended rate coefficients and photodissociation coefficients that is given at the end of the chapter.

12.2. Experimental Techniques

The two classical techniques of flash photolysis and the fast flow tube have been applied with great success to measuring the rate coefficients of reactions involving both neutral and ionised metallic species in the gas phase, at the low temperatures that characterise the upper atmosphere. In addition, a new guided-ion beam method has been developed recently to measure the absolute reaction cross-sections of charge transfer reactions involving metals. These techniques will now be discussed in turn.

12.2.1. *Flash Photolysis*

This technique has been employed in a number of configurations [Plane (1991)]. A short pulse of ultraviolet light from a flash lamp or laser is used to photolyse a metal-containing precursor in the gas phase. The resulting metal atom or metallic fragment is produced in an excess of a molecular reactant diluted in a bath gas (e.g. N_2 or He), and the subsequent reaction is observed by one of several time-resolved optical monitoring techniques, which have included laser-induced fluorescence (LIF), resonance absorption, and chemiluminescence spectroscopy.

During the past five years, the pulsed laser photolysis/time-resolved LIF (PLP/LIF) technique has been used in the author's laboratory to study the reactions of a series of metal oxides with molecules such as O_3, O_2, CO_2 and H_2O. Figure 12.1 is a schematic diagram of the experimental system employed. The stainless steel reactor consists of a central cylindrical reaction chamber where the reaction is studied. The temperature of this chamber can be varied between 190 and 1100 K. The chamber is at the intersection of two sets of horizontal side-arms which cross orthogonally. These provide the optical coupling for the lasers into and out of the central chamber. The photolysis laser is an excimer laser operating at either 193 nm or 248 nm, which can be focused into the central chamber in order to achieve the multi-photon dissociation required when using certain photolytic precursors. The probe laser is a pulsed nitrogen-pumped dye laser with a pulse energy of 10–250 μJ and bandwidth of about 0.03 nm, which can be frequency-doubled

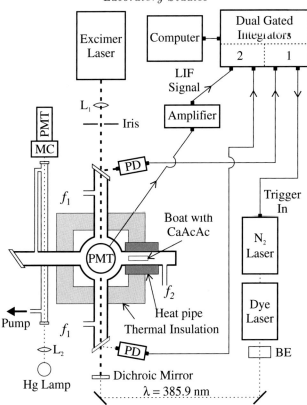

FIGURE 12.1. Block diagram of the pulsed laser photolysis/laser-induced fluorescence system: BE, beam expander; f_1, flow of reactant diluted in bath gas; f_2, flow of bath gas to entrain photolytic precursor (Ca acetyl acetonate); $L_{1,2}$, Suprasil lenses; MC, monochromator; PD, fast photodiode; PMT, photomultiplier tube.

with a β barium borate crystal when required. The side-arms also provide the means by which flows of the reactant, diluted in the bath gas, enter the chamber (flows f_2 in Figure 12.1). The bath gas dissipates heat resulting from the photodissociation, and prevents the very rapid diffusion of the metal species out of the centre of the reactor. One side-arm is independently heated to act as a heat pipe source for the metallic precursor. The opposite arm serves as an exit for the gas flows to the pump, via an absorption cell which is employed for monitoring the concentrations of a reactant such as O_3 which is prone to significant wall losses in the reactor. A fifth vertical side-arm provides the coupling for the photo-multiplier tube which monitors the LIF signal after passing through an appropriate narrow-band interference filter.

The photolytic precursors that have been employed to study a variety of metal reactions are listed in Table 12.1. Apart from ferrocene ($Fe(C_5H_5)_2$), which is quite volatile [Helmer & Plane (1994)], the other metal precursors are placed directly into the heat pipe and heated to the temperature required to generate a sufficient vapour pressure. The resulting precursor vapour is then entrained in a flow of inert bath gas (flow f_2 in Figure 12.1) and carried into the central chamber, where it is mixed with the flow of the reactant (f_1) before being photolysed. Sometimes it is advantageous for the photolytic precursor to be formed *in situ* by reaction in the central chamber. One recent example is the formation of an FeO precursor, formed from the reaction between ferrocene entrained from the side-arm, and NO_2 or O_3 [Rollason & Plane (2000)]. Pure metal vapours such as Na or

TABLE 12.1. Laser flash photolysis/laser induced fluorescence studies of metal atom, ion and oxide reactions

Metal	Photolytic precursor[a]	LIF probe (nm)[b]	Reference
Li	LiI (1×193 nm)	670.7 [2^2P_J–$2^2S_{1/2}$]	Plane & Helmer (1994)
Na	NaCl (1×193 nm)	589.0 [3^2P_J–$3^2S_{1/2}$]	Plane et al. (1993)
NaO	NaCl/NaOH/O_3 (1×193 nm)	589.0 [from NaO + O]	Cox & Plane (1999)
K	KCl, CH_3COOK (1×193 nm)	760.0 [4^2P_J–$4^2S_{1/2}$]	Plane & Helmer (1994)
KO	KCl/O_3 (1×193 nm)	760.0 [from KO + O]	Plane & Helmer (1994)
Mg	Mg($C_5H_{10}O_2$)$_2$ (2×193 nm)	285.2 [3^1P_1–3^1S_0]	Plane & Helmer (1995)
MgO	Mg($C_5H_{10}O_2$)$_2$/O_3 (2×193 nm)	499.4 [$B^1\Sigma^+$–$X^1\Sigma^+$][c]	Plane & Helmer (1995)
Ca	Ca($C_5H_{10}O_2$)$_2$ (2×193 nm)	422.7 [4^1P_1–4^1S_0]	Helmer et al. (1993)
CaO	Ca($C_5H_{10}O_2$)$_2$ (2×193 nm)	385.9 [$B^1\Pi^+$–$X^1\Sigma^+$][d]	Plane & Rollason (2001)
Fe	Fe(C_5H_5)$_2$ (2×193 nm)	248.3 [$x^5F^0_5$–a^5D_4]	Plane & Rollason (1999c)
FeO	Fe(C_5H_5)$_2$/NO_2 (2×193 nm)	582.0 [$D^5\Delta_4$–$X^5\Delta_4$]	Rollason & Plane (2000)
Fe$^+$	Fe(C_5H_5)$_2$ (4×248 nm)	260.0 [$z^6D^0_{9/2}$–$a^6D_{9/2}$]	Rollason & Plane (1998)

[a] The photolysis wavelength and number of photons required are given in parenthesis.
[b] The excitation wavelength is followed by the spectroscopic transition in parenthesis.
[c] Non-resonant emission $\lambda \geq 600$ nm MgO[$B^1\Sigma^+$–$A^1\Sigma^+$].
[d] Non-resonant emission $\lambda \geq 693$ nm CaO[$B^1\Pi^+$–$A^1\Sigma^+$].

K can also be entrained from the heat pipe, which is converted into a heat-pipe oven by inserting a spiral of stainless steel mesh. These metals then react with reagents in the central chamber to form precursors of the metal atoms or their oxides [Plane et al. (1993)].

Following the photolysis pulse, the metallic species are probed by the pulsed dye laser, and the time-resolved LIF signal is recorded using a gated integrator interfaced to a microcomputer. Alternatively, for certain reactions time-resolved chemiluminescence can be employed. For example, the reactions of Ca and Fe with O_3 are so exothermic that the metal oxides are produced in excited electronic states, giving rise to chemiluminescence which can be recorded by time-resolved photon counting with a multichannel scaler [Helmer et al. (1993); Helmer & Plane (1994)]. The reactions of NaO and KO have to be studied indirectly, because these oxides do not appear to have suitable spectroscopic transitions for detection by LIF. For example, to study the reaction NaO + $H_2O \rightarrow$ NaOH + OH, Na vapour was mixed with flows of N_2O and H_2O in the central chamber. The resulting NaOH was then photolysed to yield Na, which immediately reacted with N_2O to form NaO. A small fraction of the excess N_2O was also photolysed by the excimer laser pulse at 193 nm. The resulting O atoms reacted with NaO to produce Na(3^2P_J); this emitted a photon at 589 nm, thereby providing a spectroscopic marker for NaO as it proceeded to react with H_2O [Cox & Plane (1999)].

Note that in these experiments the reactant concentration is in excess, by at least three orders of magnitude, over the metal species concentration that typically ranges from 10^8 to 10^{10} cm^{-3}. This ensures that the reaction is pseudo-first order, so that only the *relative* concentration of the metallic species, which is proportional to the spectroscopic signal, is required for analysis. Examples of time-resolved LIF decays for the reaction between CaO and O_2 are shown in Figure 12.2 [Plane & Rollason (2001)].

The PLP/LIF technique has also been employed to measure the absolute photolysis cross-sections of the sodium-containing molecules NaCl [Silver et al. (1985)] and NaO_2 [Rajasekhar et al. (1989)]. Recently, D. E. Self and the author have measured the pho-

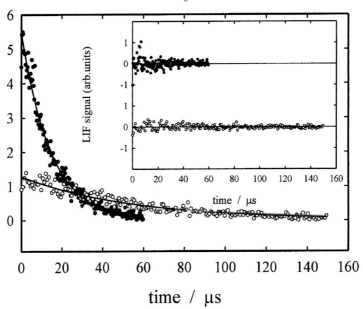

FIGURE 12.2. Main panel: time-resolved decays of the LIF signal from CaO probed at 385.5 nm [CaO(B$^1\Pi^+$–X$^1\Sigma^+$)] and detected by the non-resonant LIF signal [CaO(B$^1\Pi^+$–A$^1\Sigma^+$)] at 693 nm. The slower and faster LIF decays are with [O$_2$] = 0 and 3.48×10^{15} molecule cm^{-3}, respectively; [N$_2$] = 6.61×10^{16} molecule cm^{-3}, $T = 483$ K. The solid curves are fits of the form $A\exp(-k't)$ to the data points, yielding $k' = 20580$ and 68670 s^{-1} respectively. Note that the LIF signal is greatly enhanced by the presence of O$_2$. Inset: plots of the residuals between the data-points and the fitted curves.

tolysis cross-sections of NaO, NaO$_2$, NaO$_3$, NaOH and NaHCO$_3$, using a Raman-shifted excimer laser as the pulsed photolysis source. By operating the excimer either with ArF (193 nm) or KrF (248 nm), and using H$_2$ or D$_2$ in the Raman cell, cross-sections were measured at a large number of individual wavelengths between 190 and 400 nm. Figure 12.3 illustrates the cross-sections for NaHCO$_3$ and NaOH. Note that even though the NaHCO$_3$ cross-section is relatively small, and the onset of photolysis is below 240 nm, the photodissociation coefficient of NaHCO$_3$ in the upper mesosphere is about 10^{-4} s^{-1}, corresponding to a lifetime of only about 3 hours. The absence of a significant diurnal variation in the Na layer below 80 km [States & Gardner (2000)] is *prima facie* evidence that the molecule polymerises with other metal-containing molecules to form stable dust particles [Hunten *et al.* (1980)].

12.2.2. *Fast Flow Tubes*

In this technique the time resolution required for kinetic measurements is obtained by setting up a rapid flow (≥ 5 m s^{-1}) of a carrier gas (e.g. N$_2$, He or Ar) down a tube to an observation point, where at least one of the reactants or products is monitored. In most flow tube arrangements, one reactant is added to the carrier gas flow at the upstream end of the tube, and the second is added further downstream through a moveable injector. The rate of reaction is then measured by changing the distance between the injection and observation points, which alters the reaction time when the reactants are in contact. When studying the kinetics of metallic species, it is important to take account of the strong radial concentration gradient that develops because these species are deposited with close to unit efficiency on the flow tube walls [Helmer & Plane (1993a)]. Convention-

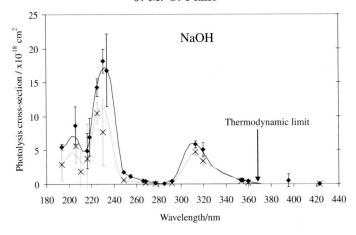

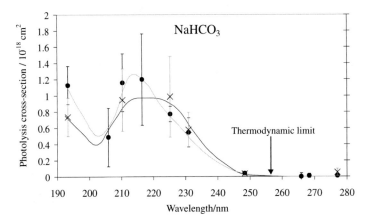

FIGURE 12.3. Absolute photolysis cross-sections for NaHCO3 and NaOH, measured by pulsed laser photolysis using a Raman-shifted excimer laser, with detection of the photofragments by laser induced fluorescence.

ally, sub-ambient reaction temperatures relevant to the upper mesosphere are obtained by enclosing the flow tube in a jacket circulating a cryogenic liquid [e.g. Cox & Plane (1997)]. An alternative method is to use a supersonic expansion of the flow through a Laval nozzle, which can provide temperatures down to a few K [Le Picard *et al.* (1997)].

Studies of ion–molecule reactions of metallic species have been reviewed by Ferguson and Fehsenfeld [1967], and are listed in a compilation of ion–molecule reactions [Ikezoe *et al.* (1987)]. The pioneering work by Fontijn [Fontijn *et al.* (1977)], Kolb [Silver *et al.* (1984a); Silver *et al.* (1984b); Silver *et al.* (1985); Silver & Kolb (1986a); Silver & Kolb (1986b); Worsnop *et al.* (1991)], Howard [Ager *et al.* (1986); Ager & Howard (1986); Ager & Howard (1987a); Ager & Howard (1987b)] and their coworkers in developing the flow tube technique for measuring reactions of *neutral* metallic species at low temperatures has been reviewed previously [Plane (1991)].

Since then, flow tubes have been used to investigate three aspects of sodium chemistry. The first concerned sporadic Na layers in the lower thermosphere (see Chapter 11), and involved a low-temperature (90–250 K) study of the kinetics of cluster formation between Na^+ and N_2, O_2 or CO_2, as well as the reactivity of the resulting cluster ions with atomic

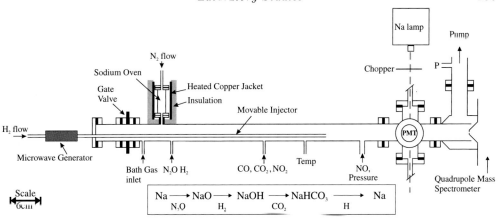

FIGURE 12.4. Schematic drawing (approximately to scale) of the fast flow tube employed to study the reaction between $NaHCO_3$ and H (PMT = photomultiplier tube). Reproduced with permission from the American Geophysical Union [Cox *et al.* (2001)].

O [Cox & Plane (1997)]. The mechanism that was constructed from the laboratory work appears to provide a convincing explanation of this phenomenon [Cox & Plane (1998)].

The second investigation concerned the reactions of reservoir Na-containing species with atomic O and H [Helmer & Plane (1993a); Cox *et al.* (2001)]. For example, the reaction $NaHCO_3 + H \rightarrow Na + H_2CO_3$, which provides chemical closure in the cycling of sodium between atomic Na and its major reservoir (see Chapter 11), was investigated using the stainless steel flow tube illustrated in Figure 12.4 [Cox *et al.* (2001)]. The reaction was studied by reacting $NaHCO_3$ under pseudo first-order conditions in a known excess of atomic H. $NaHCO_3$ was produced *in situ* via a sequence of reactions: atomic Na vapour was first entrained in a flow of N_2 from the heat-pipe oven, mixed with the main N_2 carrier flow, and then converted to $NaHCO_3$ by the sequential addition of N_2O (analogous to O_3 in the mesosphere), H_2 and CO_2. Atomic H, produced by the microwave discharge of H_2, was then introduced further downstream through a moveable injector.

The relative Na atom concentration was monitored at the downstream end of the tube (Figure 12.4) by phase sensitive detection of the modulated resonance fluorescence signal produced by chopping the output from a sodium discharge lamp. The *absolute* H atom concentration was measured by titration with NO_2, using the quadrupole mass spectrometer which sampled along the cylindrical axis of the tube. Because the reaction between $NaHCO_3$ and H is slow at room temperature and below, it was important to prevent any Na produced by reaction with H reforming $NaHCO_3$ via NaO. This was achieved by adding a large excess of CO or NO just above the H atom injection point: CO and NO react very rapidly with NaO to yield Na. In practice, a kinetic experiment of this type is quite complex, because each of the metallic species and the atomic H are lost on the flow tube walls at different rates. The final determination of the target rate coefficient therefore requires a full-scale model treating gas-phase reactions, diffusion and wall losses [Helmer & Plane (1993a); Cox *et al.* (2001)]. Nevertheless, this experiment illustrates one of the major advantages of the flow tube over flash photolysis: a species such as $NaHCO_3$ can be produced by a sequence of reactions in one section of the tube, and then have its chemistry studied further downstream. By contrast, in a flash photolysis study it may not be possible to find a photolytic precursor for such a species, or indeed a spectroscopic transition that permits rapid time-resolved detection.

The third aspect of sodium chemistry involving a recent flow-tube study has been experimental confirmation of the Chapman mechanism for the production of Na(^{2}P) and hence the Na nightglow at 589 nm in the upper atmosphere [Griffin et al. (2001)]. Chapman proposed that Na atoms were first oxidised to NaO by O_3, followed by the reaction between NaO and O being sufficiently exothermic to produce Na(^{2}P) [Chapman (1939)]. Laboratory studies of the reactions Na + O_3 → NaO + O_2 [Ager et al. (1986); Worsnop et al. (1991); Plane et al. (1993)] and NaO + O → Na + O_2 [Plane & Husain (1986)] have shown that both are fast enough to sustain the observed D-line emission intensity of 50–200 R (1 Rayleigh = 10^6 photons cm^{-2} s^{-1}), if the branching ratio, f, for production of Na(^{2}P) in the second reaction, is about 0.1. A value of f in the range 0.05 to 0.2 was obtained from a field experiment where the atomic Na layer density was measured by lidar simultaneously with a rocket-borne photometric measurement of the D-line emission intensity [Clemesha et al. (1995)]. Theoretical considerations of the electronic potential surfaces correlating through an ion-pair intermediate – NaO + O → NaO$^+$ ... O$^-$ → Na(^{2}P) + O_2 – showed that f could be as high as 0.3 [Bates & Ohja (1980)].

Experimentally, f has proven very difficult to measure. The first estimate was obtained from an experiment employing the flash photolysis of N_2O to produce atomic O in an excess of NaO [Plane & Husain (1986)]. This yielded an upper limit to f of 0.01, much smaller than required to match the airglow intensity. However, studies of the Na + O_3 reaction products by photoelectron spectroscopy [Wright et al. (1993)], and in a molecular beam with magnetic deflection analysis [Shi et al. (1993)], have shown that NaO is produced almost entirely in the first excited $^2\Sigma^+$ state, rather than the ground $^2\Pi$ state. Electronic symmetry correlations show that the reaction NaO($^2\Sigma^+$) + O → Na(^{2}P) + O_2 could have a branching ratio between 0.5 and 0.67 [Herschbach et al. (1992)]. Very recently, a low-pressure flow tube study, using an apparatus similar to that illustrated in Figure 12.4, has shown that the reaction between NaO($^2\Sigma^+$) and O is very fast, and has a branching ratio f of 0.14±0.04. Thus, if the loss of NaO($^2\Sigma^+$) by spontaneous emission or by quenching with N_2 or O_2 is slow compared with reaction with atomic O at 88 km (which seems to be the case), then the observed airglow intensities are accounted for. This also reconciles the apparently low estimate of f by Plane and Husain [1986], since the author's experiment would have observed the reaction of atomic O with predominantly ground-state NaO($^2\Pi$). Interestingly, a very recent combined lidar/rocket photometry experiment [Hecht et al. (2000)] has reported a value for f between 0.01 and 0.05. This field measurement was made under conditions of anomolously low atomic O in the upper mesosphere, so that the low f is probably explained by significant quenching or spontaneous emission from NaO($^2\Sigma^+$). Further rocket experiments should be carried out to confirm this.

12.2.3. Guided Ion Beams

The topsides of the metal atom layers in the upper mesosphere are controlled by their conversion to ions (see Chapter 11). Although photo-ionization by solar radiation makes some contribution, metal ions are formed predominantly by charge transfer with the major E region ions, NO$^+$ and $O_2{}^+$. These charge transfer reactions occur over a range of collision energies, from thermal collisions in the background layer to hyperthermal collisions during the ablation process itself. Although charge transfer rate coefficients can be measured in a flowing afterglow experiment using a fast flow tube [e.g. Farragher et al. (1969)], this only provides data averaged over thermal energies. In contrast, crossed molecular beam experiments have been used to measure the charge transfer cross section at high collision energies. However, these instruments cannot operate efficiently below

FIGURE 12.5. Schematic of the high-temperature octopole ion-guide instrument [Levandier *et al.* (1997)]. The labelled components are as follows: electron impact source, (A); 90° DC quadrupole bender, (B); Wien velocity filter, (C); high-temperature octopole assembly, (D); heated oven-collision cell, (E); quadrupole mass filter, (F); off-axis twin-microchannel plate detector, (G). The unlabelled elements are electrostatic ion transport lenses. The "upstream" ends of the octopole rod heaters are curved away from the instrument centreline to accommodate the heater lead connections. The injection lens has a conical shape of dimensions that preclude the ion beam being affected by the fields in the region of the leads in the poles. Reproduced with permission from The Royal Society of Chemistry, Faraday Transactions.

collision energies of about 2 eV, so that significant extrapolation to thermal collision energies (ca. 0.03 eV) is required in order to calculate a thermal rate coefficient [e.g. Rutherford *et al.* (1971); Rutherford *et al.* (1972)].

During the last 4 years, a new technique has been developed to bridge the gap between thermal and hyperthermal energies. This is the guided ion beam experiment, shown schematically in Figure 12.5 [Levandier *et al.* (1997)]. The ion beam is generated in a continuous electron impact source (A) and directed onto the main instrument axis by the quadrupole bender (B). After mass selection by a velocity filter (C), the primary ion beam is decelerated to the desired energy and injected into the high-temperature octopole assembly (D). The octopole transports the ions through the target vapour in the heated oven-collision cell (E), and the resulting secondary ions as wells as the transmitted primary ions are extracted into a quadrupole mass filter (F) and detected with a microchannel plate (G). Beam energies can be varied from 0.1 to 10 eV, with a precision of ±0.05 eV. A novel feature of this apparatus is that the octopole rods are also coaxial sheath heaters, which provide a heat source for the metal vapour cell (E). In order to provide *absolute* cross-sections, the metal vapour concentration can be measured in the cell by atomic resonance absorption, and calculated assuming that the vapour is in equilibrium with the coldest part of the cell wall. So far, the guided ion beam technique has produced much improved charge transfer cross-section for Na with NO^+ and O_2^+.

12.3. Theoretical Methods

Advances in computing power and the development of new theoretical tools mean that *ab initio* quantum calculations play an increasingly important role in understanding metal chemistry. These calculations provide a framework for interpreting experimental measurements, and then for extrapolating to the temperature and pressure regimes of the upper mesosphere that are often not experimentally accessible. Quantum calculations on reaction intermediates and transition states also enable reaction pathways and hence mechanisms to be elucidated. Examples of these types of applications are given below.

12.3.1. *Ab Initio* Quantum Theory

An *ab initio* quantum calculation provides an approximate solution to the Schrödinger equation within the Born–Oppenheimer approximation. From this solution various molec-

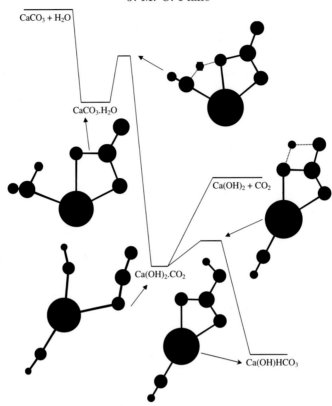

FIGURE 12.6. *Ab initio* quantum calculations at the B3LYP/6-311+G(2d,p) level for the potential energy surface of the reaction $CaCO_3 + H_2O$. The eventual products are either $Ca(OH)_2 + CO_2$ ($\Delta H_0 = -144$ kJ mol^{-1}), or stabilisation to $Ca(OH)HCO_3$, which is 237 kJ mol^{-1} lower in energy than the reactants. Note that the unlabelled atom is carbon.

ular properties are determined, including the optimised geometry, vibrational frequencies, dipole moment and bond energy of the molecule. For metal-containing species such theoretical information is particularly important, because of the difficulty of measuring these quantities in the gas phase. It is worth noting, however, that some data have been obtained from infra-red spectroscopy on metallic species deposited into an inert gas matrix [e.g. Andrews *et al.* (1996)]. Also, high-temperature thermochemical data has been obtained from flame photometry [e.g. Schofield (1992)] and the technique of coupling a Knudsen cell to a mass spectrometer [e.g. Murad (1981)].

One of the difficulties in carrying out *ab initio* calculations on reactions involving metal atoms is to find a single basis set which adequately describes both the covalent reactants and the ionic saddle points and products. Fortunately, this problem has been considerably reduced by using new hybrid theoretical formalisms which include some description of electron correlation, together with large basis sets describing the atomic orbitals. These methods are routinely available within commercial software packages, such as *Gaussian98*.

Recently, the author has used hybrid density functional/Hartree Fock B3LYP theory together with the large 6-311+G(2d,p) basis set, which includes both polarisation and diffuse functions, to carry out a series of calculations on various Na-, Ca- and Fe-containing species. The results compare well with the limited set of available experimental data [Rollason & Plane (2000); Cox *et al.* (2001); Plane & Rollason (2001)]. Figure 12.6 illustrates

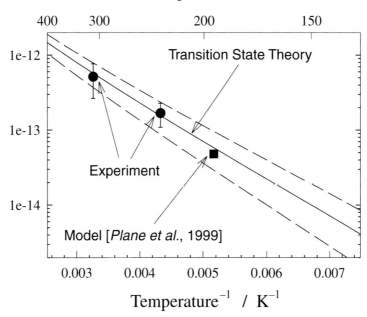

FIGURE 12.7. Arrhenius plot for the reaction NaHCO₃ + H, illustrating the experimental points obtained from the flow tube illustrated in Figure 12.4, transition state theory fitted to these experimental points (solid line) with upper and lower limits (broken lines), and model predictions of the rate coefficient from the seasonal behaviour of the sodium layer at 40° N [Plane *et al.* (1999a)] Reproduced with permission from the American Geophysical Union [Cox *et al.* (2001)].

the reaction pathways for the reaction between $CaCO_3$ and H_2O. Initially, the H_2O is strongly attracted by the extremely large dipole of $CaCO_3$, calculated to be 13.5 Debye. The resulting complex then rearranges via a saddle point to form the $Ca(OH)_2CO_2$ intermediate. This can then either decompose, which is more likely at low pressures, or rearrange via another saddle point to form the very stable $Ca(OH)HCO_3$ molecule. This is an example of where calculations elucidate the probable reaction pathways, by showing that they are exothermic and that there are no high barriers on the potential energy surface. Quantum calculations were also used recently to demonstrate that the very large dipole moment of $NaHCO_3$ (6.9 Debye) enables it to cluster strongly with H_2O, hence providing a nucleating agent for noctilucent cloud formation [Plane (2000)].

12.3.2. *Statistical Rate Theories*

The results of *ab initio* calculations can also be used in statistical rate theories to predict rate coefficients, such as Transition State Theory and Rice–Ramsberger–Kassel–Marcus (RRKM) theory [e.g. Steinfeld *et al.* (1989)]. This is particularly useful when combined with experimental data: the theory is constrained by experiment within the experimental temperature and pressure range, and is then used to extrapolate with confidence outside it. Two examples are presented here. The first involves the reaction between $NaHCO_3$ + H; the experimental study of this reaction is described in Section 12.2.2. Figure 12.7 is an Arrhenius plot for the reaction, which shows that the experimental measurements could only be made down to about 240 K. The reaction clearly has a significant temperature dependence (or activation energy), but extrapolation to lower temperature using just two data-points, with unavoidably large uncertainties, would be rather unsafe. Figure 12.8

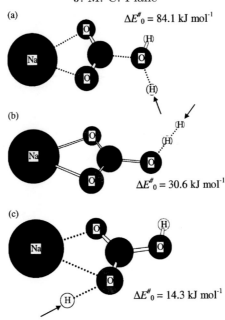

(a) $\Delta E^{\#}_0 = 84.1 \text{ kJ mol}^{-1}$

(b) $\Delta E^{\#}_0 = 30.6 \text{ kJ mol}^{-1}$

(c) $\Delta E^{\#}_0 = 14.3 \text{ kJ mol}^{-1}$

FIGURE 12.8. Transition states for the reaction between $NaHCO_3$ and H to yield (a) Na + H_2O + CO_2, (b) $NaCO_3$ + H_2, and (c) Na + H_2CO_3. The arrow indicates the reactant H; breaking and forming bonds are shown as dotted lines. The saddle point energies (including zero point energy) with respect to the reactants are calculated at the B3LYP/6-311+G(2d,p) level of theory. Reproduced with permission from the American Geophysical Union [Cox *et al.* (2001)].

illustrates the three ways in which atomic H can attack $NaHCO_3$. In each case, the geometry of the saddle point is illustrated, together with the energy barrier ΔE_0.[‡] The lowest barrier involves formation of the products Na + H_2CO_3. Taking the calculated vibrational frequencies, geometry and height of the saddle point, Transition State Theory (including tunnelling through the barrier by the light H atom [Steinfeld *et al.* (1989)]) produces an excellent fit through the experimental points, as shown in Figure 12.7. Note that the extrapolated rate coefficient agrees well with the requirements of an atmospheric model constrained to lidar observations [Plane *et al.* (1999a)], thereby establishing probable closure of the gas-phase Na chemistry [Cox *et al.* (2001)].

The second example is where quantum calculations provide the input required for carrying out RRKM calculations on recombination reactions [e.g. Cox & Plane (1999); Rollason & Plane (2000)]; Campbell & Plane (2001); Plane & Rollason (2001)]. The interested reader is referred to two fairly recent texts on RRKM theory [Steinfeld *et al.* (1989); Gilbert & Smith (1990)]. An RRKM calculation solves the Master Equation describing the time-evolution of the populations of the rovibrational states of the newly formed adduct. The microcanonical rate coefficients governing the dissociation of each rovibrational state of the adduct back to the reactants can be computed efficiently by using inverse Laplace transformation [De Avillez Pereira *et al.* (1997)] to link them directly to $k_{\text{rec},\infty}$, the high pressure limiting recombination coefficient. For the recombination reactions of many metallic species, $k_{\text{rec},\infty}$ is governed by long-range dipole-induced dipole interactions, and hence can be calculated analytically [Rollason & Plane (2000)]. Figure 12.9 illustrates the results of an RRKM calculation on the recombination reaction Ca + O_2 (+ N_2) $\rightarrow$ CaO_2. This reaction has an unusual positive temperature depen-

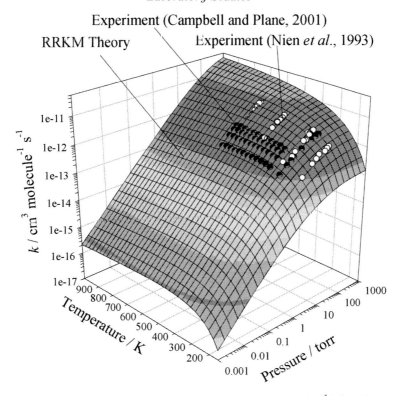

FIGURE 12.9. A mesh plot of the recombination rate coefficient for $Ca(^1S_0) + O_2$ in N_2, calculated from RRKM theory as a function of temperature and pressure. Experimental points from Campbell and Plane [2001] (dark circles) and Nien *et al.* [1993] (white circles) are also shown.

dence, as well as being in the fall-off region between third and second order over the experimental pressure range. Figure 12.9 is a 3D mesh plot, showing the recombination rate coefficient as a function of pressure and temperature. The mesh surface is the RRKM calculation, which fits extremely well to a recent experimental study [Campbell & Plane (2001)]. The rather complex temperature–pressure dependence of this reaction can now be extrapolated reliably to mesospheric conditions.

12.4. A Compilation of Recommended Rate Coefficients

This section lists the measured rate coefficients for reactions of metallic species that are of atmospheric relevance. The list is divided into four tables, which contain reactions of Na, Fe, Ca, and other metal species (Li, K, Al and Ni), respectively. The experimental temperature range for each reaction is given. In order for the compilation to be useful for atmospheric modelling at temperatures below those reached in most experimental studies, the rate coefficients need to be extrapolated. Where a temperature-dependent expression for a rate coefficient was reported in the original paper, this is listed in the relevant table. However, a number of reactions have only been studied at room temperature. In these cases, temperature-dependent expressions have been derived using the typical mesospheric temperature of 200 K as a reference: bimolecular reactions that proceed essentially at the collision frequency are assigned a $T^{1/2}$ dependence, and termolecular reactions are assigned a T^{-1} dependence. Where recombination reactions were found experimentally to be in the fall-off region, the low-pressure limiting rate coefficient ($k_{rec,0}$),

TABLE 12.2. Reactions of Na-containing species

Reaction	T range (K)	$k(T)$[a]	Reference
$Na + O_3 \rightarrow NaO + O_2$	207–377	$1.05 \times 10^{-9} \exp(-116/T)$	Plane et al. (1993)
$Na + HCl \rightarrow NaCl + H$	591–791	$2.1 \times 10^{-9} \exp(-5027/T)$	Plane et al. (1989)
$Na + O_2 (+ N_2) \rightarrow NaO_2$	233–1118	$5.0 \times 10^{-30} (T/200 \text{ K})^{-1.22}$	Plane & Rajasekhar (1989)
$Na + O_2^+ \rightarrow Na^+ + O_2$	[b]	2.7×10^{-9}	Levandier et al. (1997)
$Na + NO^+ \rightarrow Na^+ + NO$	[b]	8.0×10^{-10}	Levandier et al. (1997)
$Na^+ + N_2 (+ N_2) \rightarrow Na^+ \cdot N_2$	93–255	$4.8 \times 10^{-30} (T/200 \text{ K})^{-2.20}$	Cox & Plane (1997)
$Na^+ + O_2 (+ N_2) \rightarrow Na^+ \cdot O_2$	93–130	$2.1 \times 10^{-30} (T/200 \text{ K})^{-2.64}$	Cox & Plane (1997)
$Na^+ + CO_2 (+ N_2) \rightarrow Na^+ \cdot CO_2$	200–300	$3.6 \times 10^{-29} (T/200 \text{ K})^{-2.84}$	Cox & Plane (1997)
$Na^+ + H_2O (+ N_2) \rightarrow Na^+ \cdot H_2O$	300 K	$1.5 \times 10^{-28} (T/200 \text{ K})^{-1}$	Johnsen et al. (1971)
$Na^+ \cdot N_2 + O \rightarrow NaO^+ + N_2$	140 K	4×10^{-10}	Cox & Plane (1998)
$NaO(^2\Pi) + O \rightarrow Na + O_2$	573 K	$2.2 \times 10^{-10} (T/200 \text{ K})^{1/2}$	Plane & Husain (1986)
$NaO(^2\Sigma^+) + O \rightarrow Na + O_2$	290 K	$4.2 \times 10^{-10} (T/200 \text{ K})^{1/2}$	Griffin et al. (2001)
$NaO + O_3 \rightarrow NaO_2 + O_2$	206–378	$1.11 \times 10^{-9} \exp(-568/T)$	Plane et al. (1993)
$NaO + O_3 \rightarrow Na + 2O_2$	290 K	$3.2 \times 10^{-10} \exp(-550/T)$	Ager et al. (1986)
$NaO + H_2 \rightarrow NaOH + H$	296 K	$1.1 \times 10^{-9} \exp(-1100/T)$	Ager & Howard (1987a)
$NaO + H_2 \rightarrow Na + H_2O$	296 K	$1.1 \times 10^{-9} \exp(-1400/T)$	Ager & Howard (1987a)
$NaO + H_2O \rightarrow NaOH + OH$	260–716	$4.4 \times 10^{-10} \exp(-507/T)$	Cox & Plane (1999)
$NaO + HCl \rightarrow NaCl + OH$	308 K	$2.3 \times 10^{-10} (T/200 \text{ K})^{1/2}$	Silver et al. (1984b)
$NaO + O_2 (+ N_2) \rightarrow NaO_3$	297 K	$5.3 \times 10^{-30} (T/200 \text{ K})^{-1}$	Ager & Howard (1986)
$NaO + CO_2 (+ N_2) \rightarrow NaCO_3$	297 K	$1.3 \times 10^{-27} (T/200 \text{ K})^{-1}$	Ager & Howard (1986)
$NaO + h\nu \rightarrow Na + O$	200 K	5.5×10^{-2}	Self and Plane[c]
$NaO_2 + O \rightarrow NaO + O_2$	300 K	$5.0 \times 10^{-10} \exp(-940 \text{ K}/T)$	Helmer & Plane (1993a)
$NaO_2 + HCl \rightarrow NaCl + HO_2$	295 K	$1.9 \times 10^{-10} (T/200 \text{ K})^{1/2}$	Silver & Kolb (1986b)
$NaO_2 + h\nu \rightarrow Na + O_2$	200 K	1.9×10^{-2}	Self and Plane[c]
$NaOH + H \rightarrow Na + H_2O$	300 K	3×10^{-11}	Cox et al.[d]
$NaOH + HCl \rightarrow NaCl + H_2O$	308 K	$2.3 \times 10^{-10} (T/200 \text{ K})^{1/2}$	Silver et al. (1984b)
$NaOH + CO_2 \rightarrow NaHCO_3$	290 K	$1.9 \times 10^{-28} (T/200 \text{ K})^{-1}$	Ager & Howard (1987b)
$NaOH + h\nu \rightarrow Na + OH$	200 K	1.8×10^{-2}	Self and Plane[c]
$NaHCO_3 + H \rightarrow Na + H_2CO_3$	227–307	$1.84 \times 10^{-13} T^{0.78}$ $\times \exp(-1014/T)$	Cox et al. (2001)
$NaHCO_3 + h\nu \rightarrow Na + HCO_3$	200 K	1.2×10^{-4}	Self and Plane[c]

[a]Units: photodissociation rate coefficients, s^{-1}; bimolecular rate coefficients, cm^3 molecule^{-1} s^{-1}; termolecular rate coefficient, cm^6 molecule^{-2} s^{-1}.
[b]Derived from the excitation function at low collision energies.
[c]Unpublished data from D. E. Self & J. M. C. Plane, University of East Anglia.
[d]Unpublished data from R. M. Cox, D. E. Self & J. M. C. Plane, University of East Anglia.

the high-pressure limiting rate coefficient $(k_{rec,\infty})$ and the broadening factor (F_c) are provided so that fall-off calculations can be carried out.

Only a few of the reactions in these tables have been investigated by more than one research group. There is excellent agreement between the recent studies of the reactions $Na + O_3$ and $NaO + O_3$ [Ager et al. (1986); Silver & Kolb (1986a); Plane et al. (1993)]. In the case of $Na + O_2 + N_2$, the flow tube study [Helmer & Plane (1993a)] has resolved the discrepancy between the previous flash photolysis [Plane & Rajasekhar (1989)] and flow tube [Silver et al. (1984a)] investigations. Likewise, the more recent flash photolysis study [Plane et al. (1990)] of the reaction $K + O_2 + N_2$ is recommended over the earlier flow tube study [Silver et al. (1984a)]. For the reaction $NaO + H_2O$, the result of a recent temperature-dependent study [Cox & Plane (1999)] is listed in preference to an earlier single-temperature (296 K) study [Ager & Howard (1987a)], for reasons given in the later paper. For the reaction $Al + O_2$, the early study by Fontijn et al. [1977]

TABLE 12.3. Reactions of Fe-containing species

Reaction	T range (K)	$k(T)^a$	Reference
$Fe + O_3 \rightarrow FeO + O_2$	190–358	$3.44 \times 10^{-10}\ exp(-146/T)$	Helmer & Plane (1994)
$Fe + O_2\ (+ N_2) \rightarrow FeO_2$	288–592	$3.7 \times 10^{-30}\ exp(-2107/T)$	Helmer & Plane (1994)
$Fe + H_2O \rightarrow products$	296	$\leq 5 \times 10^{-15}$	Mitchell & Hackett (1990)
$Fe + O_2^+ \rightarrow Fe^+ + O_2$	b	1.1×10^{-9}	Rutherford & Vroom (1972)
$Fe + NO^+ \rightarrow Fe^+ + NO$	b	9.2×10^{-10}	Rutherford & Vroom (1972)
$Fe^+ + O_3 \rightarrow FeO^+ + O_2$	200–383	$7.07 \times 10^{-10}\ exp(-130/T)$	Rollason & Plane (1998)
$Fe^+ + N_2\ (+ N_2) \rightarrow Fe^+.N_2$	196–482	$1.5 \times 10^{-29}\ (T/200\ K)^{-1.52}$	Rollason & Plane (1998)
$Fe^+ + O_2\ (+ N_2) \rightarrow Fe^+.O_2$	194–479	$3.5 \times 10^{-29}\ (T/200\ K)^{-1.86}$	Rollason & Plane (1998)
$FeO + O \rightarrow Fe + O_2$	249–381	$5.0 \times 10^{-10}\ exp(-171/T)$	Self and Plane[c]
$FeO + O_3 \rightarrow FeO_2 + O_2$	194–341	$2.9 \times 10^{-10}\ exp(-174/T)$	Rollason & Plane (2000)
$FeO + H_2 \rightarrow Fe + H_2O$	320 K	$< 7 \times 10^{-14}$	Rollason & Plane (2000)
$FeO + H_2O\ (+ N_2) \rightarrow Fe(OH)_2$	280–550	RRKM fit[d]	Rollason & Plane (2000)
$FeO + O_2\ (+ N_2) \rightarrow FeO_3$	190–330	RRKM fit[e]	Rollason & Plane (2000)
$FeO + CO_2\ (+ N_2) \rightarrow FeCO_3$	190–550	RRKM fit[f]	Rollason & Plane (2000)
$FeO_2 + O \rightarrow FeO + O_2$	249–381	$2.2 \times 10^{-10}\ (T/200\ K)^{1/2}$	Self & Plane[c]
$FeO_2 + O_3 \rightarrow FeO_3 + O_2$	224–299	$7.1 \times 10^{-11}\ exp(-74/T)$	Self & Plane[c]

[a]Units: photodissociation rate coefficients, s^{-1}; bimolecular rate coefficients, $cm^3\ molecule^{-1}\ s^{-1}$; termolecular rate coefficient, $cm^6\ molecule^{-2}\ s^{-1}$.
[b]Derived from the excitation function at low collision energies.
[c]Unpublished data from D. E. Self and J. M. C. Plane, University of East Anglia.
[d]$log_{10}(k_{rec,0}) = -31.05 + 4.44\ log_{10} T - 1.22 log_{10}^2 T$, $k_{rec,\infty} = 5.35 \times 10^{-10} exp(-611/T)$, $F_c = 0.28$.
[e]$log_{10}(k_{rec,0}) = -41.66 + 10.17\ log_{10} T - 2.09 log_{10}^2 T$, $k_{rec,\infty} = 2.37 \times 10^{-10} exp(-454/T)$, $F_c = 0.44$.
[f]$log_{10}(k_{rec,0}) = -37.87 + 7.07\ log_{10} T - 1.649 log_{10}^2 T$, $k_{rec,\infty} = 4.81 \times 10^{-10} exp(-430/T)$, $F_c = 0.34$.

yielded rate coefficients that were erroneously small, and two more recent studies, which are in excellent agreement, are recommended [Garland & Nelson (1992); Le Picard *et al.* (1997)]. Similarly, in the case of $Al + CO_2$, the two most recent studies [Parnis *et al.* (1988); Garland *et al.* (1992)] are in very good agreement and are recommended.

In the case of the ion–molecule reactions of Na^+, the most recent study by Cox and Plane [1997], which in general agreed well with earlier work, is preferred because of the very extensive temperature range of the measurements. For the charge transfer reactions of Na with NO^+ and O_2^+, the recent results of Levandier *et al.* [1997] are recommended over an earlier study [Rutherford *et al.* (1972)] because of the ability of the guided ion beam experiment to operate at low collision energies. For the reaction of Fe^+ with O_3, the recent PLP/LIF study by Rollason and Plane [1998] is preferred over the early flowing afterglow experiment of Ferguson and Fehsenfeld [1968], since the analogous reaction for Mg^+ was subsequently found by those authors to have been slow by about a factor of 3 [Rowe *et al.* (1981)]. The $Ca^+ + O_3$ rate constant in Table 12.4 has also been corrected accordingly.

12.5. Conclusions

The aim of this chapter has been to provide a selective overview of novel experimental and theoretical techniques for studying the reaction kinetics of metallic species. It is clear that much has been accomplished in the past five years, and that a relatively large kinetic data-base now exists. Most of the important reactions involving metal atoms

TABLE 12.4. Reactions of Ca-containing species

Reaction	T range (K)	$k(T)$[a]	Reference
$Ca + O_3 \rightarrow CaO + O_2$	213–383	$8.23 \times 10^{-10} \exp(-192/T)$	Helmer *et al.* (1993)
$Ca + O_2 (+ N_2) \rightarrow CaO_2$	296–623	RRKM fit[b]	Campbell & Plane (2001)
$Ca + O_2^+ \rightarrow Ca^+ + O_2$	[c]	1.8×10^{-9}	Rutherford *et al.* (1972)
$Ca + NO^+ \rightarrow Ca^+ + NO$	[c]	4.0×10^{-9}	Rutherford *et al.* (1972)
$Ca^+ + O_3 \rightarrow CaO^+ + O_2$	300	$4.0 \times 10^{-10} (T/200 \text{ K})^{1/2}$[d]	Ferguson & Fehsenfeld (1968)
$Ca^+ + O_2 (+ N_2) \rightarrow CaO_2^+$	300	$5.4 \times 10^{-29} (T/200 \text{ K})^{-1.8}$	Ferguson & Fehsenfeld (1968)
$CaO + O \rightarrow Ca + O_2$	805	$3.2 \times 10^{-10} (T/200 \text{ K})^{1/2}$	Plane & Nien (1990)
$CaO + O_3 \rightarrow CaO_2 + O_2$	204–318	$5.7 \times 10^{-10} \exp(-267/T)$	Plane & Rollason (2001)
$CaO + H_2O (+ N_2) \rightarrow Ca(OH)_2$	278–513	RRKM fit[e]	Plane & Rollason (2001)
$CaO + O_2 (+ N_2) \rightarrow CaO_3$	199–505	RRKM fit[f]	Plane & Rollason (2001)
$CaO + CO_2 (+ N_2) \rightarrow CaCO_3$	213–480	RRKM fit[g]	Plane & Rollason (2001)

[a] Units: photodissociation rate coefficients, s^{-1}; bimolecular rate coefficients, cm^3 molecule^{-1} s^{-1}; termolecular rate coefficient, cm^6 molecule^{-2} s^{-1}.
[b] $\log_{10}(k_{rec,0}) = -57.10 + 19.70 \log_{10} T - 3.410 \log_{10}^2 T$, $k_{rec,\infty} = 1.36 \times 10^{-10} \exp(-1020/T)$, $F_c = 0.67$.
[c] Derived from the excitation function at low collision energies.
[d] Increased by a factor of 3.0 [Rowe *et al.* (1981)].
[e] $\log_{10}(k_{rec,0}) = -23.39 + 1.41 \log_{10} T - 0.751 \log_{10}^2 T$, $k_{rec,\infty} = 7.02 \times 10^{-10} \exp(-38/T)$, $F_c = 0.31$.
[f] $\log_{10}(k_{rec,0}) = -42.19 + 13.15 \log_{10} T - 2.87 \log_{10}^2 T$, $k_{rec,\infty} = 9.90 \times 10^{-10} \exp(-195/T)$, $F_c = 0.43$.
[g] $\log_{10}(k_{rec,0}) = -36.14 + 9.24 \log_{10} T - 2.19 \log_{10}^2 T$, $k_{rec,\infty} = 7.97 \times 10^{-10} \exp(-190/T)$, $F_c = 0.36$.

have been studied. However, much less is known about the reactions and photochemistry of metallic compounds (higher oxides, hydroxides, carbonates and bicarbonates), and much of the available data has come from complex experiments where a compound is detected indirectly by reduction to the metal atom or oxide. Direct detection is not possible because the spectroscopy of many of these metallic compounds has yet to be investigated. Indeed, this lack of electronic spectroscopic data on metallic compounds is a major obstacle to future kinetic studies.

In terms of ion–molecule chemistry, most of the important atomic metal ion reactions have now been studied. However, once again very little is known about the reactions of metal oxide ions with species such as O, H and O_3. Another major deficiency is that there do not appear to have been any studies of the dissociative recombination of metallic molecular ions with electrons – the final step in neutralising charged species.

Acknowledgments: Laboratory studies of meteoric metal chemistry at the University of East Anglia are supported by the U.K. Natural Environment Research Council, the Royal Society and the U.S. Air Force Office of Scientific Research.

TABLE 12.5. Reactions of species containing Li, K, Mg, Al and Ni

Reaction	T range (K)	$k\,(T)$	Reference
$Li + O_3 \rightarrow LiO + O_2$	199–306	$1.06 \times 10^{-9}\,\exp(-213/T)$	Plane & Helmer[a]
$Li + O_2\,(+\,N_2) \rightarrow LiO_2$	267–1100	$6.3 \times 10^{-30}\,(T/200\,\mathrm{K})^{-0.93}$	Plane & Rajasekhar (1988)
$Li + HCl \rightarrow LiCl + H$	700–1000	$3.8 \times 10^{-10}\,\exp(-883/T)$	Plane & Saltzman (1987)
$K + O_3 \rightarrow KO + O_2$	200–356	$1.15 \times 10^{-9}\,\exp(-120/T)$	Plane & Helmer[a]
$K + O_2\,(+\,N_2) \rightarrow KO_2$	267–1100	$1.3 \times 10^{-29}\,(T/200\,\mathrm{K})^{-1.23}$	Plane et al. (1990)
$K + HCl \rightarrow KCl + H$	252–780	Complex $k(T)$[b]	Helmer & Plane (1993b)
$KO + O_3 \rightarrow KO_2 + O_2$	209–355	$6.86 \times 10^{-10}\,\exp(-382/T)$	Plane & Helmer[a]
$Mg + O_3 \rightarrow MgO + O_2$	196–368	$2.28 \times 10^{-10}\,\exp(-139/T)$	Plane & Helmer (1995)
$Mg + O_2\,(+\,N_2) \rightarrow MgO_2$	350–624	$4.9 \times 10^{-30}\,\exp(-2790/T)$	Nien et al. (1993)
$Mg + O_2{}^+ \rightarrow Mg^+ + O_2$	c	1.2×10^{-9}	Rutherford et al. (1971)
$Mg + NO^+ \rightarrow Mg^+ + NO$	c	8.1×10^{-10}	Rutherford et al. (1971)
$Mg^+ + O_3 \rightarrow MgO^+ + O_2$	300	$5.8 \times 10^{-10}\,(T/200\,\mathrm{K})^{1/2}$	Rowe et al. (1981)
$Mg^+ + O_2\,(+\,N_2) \rightarrow MgO_2{}^+$	300	$6.0 \times 10^{-30}\,(T/200\,\mathrm{K})^{-1}$	Rowe et al. (1981)
$MgO^+ + O_3 \rightarrow Mg^+ + 2O_2$	300	$6.5 \times 10^{-10}\,(T/200\,\mathrm{K})^{1/2}$	Rowe et al. (1981)
$MgO^+ + O \rightarrow Mg^+ + O_2$	300	$8 \times 10^{-11}\,(T/200\,\mathrm{K})^{1/2}$	Rowe et al. (1981)
$MgO + O_3 \rightarrow MgO_2 + O_2$	217–366	$2.19 \times 10^{-10}\,\exp(-548/T)$	Plane & Helmer (1995)
$MgO + H_2O\,(+\,N_2) \rightarrow Mg(OH)_2$	278–513	RRKM fit[d]	e
$MgO + O_2\,(+\,N_2) \rightarrow MgO_3$	199–505	RRKM fit[f]	e
$MgO + CO_2\,(+\,N_2) \rightarrow MgCO_3$	213–480	RRKM fit[g]	e
$Al + O_2 \rightarrow AlO + O$	23–295	$1.75 \times 10^{-10}(T/298\,\mathrm{K})^{-0.25}$ $\times \exp(7/T)$	Le Picard et al. (1997)
$Al + CO_2 \rightarrow AlO + CO$	298–483	$5.6 \times 10^{-12}\,\exp(-241/T)$	Garland et al. (1992)
$Al + H_2O \rightarrow$ products	298–1174	Complex $k(T)$[h]	McClean et al. (1993)
$Al + HCl \rightarrow$ products	475–1275	$1.52 \times 10^{-10}\,\exp(-803/T)$	Rogowski et al. (1989)
$AlO + O_2 \rightarrow AlO_2 + O$	300–1400	4.8×10^{-13}	Fontijn et al. (1977)
$AlO + CO_2 \rightarrow AlO_2 + CO$	296	1.4×10^{-11}	Parnis et al. (1989)
$AlO + H_2 \rightarrow$ products	296	$\leq 5 \times 10^{-14}$	Parnis et al. (1989)
$AlO + HCl \rightarrow$ products	440–1590	$5.6 \times 10^{-11}\,\exp(-139/T)$	Slavejkov et al. (1990)
$Ni + O_2\,(+N_2) \rightarrow NiO_2$	296	$2.6 \times 10^{-30}(T/200\,\mathrm{K})^{-1}$	Brown et al. (1991)
$Ni + H_2O \rightarrow$ products	298	3×10^{-10}	Mitchell et al. (1994)

[a]Unpublished data from J. M. C. Plane & M. Helmer, University of East Anglia.
[b]$k(T) = 1.69 \times 10^{-10}\,\exp(-1829/T) + 1.51 \times 10^{-11}\,\exp(-594/T)$.
[c]Derived from the excitation function at low collision energies.
[d]$\log_{10}(k_{\mathrm{rec},0}) = -34.61 + 7.27\,\log_{10} T - 1.61\log_{10}^2 T$, $k_{\mathrm{rec},\infty} = 6.44 \times 10^{-10}\exp(-455/T)$, $F_c = 0.34$.
[e]Unpublished data from J. M. C. Plane & R. J. Rollason, University of East Anglia.
[f]$\log_{10}(k_{\mathrm{rec},0}) = -28.05 + 1.423\,\log_{10} T - 0.683\log_{10}^2 T$, $k_{\mathrm{rec},\infty} = 1.16 \times 10^{-10}\exp(-219/T)$, $F_c = 0.34$.
[g]$\log_{10}(k_{\mathrm{rec},0}) = -33.67 + 5.83\,\log_{10} T - 1.49\log_{10}^2 T$, $k_{\mathrm{rec},\infty} = 6.79 \times 10^{-10}\exp(-310/T)$, $F_c = 0.36$.
[h]$k(T) = 1.9 \times 10^{-12}\,\exp(-442/T) + 1.6 \times 10^{-10}\,\exp(-2870/T)$.

REFERENCES

AGER J. W., TALCOTT C. L. & HOWARD C. J. 1986. Gas-phase kinetics of the reactions of Na and NaO with O_3 and N_2O, *J. Chem. Phys.*, **85**, 5584–5590.

AGER J. W. & HOWARD C. J. 1986. The kinetics of NaO + O_2 + M and NaO + CO_2 + M and their role in atmospheric sodium chemistry, *Geophys. Res. Lett.*, **13**, 1395–1398.

AGER J. W. & HOWARD C. J. 1987. Gas-phase kinetics of the reactions of NaO with H_2, D_2, H_2O, and D_2O, *J. Chem. Phys.*, **87**, 921–925.

AGER J. W. & HOWARD C. J. 1987. Rate coefficient for the gas-phase reaction of NaOH with CO₂, *J. Geophys. Res.*, **92**, 6675–6678.

ANDREWS L., CHERTIHIN G V., THOMPSON C. A., DILLON J., BYRNE S. & BAUSCHLICHER JR. C. W. 1996. Infrared spectra and quantum chemical calculations of Group 2 MO₂, O₂MO₂, and related molecules, *J. Phys. Chem.*, **100**, 10088–10099.

BATES D. R. & OHJA P. C. 1980. Excitation of the Na D-doublet of the nightglow, *Nature*, **286**, 790–791.

BERNARD R. 1938. *Z. Phys.*, **110**, 291–296.

BOWMAN M. R., GIBSON A. J. & SANDFORD M. C. W. 1969. Atmospheric sodium measured by a tuned laser radar, *Nature*, **221**, 456–457.

BROWN C. E., MITCHELL S. A. & HACKETT P. A. 1991. Dioxygen complexes of 3d transition-metal atoms: formation reactions in the gas phase, *J. Phys. Chem.*, **95**, 1062–1066.

CAMPBELL M. L. & PLANE J. M. C. 2001. Kinetic Study of Gas Phase Ca(^{1}S$_0$) with O₂ from 296–623 K, *J. Phys. Chem. A*, **105**, 3515–3520.

CHAPMAN S. 1939. Notes on atmospheric sodium, *J. Astrophys.*, **90**, 309–316.

CLEMESHA B. R., SIMONICH D. M., TAKAHASHI H. & MELO S. M. L. 1995. Experimental evidence for photochemical control of the atmospheric sodium layer, *J. Geophys. Res.*, **100**, 18909–18916.

COX R. M. & PLANE J. M. C. 1997. An experimental and theoretical study of the clustering reactions between Na⁺ ions and N₂, O₂ and CO₂, *J. Chem. Soc., Faraday Trans.*, **93**, 2619–2629.

COX R. M. & PLANE J. M. C. 1998. An ion–molecule mechanism for the formation of neutral sporadic Na layers, *J. Geophys. Res.*, **103**, 6349–6360.

COX R. M. & PLANE J. M. C. 1999. An experimental and theoretical study of the reactions NaO + H₂O(D₂) → NaOH(D) + OH(OD), *Phys. Chem. Chem. Phys.*, **1**, 4713–4720.

COX R. M., SELF D. E. & PLANE J. M. C. 2001. A study of the reaction between NaHCO₃ and H: apparent closure on the neutral chemistry of sodium in the upper mesosphere, *J. Geophys. Res.*, **106**, 1733–1739.

DE AVILLEZ PEREIRA R., BAULCH D. L., PILLING M. J., ROBERTSON S. H. & ZENG G. 1997. Temperature and pressure dependence of the multichannel rate coefficients for the CH₃ + OH system, *J. Phys. Chem.*, **101**, 9681–9693.

ESKA V., VON ZAHN U. & PLANE J. M. C. 1999. The terrestrial potassium layer (75–110 km) between 71° S and 54° N: Observations and modeling, *J. Geophys. Res.*, **104**, 17173–17186.

FARRAGHER A. L., PEDEN J. A. & FITE W. L. 1969. *J. Chem. Phys.*, **50**, 287–294.

FERGUSON E. E. & FEHSENFELD F. C. 1968. Some aspects of the metal ion chemistry of the Earth's atmosphere, *J. Geophys. Res.*, **73**, 6215–6223.

FONTIJN A., FELDER W. & HOUGHTON J. J. 1977. *Proceedings of 16th International Symposium on Combustion*, Pittsburgh: The Combustion Institute, pp. 871–876.

GARLAND N. L. & NELSON H. H. 1992. Temperature dependence of the kinetics of the reaction Al + O₂ → AlO + O, *Chem. Phys. Lett.*, **191**, 269–272.

GARLAND N. L., DOUGLASS C. H. & NELSON H. H. 1992. Pressure and temperature dependence of the kinetics of the reaction Al + CO₂, *J. Phys. Chem.*, **96**, 8390–8394.

GERDING M., ALPERS M., VON ZAHN U., ROLLASON R. J. & PLANE J. M. C. 2000. The atmospheric Ca and Ca⁺ layers: Midlatitude observations and modeling, *J. Geophys. Res.*, **105**, 27131–27146.

GILBERT R. C. & SMITH S. C. 1990. *Theory of Unimolecular and Recombination Reactions*, Blackwell, Oxford.

GRIFFIN J., WORSNOP D. R., BROWN R. C., KOLB C. E. & HERSCHBACH D. R. 2001. Chemical kinetics of the NaO($A^2\Sigma^+$)+O(^{3}P) reaction, *J. Phys. Chem.*, **105**, 1643–1648.

HECHT J. H., COLLINS R. L., KRUSCHWITZ C., KELLEY M. C., ROBLE R. G. & WALTERSCHEID R. L. 2000. The excitation of the Na airglow from Coqui Dos rocket and ground-based observations, *Geophys. Res. Lett.*, **27**, 453–456.

HELMER M. & PLANE J. M. C. 1993. A study of the reaction NaO₂ + O → NaO + O₂: implications for the chemistry of sodium in the upper atmosphere, *J. Geophys. Res.*, **98**, 23207–23222.

HELMER M. & PLANE J. M. C. 1993. Experimental and theoretical study of the reaction K + HCl, *J. Chem. Phys.*, **99**, 7696–7702.

HELMER M., PLANE J. M. C. & ALLEN M. 1993. A kinetic investigation of the reaction Ca + O_3 over the temperature range 213–383 K, *J. Chem. Soc., Faraday Trans.*, **89**, 763–769.

HELMER M. & PLANE J. M. C. 1994. Kinetic study of the reaction between Fe and O_3 under mesospheric conditions, *J. Chem. Soc., Faraday Trans.*, **90**, 31–37.

HELMER M. & PLANE J. M. C. 1994b. Experimental and theoreical study of the reaction Fe + O_2 + N_2 → FeO_2 + N_2, *J. Chem. Soc., Faraday Trans.*, **90**, 395–401.

HELMER M., PLANE J. M. C., QIAN J. & GARDNER C. S. 1998. A model of meteoric iron in the upper atmosphere, *J. Geophys. Res.*, **103**, 10913–10925.

HERSCHBACH D. R., KOLB C. E., WORSNOP D. R. & SHI X. 1992. Excitation mechanism of the mesospheric sodium nightglow, *Nature*, **356**, 414–416.

HUNTEN D. M. 1967. *Space Sci. Rev.*, **6**, 493.

HUNTEN D. M., TURCO R. P. & TOON O. B. 1980. Smoke and dust particles of meteoric origin in the mesosphere and stratosphere, *J.Atmos. Sci.*, **37**, 1342–1357.

IKEZOE Y., MATSUOKA S., TAKEBE M. & VIGGIANO A. 1987. Gas phase ion–molecule reaction rate constants through 1986, *Ion Reaction Research Group of The Mass Spectroscopy Society of Japan*, Daito Bldg., 18–10, Koraku 2-chome, Bunkyo-ku, Tokyo, 112 Japan.

JOHNSEN R., BROWN H. L. & BIONDI M. A. 1971. Reactions of Na^+, K^+, and Ba^+ ions with O_2, NO, and H_2 molecules, *J. Chem. Phys.*, **55**, 186–188.

KOPP E. 1997. On the abundance of metal ions in the lower ionosphere, *J. Geophys. Res.*, **102**, 9667–9674.

LE PICARD S. D., CANOSA A., TRAVERS D., CHASTAING D. & ROWE B. R. 1997. Experimental and theoretical kinetics for the reaction of Al with O_2 at temperatures between 23 and 295 K, *J. Phys. Chem.*, **101**, 9988–9992.

LEVANDIER D. J., DRESSLER R. A., WILLIAMS S. & MURAD E. 1997. A high-temperature guided-ion beam study of Na + X^+ (X = O_2, NO, N_2) charge-transfer reactions, *J. Chem. Soc., Faraday Trans.*, **93**, 2611–2617.

LÜBKEN F.-J. 1999. Thermal structure of the Arctic summer mesosphere, *J. Geophys. Res.*, **104**, 9135–9149.

MCCLEAN R. E., NELSON H. H. & CAMPBELL M. L. 1993. Kinetics of the reaction $Al(^2P_0)$ + H_2O over an extended temperature range, *J. Phys. Chem.*, **97**, 9673–9676.

MCNEIL W. J., MURAD E. & LAI S. T. 1995. Comprehensive model for the atmospheric sodium layer, *J. Geophys. Res.*, **100**, 16847–16855.

MCNEIL W. J., LAI S. T. & MURAD E. 1996. A model for meteoric magnesium in the ionosphere, *J. Geophys. Res.*, **101**, 5251–5259.

MEGIE G. & BLAMONT J. E. 1969. Laser sounding of atmospheric sodium: Interpretation in terms of global atmospheric parameters, *Planet. Space Sci.*, **25**, 1093–1109.

MITCHELL S. A. & HACKETT P. A. 1990. Chemical reactivity of iron atoms near room temperature, *J. Chem. Phys.*, **93**, 7822–7829.

MITCHELL S. A., BLITZ M. Z., SIEGBAHN E. M. & SVENSSON M. 1994. Experimental and theoretical study of oxidative addition reaction of nickel atom to O–H bond of water, *J. Chem. Phys.*, **100**, 423–433.

MURAD E. 1981. Thermochemical properties of the gaseous alkaline earth monohydroxides, *J. Chem. Phys.*, **75**, 4080–4085.

NARCISI R. S. 1968. Processes associated with metal-ion layers in the E-region of the ionosphere, *Space Res.*, **8**, 360–369.

NIEN C.-F., RAJASEKHAR B. & PLANE J. M. C. 1993. The unusual kinetic behavior of the reactions Mg + O_2 + M and Ca + O_2 + M (M = N_2, He) over extended temperature ranges, *J. Phys. Chem.*, **94**, 4161–4167.

PARNIS J. M., MITCHELL S. A. & HACKETT P. A. 1988. Complexation and abstraction channels in the Al + CO_2 reaction, *Chem. Phys. Lett.*, **151**, 485–489.

PARNIS J. M., MITCHELL S. A., KANIGAN T. S. & HACKETT P. A. 1989. Gas-phase reactions of AlO with small molecules, *J. Phys. Chem.*, **93**, 8045–8051.

PLANE J. M. C. & HUSAIN D. 1986. The absolute rate of the reaction O + NaO → Na + O_2 and its effect on atomic sodium in the mesosphere, *J. Chem. Soc., Faraday Trans. 2*, **82**, 2047–2052.

PLANE J. M. C. & SALTZMAN E. S. 1987. A study of the reaction Li + HCl by the technique of time-resolved laser induced fluorescence of Li (2^2P_J–$2^2S_{1/2}$, $\lambda = 670.7$ nm) between 700 and 1000 K, *J. Chem. Phys.*, **87**, 4606–4611.

PLANE J. M. C. & RAJASEKHAR B. 1988. A study of the reaction Li + O_2 + M (M = N_2, He over the temperature range 267–1100 K by time-resolved laser induced fluorescence of Li (2^2P_J–$2^2S_{1/2}$), *J. Phys. Chem.*, **92**, 3884–3890.

PLANE J. M. C. & RAJASEKHAR B. 1989. A kinetic study of the reactions Na + O_2 + N_2 and Na + N_2O over an extended temperature range, *J. Phys. Chem.*, **93**, 3135–3140.

PLANE J. M. C., RAJASEKHAR B. & BARTOLOTTI L. 1989. An experimental and theoretical study of the reactions Na + HCl and Na + DCl, *J. Chem. Phys.*, **91**, 6177–6186.

PLANE J. M. C., RAJASEKHAR B. & BARTOLOTTI L. 1990. A kinetic study of the reaction K + O_2 + M (M = N_2 or He) from 250–1100 K, *J. Phys. Chem.*, **94**, 4161–4167.

PLANE J. M. C. & NIEN C.-F. 1990. A kinetic investigation of the Ca/CaO system: non-Arrhenius behavior of the reaction Ca(^{1}S) + N_2O over the temperature range 250–898 K, and a study of the reaction CaO + O, *J. Phys. Chem.*, **94**, 5255–5261.

PLANE J. M. C. 1991. The chemistry of meteoritic metals in the upper atmosphere, *Int. Rev. Phys. Chem.*, **10**, 55–106.

PLANE J. M. C., NIEN C.-F., ALLEN M. R. & HELMER M. 1993. A kinetic investigation of the reactions Na + O_3 and NaO + O_3 over the temperature range 207–377 K, *J. Phys. Chem.*, **97**, 4459–4467.

PLANE J. M. C. & HELMER M. 1994. Laboratory studies of the chemistry of meteoric metals, *Research in Chemical Kinetics*, **2**, 313–365.

PLANE J. M. C. & HELMER M. 1995. Laboratory study of the reactions Mg + O_3 and MgO + O_3: implications for the chemistry of magnesium in the upper atmosphere, *Faraday Discuss.*, **100**, 411–430.

PLANE J. M. C., GARDNER C. S., YU J., SHE C. Y., GARCIA R. R. & PUMPHREY H. C. 1999. Mesospheric Na layer at 40-deg N: modeling and observations, *J. Geophys. Res.*, **104**, 3773–3788.

PLANE J. M. C., COX R. M. & ROLLASON R. J. 1999. Metallic layers in the mesopause and lower thermosphere region, *Adv. Space Res.*, **24**, 1559–1570.

PLANE J. M. C. & ROLLASON R. J. 1999. A study of the reactions of Fe and FeO with NO_2, and the structure and bond energy of FeO_2, *Phys. Chem. Chem. Phys.*, **1**, 1843–1849.

PLANE J. M. C. 2000. The chemistry of the mesosphere and lower thermosphere region, in *ERCA – Volume 4, From Weather Forecasting to Exploring the Solar System*, Boutron, C. (Editor), EDP Sciences, Les Ulis, France, pp. 257–282.

PLANE J. M. C. 2000. The role of sodium bicarbonate in the nucleation of noctilucent clouds, *Ann. Geophys.*, **18**, 807–814.

PLANE J. M. C. & ROLLASON R. J. 2001. A kinetic study of the reactions of CaO with H_2O, CO_2, O_2 and O_3: implications for calcium chemistry in the mesosphere, *J. Phys. Chem. A*, **105**, 7047–7056.

RAJASEKHAR B., PLANE J. M. C. & BARTOLOTTI L. 1989. Determination of the absolute photolysis cross-section of sodium superoxide at 230 K: evidence for the formation of sodium tetroxide in the gas-phase, *J. Phys. Chem.*, **93**, 7399–7404.

ROGOWSKI D. F., MARSHALL P. & FONTIJN A. 1989. High-temperature fast-flow reactor kinetics studies of the reactions of Al with Cl_2, Al with HCl, and AlCl with Cl_2 over wide temperature ranges, *J. Phys. Chem.*, **93**, 1118–1124.

ROLLASON R. J. & PLANE J. M. C. 1998. A study of the reactions of Fe^+ with O_3, O_2 and N_2, *J. Chem. Soc., Faraday Trans.*, **94**, 3067–3075.

ROLLASON R. J. & PLANE J. M. C. 2000. The reactions of FeO with O_3, H_2, H_2O, O_2 and CO_2, *Phys. Chem. Chem. Phys.*, **2**, 2335–2343.

ROWE B. R., FAHEY D. W., FERGUSON E. E. & FEHSENFELD F. C. 1981. Flowing afterglow studies of gas phase magnesium ion chemistry, *J. Chem. Phys.*, **75**, 3225–3228.

RUTHERFORD J. A., MATHIS R. F., TURNER B. R. & VROOM D. A. 1971. Formation of magnesium ions by charge transfer, *J. Chem. Phys.*, **55**, 37854–3793.

RUTHERFORD J. A., MATHIS R. F., TURNER B. R. & VROOM D. A. 1972. Formation of sodium ions by charge transfer, *J. Chem. Phys.*, **56**, 4654–4658.

RUTHERFORD J. A., MATHIS R. F., TURNER B. R. & VROOM D. A. 1972. Formation of calcium ions by charge transfer, *J. Chem. Phys.*, **57**, 3087–3090.

RUTHERFORD J. A. & VROOM D. A. 1972. Formation of iron ions by charge transfer, *J. Chem. Phys.*, **57**, 3091–3093.

SCHOFIELD K. S. 1992. The flame chemistry of alkali and alkaline earth metals, in *Gas Phase Metal Reactions*, Fontijn A. (Editor), Elsevier, Amsterdam, pp. 529–571.

SHI X., HERSCHBACH D. R., WORSNOP D. R. & KOLB C. E. 1993. Molecular beam chemistry: magnetic deflection analysis of monoxide electronic states from the alkali-metal atom + ozone reactions, *J. Phys. Chem.*, **97**, 2113–2122.

SILVER J. A., ZAHNISER M. S., STANTON A. C. & KOLB C. E. 1984. *Proceedings of 20th International Symposium on Combustion*, Pittsburgh: The Combustion Institute, pp. 605–612.

SILVER J. A., STANTON A. C., ZAHNISER M. S. & KOLB C. E. 1984. Gas-phase reaction rate of sodium hydroxide with hydrochloric acid, *J. Phys. Chem.*, **88**, 3123–3129.

SILVER J. A., WORSNOP D. R., FREEDMAN A. & KOLB C. E. 1985. Absolute photodissociation cross-sections of gas-phase sodium chloride at room temperature, *J. Chem. Phys.*, **84**, 4378–4384.

SILVER J. A. & KOLB C. E. 1986. Determination of the absolute rate constants for the room-temperature reactions of atomic sodium with ozone and nitrous oxide, *J. Phys. Chem.*, **90**, 3263–3266.

SILVER J. A. & KOLB C. E. 1986. Gas-phase reaction rate of sodium superoxide with hydrochloric acid, *J. Phys. Chem.*, **90**, 3267–3269.

SLAVEJKOV A. G., STANTON C. T. & FONTIJN A. 1990. High-temperature fast-flow reactor kinetics studies of the reactions of AlO with Cl_2 and HCl over wide temperature ranges, *J. Phys. Chem.*, **94**, 3347–3352.

SLIPHER V. M. 1929. Emission in the spectrum of the light of the night sky, *Z. Phys.*, **110**, 262.

STATES R. J. & GARDNER C. S. 2000. Thermal structure of the mesopause region (80–105 km) at 40 degrees N latitude. Part II: diurnal variations, *J. Atmos. Sci.*, **57**, 78–92.

STEINFELD J. I., FRANCISCO J. S. & HASE W. L. 1989. *Chemical Kinetics and Dynamics*, Prentice Hall, New Jersey.

WORSNOP D. R., ZAHNISER M. S. & KOLB C. E. 1991. Low-temperature absolute rate constants for the reaction of atomic sodium with ozone and nitrous oxide, *J. Phys. Chem.*, **95**, 3960–3964.

WRIGHT T. G., ELLIS A. M. & DYKE J. M. 1993. A study of the products of the gas-phase reaction M + N_2O and M + O_3, with ultraviolet photoelectron spectroscopy, *J. Chem. Phys.*, **98**, 2891–2907.

13
SUMMARY AND FUTURE OUTLOOK

By EDMOND MURAD[1†] AND IWAN P. WILLIAMS[2‡]

[1]Space Weather Center of Excellence, Space Vehicles Directorate, Air Force Research
Laboratory, Hanscom AFB, MA 01731, USA

[2]Astronomy Unit, Queen Mary, University of London, London, E1 4NS, UK

13.1. Summary

The preceding chapters have presented a comprehensive summary of recent work in
an exciting and dynamic area of research, namely meteoric phenomena. One aim was to
use this monograph as a vehicle to update the excellent, but now dated, works of Öpik
(1958), McKinley (1961), and Bronshten (1983). For example, when these monographs
were written, lidar was in its infancy and had not been used to study meteoric phenomena,
whereas in the last few years its application to the study of meteor trails has lead to many
advances and is described in Chapter 7. Sound understanding of meteoroid–atmosphere
interaction requires a knowledge of the meteoroid velocity immediately before encountering the atmosphere and this in turn requires a knowledge of the velocities of meteoroids
immediately after ejection from comets and other bodies. The previous standard work
on this subject [Porter (1952)] is brought up to date in the discussion given in Chapter
2. Likewise, in discussing models for the interaction with the atmosphere of metal atoms
from meteor ablation (Chapter 11), an attempt was made for the first time to include the
vaporization of meteoroids due to heating by collisional energy transfer, the chemical processes that lead to the formation of molecular species that eventually condense and the
effects of important geophysical processes (e.g. winds, electric fields) that are responsible
for the transportation of these species – eventually to ground. The fundamental chemical
processes that take place, of course, had to be measured, and in Chapter 12 the latest
kinetic data are summarized within the framework of meteor physics and chemistry. Another aim was to provide a forum for bringing together work that is not traditionally
regarded as meteor research, but that has significant impact on the field. An example
of this cross-fertilization is the chapters on extraterrestrial dust in deep space (Chapter
3), interplanetary dust particles (Chapter 4), and satellite impact studies (Chapter 10).
These are important topics that can provide much information about the composition
and velocities of dust particles that are captured by Earth. Two major uncertainties that
hinder the development of a comprehensive model that explains and predicts the results
of meteoroids entering the Earth's atmosphere are: the composition of meteoroids and
the total flux of extraterrestrial material. The uncertainty in composition arises from
the fact that much of what we know in that respect comes from meteorites that are
collected on the ground. In turn, these meteorites arise mostly from the fragmentation
of asteroids. In contrast, except for the Geminids, all major meteor showers arise from
comets. Because of this and because we do not know the difference in composition between asteroids and comets, there is some uncertainty about the starting point of the
components. This controversy is illustrated in this book by different opinions stated in
Chapters 7 and 9 with regards to composition. Fortunately, recent advances in the use of

† Email:ed.murad@hanscom.af.mil
‡ Email:I.P.Williams@qmul.ac.uk

CCDs for spectroscopic observations of meteor trails [Borovička *et al.* (1996); Borovička *et al.* (1999); Abe *et al.* (2000)] seem to support the use of chondritic composition [Fegley & Cameron (1987)] for meteoroids. Some of the spectroscopic observations are discussed in Chapters 5 and 7; space limitations precluded a more detailed discussions of this interesting topic. Uncertainty in the total flux arises from attempts to merge visual, radar, space data, and landfall collections. Radar has been very important in the development of an observational database for meteors, as shown in Chapter 6. Space limitations precluded the inclusion of work using wide aperture radar [Pellinen-Wannberg & Wannberg (1994); Pellinen-Wannberg & Wannberg (1996); Pellinen-Wannberg *et al.* (1998); Close *et al.* (2000)]. Likewise, the use of radar to study geophysical phenomena such as winds, temperatures, and atmospheric dynamics [Elford (2001); Hocking & Thayaparan (1997); Hocking (1999); Kobayashi *et al.* (1999); Portnyagin *et al.* (1999)] was not covered. As demonstrated in Chapter 7, lidar has recently played a very important role in the analysis and understanding of meteor trails. This new development was preceded by a great deal of work on the use of lidar to study the metal atoms generated in the ablation of those meteoroids that form part of the sporadic background. Again, due to limitations of space, this work was not covered in this book, although citation may be made to reviews by Gardner (1989), Gardner (1995), and Kane & Gardner (1993). Additionally, lidar has been used to study a number of geophysical parameters, such as winds and temperatures [She *et al.* (1992); She *et al.* (1995)]. Since the 1980s, great strides have been made in following the dynamical evolution of meteoroid streams through developments in larger and faster computers and some of these are described in Chapter 2. However, there has recently been a realization that more complex dynamical phenomena such as resonance and chaos can also play a part in stream evolution, the most noteworthy being the work of Asher *et al.* (1999) pointing out that meteoroids ejected from comet 55P/Tempel–Tuttle in 1333 could be trapped on a resonant orbit (hence not evolve) and could be encountered by Earth in 1998, producing an unexpected outburst.

13.2. Outlook

A major advance that will surely come is a more exact determination of the composition and flux of extraterrestrial matter. In this respect, space probes such as the *NEAR – Near Earth Asteroid Mission* and *Rosetta* missions might be very helpful in providing a firmer footing for determining the composition of comets and asteroids. The increasing use of wide aperture radar, such as the work mentioned above and others [Janches *et al.* (2000); Mathews *et al.* (1997); Zhou *et al.* (1995); Zhou & Kelley (1997)] will undoubtedly make possible a more precise determination of the absolute flux of extraterrestrial matter to Earth as well as a determination of the role of fragmentation and recondensation in the eventual transport of meteoroid matter to Earth surface. An interesting, and, perhaps, important phenomenon associated with meteoroid entry into the Earth's atmosphere is that of persistent trains. The phenomenon refers to the observation of long-lived luminescent trains that have odd shapes and whose lifetimes can sometimes exceed an hour. The phenomenon was first documented by Trowbridge (1907), and has since been studied episodically [Hawkins (1957); Baggaley (1975); Baggaley (1976); Baggaley (1981)]. Recent advances in spectroscopic observations [Borovička & Jenniskens (2000); Jenniskens *et al.* (2000); Kelley *et al.* (2000)] and analysis [Murad (1901)] suggest that this phenomenon may be ripe for explanation during the next few years.

REFERENCES

ABE S, YANO H., EDIZUKA N. & WATANABE J. 2000. First results of high-definition TV spectroscopic observations of the 1999 Leonid meteor shower, *Earth, Moon, and Planets*, **82–83**, 369–377.

ASHER D. J., BAILEY M. E. & EMEL'YANENKO V. V. 1999. Resonant meteoroids from comet Tempel–Tuttle in 1333: the cause of the unexpected Leonid outburst in 1998, *Mon. Not. R. Astr. Soc.*, **304**, L53–L57.

BAGGALEY W. J. 1975. Sodium emission from long enduring meteor trains, *Nature*, **257**, 567–568.

BAGGALEY W. J. 1976. The chemical reduction of meteoric metal oxides as a source of meteor train emission, *Bull. Astron. Inst. Czechoslovakia*, **27**, 244–246.

BAGGALEY W. J. 1981. The source of enduring meteor train luminosity, *Nature*, **289**, 530–531.

BOROVIČKA J. & SPURNÝ P. 1996. Radiation study of two very bright terrestrial bolides, *Astron. Instit. Czech.*, **1996**, 1–25

BOROVIČKA J., STORK R. & BOČEK J. 1999. First results from video spectroscopy of 1998 Leonid meteors, *Meteorit. Planet. Sci.*, **34**, 987–994.

BOROVIČKA J. & JENNISKENS P. 2000. Time-resolved spectroscopy of a Leonid fireball afterglow, *Earth Moon Planets*, **82–83**, 399–428.

BRONSHTEN V. A. 1983. *Physics of Meteoric Phenomena*, Kluwer, Dordrecht, Holland, 356 pp.

CLOSE S., HUNT S. M., MINARDI M. J. & McKEEN F. M. 2000. Analysis of Perseid meteor head echo data collected using the Advance Research Projects Agency Long-Range Tracking and Instrumentation Radar (ALTAIR), *Radio Sci.*, **25**, 1233–1240.

ELFORD W. G. 2001. Novel applications of MST radars in meteor studies, *J. Atm. Solar Terr. Phys.*, **63**, 143–153

FEGLEY JR., B. & CAMERON A. G. W. 1987. A vaporization model for iron/silicate fractionation in the Mercury protoplanet, *Earth Planet. Sci. Lett.*, **82**, 207–222.

GARDNER C. S. 1989. Sodium resonance fluorescence lidar applications in atmospheric science and astronomy, *Proc. IEEE*, **77**, 408–418.

GARDNER C. S. 1995. Sporadic metal layers in the upper mesosphere, *Faraday Disc.*, **100**, 431–439.

HAWKINS G. S. 1957. A hollow meteor train, *Sky & Telescope*, **February 1957**, 168–169.

HOCKING W. K. & THAYAPARAN T. 1997. Simultaneous and colocated observation of winds and tides by MF and meteor radars over London, Canada (43°, 81°W), during 1994–1996, *Radio Sci.*, **32**, 833–865.

HOCKING W. K. 1999. Temperatures using radar-meteor decay times, *Geophys. Res. Lett.*, **26**, 3297–3300.

JANCHES D., MATHEWS J. D., MEISEL D. D. & ZHOU Q. 2000. Micrometeor observations using the Arecibo 430 MHz Radar: I. Determination of the ballistic parameters from measured Doppler velocity and deceleration results, *Icarus*, **145**, 53–63.

JENNISKENS P., LACEY M., ALLAN B. J., SELF D. E., & PLANE J. M. C. 2000. FeO "Orange Arc" emission detected in optical spectrum of Leonid persistent train, *Earth Moon Planets*, **82–83**, 429–438.

KANE T. J. & GARDNER C. S. 1993. Lidar observations of the meteoric deposition of mesospheric metals, *Science*, **259**, 1297–1300.

KELLEY M. C., GARDNER C. S., DRUMMOND J., ARMSTRONG T., LIU A., CHU X., PAPEN G., KRUSCHWITZ C., LOUGHMILLER P., GRIME B. & ENGELMAN J. 2000. First observations of long-lived meteor trains with resonance lidar and other optical instruments, *Geophys. Res. Lett.*, **27**, 1811–1814.

KOBAYASHI K., KITAHARA T., KAWAHARA T. D., SAITO Y., NOMURA A., NAKAMURA T., TSUDA T., ABO M., NAGASAWA C. & TSUTSUMI M. 1999. Simultaneous measurements of dynamical structure in the mesopause region with lidars and MU radar, *Earth Planets Space*, **51**, 731–739.

MATHEWS J. D., HUNTER K. P., MEISEL D. D., GETMAN V. S. & ZHOU Q. 1997. Very high resolution studies of micrometeors using the Arecibo 430 MHz Radar, *Icarus*, **126**, 157–169.

MATHEWS J. D., JANCHES D., MEISEL D. D. & ZHOU Q. H. 2001. The micrometeoroid mass flux into the upper atmosphere: Arecibo results and a comparison with prior estimates, *Geophys. Res. Lett.*, **28** 1929–1932.

McKINLEY D. W. R. 1961. *Meteor Science and Engineering*, McGraw-Hill, New York, 309 pp.

MURAD E. 2001. Heterogeneous chemical processes as ssource of persistent meteor trains, *Meteorit. Planet. Sci.*, **36**, 1217–1224.

ÖPIK, E. J. 1958. *Physics of Meteor Flight in the Atmosphere*, Interscience Publishers, New York, 174 pp.

PELLINEN-WANNBERG A. & WANNBERG G. 1994. Meteor observations with the EISCAT UHF incoherent scatter radar, *J. Geophys. Res.*, **99**, 11397 11390.

PELLINEN-WANNBERG, A. & WANNBERG G. 1996. Enhanced ion-acoustic echoes from meteor trails, *J. Atmos. Terr. Phys.*, **58**, 1–4, 495–506.

PELLINEN-WANNBERG A., WESTMAN A., WANNBERG G. & KAILA K. 1998. Meteor fluxes and visual magnitudes from EISCAT radar event rates: a comparison with cross-section based magnitude estimates and optical data, *Ann. Geophysicae*, **16**, 1475–1485.

PORTNYAGIN Y. I., SOLOVJOVA T. V. & WANG D. Y.1999. Some results of comparison between the lower thermosphere zonal winds as seen by ground-based radars and WINDII on UARS, *Earth Planets Space*, **51**, 701–709.

PORTER J. G. 1952. *Comets and Meteor Streams*, Wiley, New York, 123 pp.

SHE C. Y., YU J. R., LATIFFI H. & BILLS R. E. 1992. High-spectral-resolution fluorescence light detection and ranging for mesospheric sodium temperature measurements, *Appl. Optics*, **31**, 2095–2106.

SHE C. Y., YU J. R., KRUEGER D. A., ROBLE R. G., KECKHUT P., HAUCHECORNE A. & CHANIN M.-L. 1995. Vertical structure of the midlatitude temperature from stratosphere to mesopause (30–105 km), *Geophys. Res. Lett.*, **22**, 377–380.

TROWBRIDGE C. C. 1907. Physical nature of meteor trains, *Ap. J.*, **26**, 95–116.

WANNBERG G., PELLINEN-WANNBERG A. & WESTMAN A. 1996. An ambiguity-function-based method for analysis of Doppler decompressed radar signals applied to EISCAT measurements of oblique UHF–VHF meteor echoes, *Radio Sci.*, **31**, 497–518.

ZHOU Q., TEPLEY C. A. & SULZER M. P. 1995. Meteor observations by the Arecibo 430 MHz incoherent scatter radar I. Results from time-integrated observations, *J. Atmos. Terr. Phys.*, **57**, 421–431.

ZHOU Q. & KELLEY M. C. 1997. Meteor observations by the Arecibo 430 MHz incoherent scatter radar II. Results from time-resolved observations, *J. Atmos. Terr. Phys.*, **59**, 513–521

Glossary

Terms

ablation – evaporation of meteoroid materials caused by frictional heating in the atmosphere

blooming – spread of a signal from one pixel to an adjacent one because of leakage of electrons

body echo – reflection of radar signals from the linear plasma generated by the meteoroid as it traverses the atmosphere

chondrules – very small particles (spherules) that are incorporated into many types of small meteorites

dustball – term introduced by Öpik to denote a weakly held collection of dust particles

ejecta – material ejected when a meteoroid collides with a surface

EKBO – Edgeworth–Kuiper belt object

fireball – larger meteoroids that survive to lower altitudes where heating becomes quite intense and where shock fronts are formed so that he have very bright heads and trails at altitudes near 50 or 60 km

head echo – reflection of radar signals from the plasma and ablating meteoroid material at the entry point into the atmosphere

ionosphere – region of the atmosphere where the plasma density is sufficiently high and the neutral density is sufficiently low that the ions and electrons play an important role in the dynamics of the atmosphere. Generally this occurs at altitudes greater than 80 km

Keplerian orbit – the elliptical orbit under the gravitational field of the Sun alone that has the same energy and angular momentum as the meteroid

lidar – analogous to radar: LIght Detecting And Ranging

lifetime – time required for a signal to drop to $1/e$ of its initial value

light curve – track of meteor as it traverses the atmosphere

luminous intensity – absolute intensity in the visible region of the spectrum of a meteoroid entering the atmosphere

magnetite – iron oxide (Fe_3O_4) mineral

mesosphere – the region of the neutral atmosphere where the there is a sharp drop in temperature. It occurs at approximately 60–90 km altitude

meteor – luminosity associated with extraterrestrial matter entering the Earth's atmosphere

meteorite – a meteoroid where some of it survives its transit through the atmosphere and reaches the Earth's surface

meteoroid – solid extraterrestrial matter

meteroid stream – a set of meteoroids on very similar orbits, usually caused by the ejection of dust from either a comet or an asteroid

meteor shower – a higher influx of meteoroids all apparently originating from the same locality on the sky, usually caused by the Earth passing through a meteroid stream dust cloud left by a comet or an asteroid

micrometeorite – a meteoroid so small that its radiative losses are greater that heating by collisions with the atmosphere

Mie scattering – scattering of light by particles whose dimensions are much larger than the wavelength of the light

olivine – magnesium silicate–iron silicate mineral

overdense train – case where the plasma surrounding the meteoroid is so dense that it acts almost as a solid body reflecting the radar signal

population index – usually denoted by 'r', the ratio of the numbers of meteors in one magnitude range to the next

Poynting–Robertson effect – drag effect of solar radiation on the orbit of dust grains; this effect tends to circularize the orbit

quantum efficiency – ratio of photons detected per pixel per second to the incident photons

radar frequencies – HF: 3–30 MHz; VHF: 30–300 MHz; UHF: 0.3–1 GHz

radiation pressure – direct force exerted by solar radiation on meteoroids; this effect tends to increase the semi-major axis

radiant – the area in the sky from which the meteoroids appear to originate. Meteoroids entering the Earth's atmosphere along parallel trajectories seem to emanate from a particular point in the sky

Rayleigh scattering – scattering of light by particles whose radius is much smaller than light wavelength

SI units – international system of units: mass in kg, length in meter, time in sec, temperature in K, and luminous intensity in candela

thermosphere – the neutral component of the upper atmosphere beginning at about 95 km altitude and extending upwards

underdense train – case where the plasma surrounding the meteoroid is thin enough that the electrons reflect the radar signal individually, i.e. without interaction with each other

visual magnitude – logarithmic scale of the intensity (in units of watts) of light radiated by a meteoroid

Zenithal hourly rate – usually denoted by ZHR, the hourly rate of meteors that would be seen by one observer on a clear moonless night if the radiant was directly overhead (i.e. at the zenith)

zodiacal light – light scattered by interplanetary dust

Acronyms

AE – Atmospheric Explorer satellite
AMOR – Advanced Meteor Orbit Radar
amu – atomic mass unit
BIMS – Bennett ion mass spectrometer
CCD – charge-coupled device
CDA – (Cassini) Cosmic Dust Analyzer
CIDA – Cometary and Interstellar Dust Analyzer
COBE – Cosmic Background Explorer
DE – Dynamic Explorer satellite
EISCAT – European incoherent scatter radar
EKBO – Edgeworth–Kuiper belt object
EMI – Electromagnetic Interferences
ESA – European Space Agency
EURECA – European Retrievable Carrier
GEMS – Glass with Embedded Metal and Sulfides
HEOS – Highly Eccentric Orbit Satellite
HST – Hubble Space Telescope
IAP – Institute for Atmospheric Physics at the University of Rostock, Külungsborn, Germany
IDP – Interplanetary Dust Particle

IMF – Interplanetary Meteoroid Flux

IRAS – Infrared Astronomical Satellite

ISEE – International Sun Earth Explorer

ISO – Infrared Space Observatory

LDEF – Long-Duration Exposure Facility

MAGMA – code for calculating equilibrium composition of a multi-component gas-melt

MIMS – magnetic ion mass spectrometer

MM – micrometeorite

MORP – Meteorite Observation and Recovery Network

MSX – Midcourse Space Experiment

MU radar – Middle and Upper atmosphere radar

NEA – Near Earth Asteroid

QIMS – Quadrupole Ion Mass Spectrometer

RRKM – Rice Ramsberger Kassel Marcus theory for predicting reaction rate coefficients

VeGa – Russian satellite. Acronym for Venera Gallei (Russian for Halley)

ZHR – Zenithal hourly rate

Index

Printed in the United States
By Bookmasters